PARALLEL COMPUTING USING OPTICAL INTERCONNECTIONS

THE KLUWER INTERNATIONAL SERIES
IN ENGINEERING AND COMPUTER SCIENCE

PARALLEL COMPUTING USING OPTICAL INTERCONNECTIONS

edited by

Keqin Li
State University of New York at New Paltz

Yi Pan
University of Dayton

Si Qing Zheng
University of Texas at Dallas

SPRINGER SCIENCE+BUSINESS MEDIA, LLC

Library of Congress Cataloging-in-Publication Data

A C.I.P. Catalogue record for this book is available
from the Library of Congress.

ISBN 978-1-4757-7117-6   ISBN 978-0-585-27268-9 (eBook)
DOI 10.1007/978-0-585-27268-9

Printed on acid-free paper.

Contents

Preface

Motivation

The circuit density offered by VLSI technology provides the means for implementing systems with a large number of processors. According to today's standards, a massively parallel processing (MPP) system consists of hundreds or thousands of processors. To achieve teraflops performance, it is expected that more and more processors are incorporated into a single system. An extremely important aspect of parallel computing is data communication among processors in parallel computing systems. Due to two-dimensionality, I/O constraints, and electrical properties such as resistance, capacitance, and inductance of VLSI circuits, the VLSI technology is not suitable for interconnecting communication intensive systems.

The advances in optical technologies have made it possible to implement optical interconnections in future MPP systems. Photons are non-charged particles, and do not naturally interact. Consequently, there are many desirable characteristics of optical interconnects, e.g., high speed (speed of light), increased fanout, high bandwidth, high reliability, supporting longer interconnection lengths, exhibiting low power requirements, and immunity to EMI with reduced crosstalk. Optics can utilize free-space interconnects as well as guided wave technology, neither of which has the problems mentioned for the VLSI technology. Optical interconnections can be built at various levels, providing chip-to-chip, module-to-module, board-to-board, and node-to-node communications.

Massively parallel processing using optical interconnections poses new challenges. In general, it does not worth the effort to improve performance by simply replacing the metal interconnections of a parallel system with optical interconnections in a one-to-one fashion. This is due to the fact that in doing so the inherent large bandwidth and high parallelism of optical interconnects are underutilized. The characteristics of optical interconnects have significant implications. New system configurations need to be designed due to the changes in architectural freedom and constraints when optical interconnects are incorporated. To fully utilize the bandwidth offered by the optical interconnections, scheduling and data communication schemes based on new resource metrics need to be investigated. Algorithms for a wide variety of applications need to be developed under novel computation models that are based on optical interconnections. Computations under these new models, which can be drastically different from existing theoretical PRAMs and parallel systems with electrical interconnections, may require new algorithm design techniques and performance measures.

The book is motivated by the fact that the area of massively parallel processing using optical interconnections has received increased attention in last few years, and noticeable progress has been made. A number of research groups and authors have been actively making contributions to this relatively new field. Though research in this area is still in its early stage, promising results have been demonstrated by experimental systems built in laboratories. There is need for better understanding of what optical technologies can offer for interconnections, how optical interconnections can be constructed, and what the impacts of optical interconnections are on MPP. A significant number of research articles have been published in various journals and conference proceedings, but there has not been a book that provides current and comprehensive coverage of the field, reflects the state-of-the-art from high-level architecture design and algorithmic points of view, and points out directions for further research and development. This book will serve to fill in this gap. The book is a collection of survey articles written by leading and active scientists in the area of parallel computing using optical interconnections. The primary purpose of the book is to provide an authoritative overview of the state-of-the-art of the field.

Book Organization and Chapter Overview

This book focuses on the potential of using optical interconnections in massively parallel processing systems, and their effects on system design and parallel computation. It is intended to provide readers the new perspectives of large-scale computing using optical interconnections. The emphases include the following aspects of MPP systems with optical interconnections:

- Various optical interconnects and their implementations, high-level system design, scalability, reconfigurability, partitionability, and performance evaluation;

- Data communication schemes, resource scheduling, embedding, and mapping methods;

- Models for parallel computing using optical interconnections, algorithm design and development, and applications.

The book consists of twelve chapters, which are grouped into two parts, with six chapters in each part. Part one covers optical interconnection networks and system architectures. Chapter 1, written by D.C. Hoffmeister, J. Chu, J.A. Perreault, and P. Dowd, provides a reconfigurable, hierarchical, wavelength division multiplexing (WDM) network called LIGHTNING for large-scale parallel computing in a distributed shared memory environment. Chapter 2, contributed by T.H. Szymanski, examines the class of "Intelligent Optical Networks". In Chapter 3, A. Louri and B. Weech describe a scalable interconnection topology called Spanning Multichannel Linked Hypercube (SMLH) as well as its optical implementation, and compare SMLH with many popular networks. Chapter 4, by R. Melhem, G. Gravenstreter, D. Chiarulli, and S. Levitan, is devoted to the communication capabilities of the Partitioned Optical Passive Stars (POSP) network, together with message routing and scheduling on POSP, and analytical performance evaluation and simulations results. Chapter 5, authored by C.-f. Wang and S. Sahni, deals with the characteristics of parallel computers with optical transpose interconnection systems (OTIS), and summarizes the complexities of basic operations and fundamental algorithms on these systems. In Chapter 6, H.-A. Choi and E.J. Harder address the problems of routing and wavelength assignment in efficient use of WDM optical networks.

Part two of the book focuses on models and algorithms for optical interconnections. Several models are presented that make algorithm design, specification, and performance comparison relatively easier without worrying about optical engineering details. Algorithms and applications in this part include selection and sorting, numerical computations, matrix operations, image processing, and others. In Chapter 7, S.Q. Zheng proposes the hypernetwork model for optical interconnection networks, and shows how to design hypernetworks by using theory of hypergraphs and combinatorial block design. Chapter 8, written by C. Qiao, presents the Reconfigurable Array with Spanning Optical Buses (RASOB) model, together with example algorithms and embeddings. Chapter 9, contributed by S. Rajasekaran and S. Sahni, discusses and analyzes basic and important algorithms, deterministic and randomized, on the Array with Reconfigurable Optical Buses (AROB) model. In Chapter 10, S.D. Pavel and S.G. Akl develop optimal constant time algorithms for applications in image processing on the AROB model. Chapter 11, authored by Y. Pan, gives a comprehensive presentation of a model called Linear Array with Reconfigurable Pipelined Bus System (LARPBS), and implementation details of a number of primitive and fundamental operations. Finally, in Chapter 12, K. Li demonstrates novel implementations of fast and processor efficient parallel algorithms for matrix multiplication and related matrix manipulations on the LARPBS model by using the building blocks in Chapter 11.

The reader should keep in mind that the above division of the book into two parts is based on the emphasis of the chapters. It is virtually impossible to designate some chapters into certain category, since a chapter in Part one may contain algorithms, and a chapter in Part two may include optical implementation details, due to intimate interplay of various aspects of parallel computing. Each chapter is about 20-25 pages long and focuses on a particular topic. The material are survey/tutorial type and are based on published papers; some original results are also included. The extensive bibliography at the end of the chapters provide references for further information.

Audience and Use of the Book

The book is primarily written for scientists, researchers, engineers, developers, educators, and administrators from universities, industries, research institutes and laboratories, and government agencies in the area of parallel and distributed computing. The interdisciplinary feature of the book is likely to be appealing to a wide range of audience. They will find this book a unique source of information on recent advances and future directions of parallel computation using optical interconnections. Since this is the first book addressing parallel architectures, algorithms, and applications using optical interconnection networks, we expect that the book will be an informative and useful reference in this new and fast growing research area.

The organization of this book makes it particularly useful for graduate courses. It can be used as a text for a research oriented and seminar based advanced graduate course. Graduate students will find the material covered by this book to be stimulating and inspiring. Using this book, they can identify interesting and important research topics for their Master's theses and Ph.D. dissertations. It can also serve as a supplementary book for regular courses on Parallel Architectures, Parallel Algorithms, and High-Performance Computing taught in Computer Science, Computer Engineering, and Electrical Engineering Departments.

Acknowledgments

The book is a successful result of collaboration, cooperation, and commitment among many researchers. We are very grateful to all the 22 contributing authors from the community of parallel computing with optical interconnections. Their expertise, energy, and enthusiasm have made it possible for us to edit and publish such a book in a short period of time. We also acknowledge ten anonymous reviewers of our book proposal for their encouragement and critical comments that helped greatly in shaping the scope and coverage of the book. The publication of the book originated from a lengthly discussion between the first editor and Mr. Scott E. Delman, a senior publishing editor of Computer and Information Science of Kluwer Academic Publishers, in the summer of 1997. We appreciate Scott's suggestions, guidance, and advice during the course of this endeavor. The book would not have been published without his initial push and encouragement. Our thanks extend to other staff of Kluwer Academic Publishers, in particular, Ms. Sharon Fletcher and Ms. Suzanne St. Clair, for their strong support in many aspects during various stages of the production of this book. Last but not least, we wish to express our most sincere gratitude to our wives and children, Ling, Andrew, and Charlotte of the Li's family, Hong and Marissa of the Pan's family, Lin Li, Ting An, and Lee Ming of the Zheng's family, and the families of all the contributing authors, for their support and sacrifice when all of us were working under very tight schedule.

KEQIN LI, YI PAN, SI QING ZHENG,

Contributing Authors

(The authors are listed in alphabetical order. An author marked with $[x]$ is the contact person of Chapter x, $1 \leq x \leq 12$.)

Selim G. Akl[10] is a professor of computing and information science at Queen's University, Kingston, Ontario, Canada. His current research interests are in the area of parallel algorithms, with a particular emphasis on alternative models of computation and inherently parallel paradigms. He is the author of *Parallel Sorting Algorithms* (Academic Press, 1985), *The Design and Analysis of Parallel Algorithms* (Prentice Hall, 1989), and *Parallel Computation: Models and Methods* (Prentice Hall, 1997), and co-author of *Parallel Computational Geometry* (Prentice Hall, 1992). He is an editor of *Information Processing Letters* (North-Holland), *Computational Geometry* (Elsevier), *Parallel Processing Letters* (World Scientific Publishing), and *Parallel Algorithms and Applications* (Gordon & Breach).

Donald Chiarulli received his B.S. degree (physics, 1976) from Louisiana State University, M.S. (computer science, 1979) from Virginia Polytechnic Institute, and Ph.D. (computer science, 1986) from Louisiana State University. He was an instructor/research associate at LSU from 1979 to 1986, and has been at the University of Pittsburgh since 1986. His current research includes several projects concerned with the architecture, implementation, and application of scalable parallel processing systems.

Hyeong-Ah Choi[6] is currently a professor of computer science at the George Washington University, Washington, DC. She has published over 60 technical papers in parallel processing, optical interconnects, and algorithms. She is a coauthor of the paper received an Outstanding Paper Award at the 1996 IEEE International Conference on Parallel Processing.

John Chu received the B.S. degree in electrical engineering from the State University of New York at Buffalo in 1992. He was a Presidential Fellow at SUNY at Buffalo, and received the M.S. degree in electrical engineering in 1996. He is currently pursuing the Ph.D. degree.

Patrick Dowd[1] received the B.S. in electrical engineering and computer science (Summa Cum Laude) from the State University of New York at Buffalo in 1983, M.S. (1985) and Ph.D. (1988) degrees in electrical engineering from Syracuse University. Currently he is an associate professor in the Department of Electrical Engineering and UMIACS at the University of Maryland at College Park, and also holds the position of senior research scientist at the Advanced Networking Research Department, U.S. Department of Defense. He has served in various important positions for numerous international conferences, and is currently on the executive committee of ACM SIGCOMM, and an area technical program committee chair of INFOCOM98. He also serves on the editorial board of the *International Journal in Computer Simulation*, and is an associate editor of the *SCS Transactions on Simulation*, and the editor of the electronic journal *Computer Simulation: Modeling and Analysis.* He is a member of IEEE and ACM with research interests in high speed networks, cluster-based computing, mobile and optical communication.

Greg Gravenstreter is currently involved in networking and security at the Software Engineering Institute within Carnegie Mellon University. From 1971 to 1985, he was with the Westinghouse Electric Corporation's Research and Development Center. He worked as Director of Systems and Operations for Guidance Technologies in Pittsburgh from 1988 to 1992. His interests include a wide range of network and security issues, architectural and systems issues for parallel and distributed systems.

Eric J. Harder received his D.Sc. in computer science in 1998 from the George Washington University. He is currently a senior computer scientist with the U. S. Department of Defense. His research interests include approximation and online algorithms for routing and wavelength assignment, applied graph theory, and combinatorics.

David C. Hoffmeister received the B.S. (summa cum laude) and M.S. degrees in electrical and computer engineering from the State University of New York at Buffalo in 1994 and 1996, respectively. He is currently pursuing the Ph.D. degree in that department. His research interests include WDM networks, reconfigurable networks, distributed network control, and all-optical networks.

Steven P. Levitan is the Wellington C. Carl associate professor of electrical engineering at the University of Pittsburgh. He received the B.S. degree from Case Western Reserve University in 1972. From 1972 to 1977 he worked for Xylogic Systems. He received his M.S. (1979) and Ph.D. (1984) degrees, both in computer science, from the University of Massachusetts, Amherst. During that time he also worked for Digital Equipment Corporation, and Viewlogic Systems, as a consultant in HDL simulation and synthesis. He was an assistant professor from 1984 to 1986 in the Electrical and Computer Engineering Department at the University of Massachusetts. In 1987 he joined the electrical engineering faculty at the University of Pittsburgh where he holds a joint appointment in the Department of Computer Science. He is the editor of the *SIGDA Newsletter* and Manager of the SIGDA Internet Server.

Keqin Li[12] received B.S. degree in computer science from Tsinghua University, China, in 1985, and Ph.D. degree in computer science from the University of Houston in 1990. He is currently an associate professor of computer science in State University of New York at New Paltz. He has published over 100 technical papers and received best paper awards in 1996 International Conference on Parallel and Distributed Processing Techniques and Applications, and 1997 IEEE National Aerospace and Electronics Conference. He is the Associate Editor-in-Chief of *International Journal of Parallel and Distributed Systems and Networks.* He is the program chair of the 9th International Conference on Parallel and Distributed Computing and Systems (October 1997), a general co-chair of the 10th International Conference on Parallel and Distributed Computing and Systems (October 1998), a conference co-chair of the 4th International Conference on Computer Science and Informatics (October 1998), and the vice program chair of the IPPS Workshop on Optics and Computer Science (April 1999). He is a senior member of IEEE, and a member of IEEE Computer Society, ACM, SIGACT, SIGARCH, SIAM, ISMM, IASTED, and SCS.

Ahmed Louri[3] is currently an associate professor of electrical and computer engineering at the University of Arizona and Director of the Optical Networking and Parallel Processing Laboratory. His research interests include computer architecture, parallel processing, optical computing, and optical interconnects. He has published numerous journal and conference articles on the above topics. He has served as a member of the technical program committee of several conferences including OSA Topical Meeting on Optics in Computing, OSA/IEEE Conference on Massively Parallel Processors using Optical Interconnects, among others. In 1996 he was the general chair of the Workshop on Optics in High-Performance Computing, at Euro-Par'96 Lyon France. He is a senior member of the IEEE and a member of the ACM, OSA, and SPIE.

Rami Melhem[4] is a professor of computer science at the University of Pittsburgh. He served on program committees of numerous conferences and workshops and as the general chair for the third International Conference on Massively Parallel Processing Using Optical Interconnections. He was on the editorial board of the *IEEE Transactions on Computers* and is currently on editorial board of the *IEEE Transactions on Parallel and Distributed Systems.* His research interests include fault-tolerant and real-time systems, optical interconnection networks, high performance computing, and parallel computer architectures.

Yi Pan[11] received his B.Eng. degree in computer engineering from Tsinghua University, China, in 1982, and his Ph.D. degree in computer science from the University of Pittsburgh, USA, in 1991. Currently he is an associate professor at the University of Dayton. He is an author of more than 60 research papers, and has received several awards including NSF Research Opportunity Award, AFOSR Summer Faculty Fellowship, Andrew Mellon Fellowship from Mellon Foundation, the best paper award from PDPTA '96, and Summer Research Fellowship from the Research Council of the University of Dayton. He is currently on the editorial boards of the *Journal of Supercomputing* and the *Journal of Parallel and Distributed Computing Practices*, and is an associate editor of the *International Journal of Parallel and Distributed Systems and Networks.* He is the program chair of the 10th International Conference on Parallel and Distributed Computing and Systems in 1998, the conference co-chair of the 4th International Conference on Computer Science and Informatics in 1998, and the program chair of the 1999 IPPS Workshop on Optics and Computer Science. He is a senior member of IEEE and a member of the IEEE Computer Society. He is listed in *Men of Achievement* and *Marquis Who's Who in Midwest.*

Sandy D. Pavel is currently a member of the scientific staff at Nortel (Northern Telecom) in Ottawa, Canada. He received the Ph.D. degree in computer science from Queen's University at Kingston, Ontario, Canada. His research interests include algorithms, high-speed network protocols, IP over optical networks, and all-optical networking.

James A. Perreault received the B.S., M.S., and Ph.D. degrees in electrical and computer engineering from the State University of New York at Buffalo in 1991, 1994, and 1998, respectively. Since 1997, he has been working for the Department of Defense in advanced network research. His research interests include all-optical networking, switch design, and IP quality of service.

Chunming Qiao[8] got his BS degree in 1985 from University of Science and Technology of China (USTC) in China. He received the Andrew-Mellon Distinguished doctoral fellowship award from University of Pittsburgh and obtained his Ph.D. degree in 1993. Since then, he has been with SUNY at Buffalo. He received an NSF Research Initiation Award (RIA) in 1994 for his research on fiber-optic interconnection networks, and recently, another NSF grant for research on fiber-optic wavelength division multiplexed (WDM) networks and internetworking (e.g., optical internet). His research interests cover the two converging areas of computers and communications, including parallel and distributed computing and systems, interconnection networks, and photonic switching. He has published many papers in these areas in *IEEE Trans. on Computers*, *IEEE Trans. on Parallel and Distributed Systems*, *IEEE Trans. on Communications*, and *IEEE/ACM Trans. on Networking*. He has served as a co-chair for both the 1997 and 1998 All-optical Networking Conferences sponsored by SPIE (International Society for Optical Engineering), a program vice co-chair for the 1998 International Conf. on Computer Communications and Networks (IC3N), a session organizer/chair at the IEEE MILCOM'96 (the Conference on Military Communications), and program committee members and session chairs in several other conferences and workshops. He is also an editor of the *Journal on High-Speed Networks* (JHSN) and a member of IEEE Computer Society and IEEE Communications Society.

Sanguthevar Rajasekaran[9] is an associate professor in the Department of Computer and Information Science and Engineering at the University of Florida. He received his Ph.D. from Harvard University in 1988. His research interests include parallel algorithms, randomized computing, combinatorial optimization, parsing, and image processing. He has published more than 80 papers in journals, conferences, and books. He is a co-author of the texts *Computer Algorithms* and *Computer Algorithms/C++* published by W.H. Freeman Press. He serves on the editorial boards of *IEEE Transactions on Computers* and the *Journal of Parallel and Distributed Computing.*

Sartaj Sahni[5] is a University of Florida Research Foundation Professor of computer and information sciences and engineering and a Fellow of IEEE, ACM, AAAS, and Minnesota Supercomputer Institute. He received his B.Tech. (electrical engineering) degree from the Indian Institute of Technology, Kanpur, and the M.S. and Ph.D. degrees in computer science from Cornell University. He has published over 150 research papers and written several texts. His research publications are on the design and analysis of efficient algorithms, parallel computing, interconnection networks, and design automation. In 1997, he was awarded the IEEE Taylor L. Booth Education Award "for contributions to Computer Science and Engineering education in the areas of data structures, algorithms, and parallel algorithms". He is a co-editor of the *Journal of Parallel and Distributed Computing* and is on the editorial boards of *IEEE Parallel and Distributed Technology*, and *Computer Systems: Science and Engineering.* He has served as program committee chair, general chair, and been a keynote speaker at many conferences.

Ted H. Szymanski[2] is an associate professor at McGill University. He has served as the Director of the Microelectronics and Computer Systems Laboratory at McGill, and as a project leader in the Canadian Institute for Telecommunications Research. His research interests include optical networks, telecommunication and computing architectures, and performance analysis. He has been on the program committees of the 1998 and 1997 Workshops on Optics in Computer Science, the 1998 and 1997 International Conferences on Massively Parallel Processing using Optical Interconnects, the 1998 International Conference on Optical Computing, the 1997 Innovative Systems on Silicon Conference, the 1995 Workshop on High-Speed Network Computing, and the 1998, 1995 and 1994 Canadian Conferences on Programmable Logic Devices. He is a member of the IEEE Computer and Communications Societies.

Chih-fang Wang is currently a Ph.D. candidate at the Department of Computer and Information Science and Engineering, University of Florida. He got his M.S. degree in mathematics and computer science from University of Miami in 1993. His research interests include distributed and parallel algorithms, reconfigurable networks, optical interconnection computers, and all-optical networks.

Brent Weech is an M.S. candidate at the University of Arizona. He received the B.S. degree in electrical engineering from Arizona State University in 1994. His research interests include parallel and distributed processing, communication networks, and optical interconnects. He is a student member of the IEEE.

Si Qing Zheng[7] received the Ph.D. degree in computer science from the University of California, Santa Barbara, in 1987, and joined the faculty of Louisiana State University in the same year. He was an associate professor of computer science and an adjunct associate professor of electrical and computer engineering at LSU before moving to the University of Texas at Dallas in 1998, where he is currently a professor of computer science. He has published 130 refereed papers in VLSI design, computer architectures, parallel and distributed computing, optical interconnections, computer networks, and design and analysis of algorithms. He was the program committee chairman of the 8th International Conference on Computing and Information (1996), the program committee co-chairman of the 10th ISCA International Conference on Parallel and Distributed Computing Systems (1997), and the program committee vice chairman of the 2nd International Conference on Parallel and Distributed Computing and Networks (1998). He is an associate editor of several computing journals.

I Optical Interconnection Networks and System Architectures

1 LIGHTNING NETWORK AND SYSTEMS ARCHITECTURE

David C. Hoffmeister,[1,2] John Chu,[1,2]
James A. Perreault,[1,3] and Patrick Dowd [2,3]

[1] Department of Electrical Engineering
State University of New York at Buffalo
Buffalo, New York 14260
[2] Department of Electrical Engineering
University of Maryland
College Park, Maryland 20742
[3] Department of Defense
Fort Meade, Maryland 20755

p.dowd@ieee.org

Abstract: LIGHTNING is a dynamically reconfigurable, hierarchical, WDM network testbed project designed to interconnect a large number of supercomputers and create a distributed shared memory environment. The hierarchical architecture achieves scalability yet avoid the requirement of multiple wavelength tunable devices per node. Single-hop all-optical communication is achieved: a packet remains in the optical form from source to destination and does not require intermediate routing. The wavelength multiplexed hierarchical structure features wavelength channel re-use at each level, allowing scalability to very large system sizes. It partitions the traffic between different levels of the hierarchy without electronic intervention in a combination of wavelength- and space-division multiplexing. This approach can dynamically vary the bandwidth provided to different levels of the hierarchy. Each node monitors the traffic intensities on its subset of channels and detects any temporal or spatial shift in traffic balance. Bandwidth re-allocation is completely decentralized, any node can initiate it, achieving highly fault tolerant system behavior.

1.1 INTRODUCTION

Computer systems have developed rapidly over the past few years. Specifically, processor speeds continue to increase but I/O capability continues to lag. This is not a new trend yet no general solution to this perpetual problem has emerged. This work describes an approach to support high-performance I/O - both interprocessor communication and remote file system access - with a combination of high speed optical network designed specifically for this situation.

The objective is to develop a scalable technique for clustering: a strategy that is effective (both in performance and price/performance) with both workstation-class processor interconnection and high-performance supercomputer-system level interconnection. The interconnection strategy needs to be flexible to adapt to the severe cost constraints at the low-end and performance requirements at the high-end. A general architecture for such a strategy is defined in this work. The feasibility of this architecture was proven in an experimental testbed known as "LIGHTNING," a joint project involving the State University of New York at Buffalo (Buffalo, NY), University of Maryland (College Park and Baltimore County, MD), Laboratory for Physical Sciences (College Park, MD), Center for Computational Sciences (Bowie, MD), and David Sarnoff Research Center (Princeton, NJ).

The optical network is hierarchical and based on wavelength-division multiple access (WDMA). WDM creates multiple channels, separated in wavelength, to be concurrently transmitted on a single fiber. The channels can be individually accessed, routed and switched. The architecture can be viewed as a tree, where processors reside at the leaves and the internal nodes have wavelength- and spatial-switching capabilities. This class of architecture for processor interconnection was first described in [1]. Section 1.2 briefly describes the structure.

An advantage of this architecture is that bandwidth can be dynamically re-allocated throughout the system to adapt to shifts in traffic patterns. The bandwidth assigned to different levels of the hierarchical system can be dynamically increased or decreased, based on the reference patterns and file activity of the system. However, in comparison to traditional reconfigurable architectures, the *reconfiguration of the system is automatic and hidden from the user.* The user does not have to logically map an application to a specific topology or inform the operating system at compile- or run-time of its intended communication patterns. LIGHTNING senses the traffic patterns inherent to an application and adapts itself accordingly. This function is hidden from the user and the operating system and avoids placing the burden of understanding the specifics of the system on the programmer. The programmer views the system as a pool of processors and is not involved with process placement. The operating system, during initial process placement and during migration, favors placing a process within the level-1 or level-2 cluster depending on size.

The goal of the testbed currently being developed is to support a (weakly coherent) distributed shared memory environment between the interconnected processors. This is accomplished through a combination of the network architecture, operating system, and network/memory interface design. The goal is

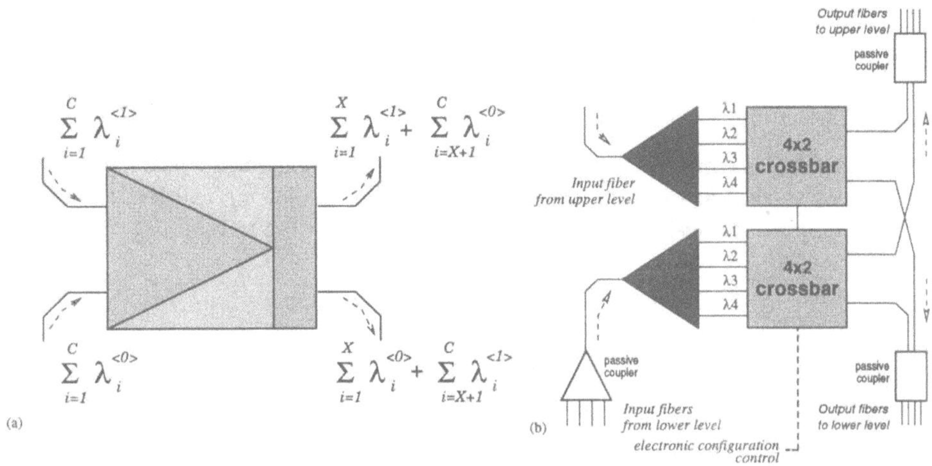

Figure 1.1 Wavelength partitioning device used to achieve spatial separation of wavelengths. (a) Functional behavior of Lambda Partitioner. Wavelengths above and below the partition point are cross-routed and bar-routed, respectively. Selection of the partition point is electronically controllable by the LIM interface. (b) Structure of lambda partitioner.

not just to provide a high-bandwidth network, but to deliver that bandwidth to the application rather than wasting it in operating system overhead.

This chapter briefly describes the basic architecture and media access protocol in Section 1.2. Section 1.3 describes the support for reconfiguration in greater detail and describes the interaction with the media access protocol. Section 1.4 gives some performance results for the reconfiguration algorithms. Finally, some conclusions are given in Section 1.5.

1.2 LIGHTNING ARCHITECTURE

This section describes the architecture of the LIGHTNING network. First background information is discussed providing a framework for discussion, and then the architecture is defined. The media access protocol is then presented.

The LIGHTNING architecture was influenced by some of the strengths and weaknesses of the Fat-Tree [2]. LIGHTNING retains the advantages of the Fat-Tree structure – a low latency, scalable interconnection – while avoiding some of the limitations through optical interconnects. Optical networks have many advantages over metallic interconnections such as a relaxed bandwidth distance product, large fan out, low power consumption, reduced crosstalk and immunity to electromagnetic noise [3]. Major considerations in determining the best placement for optical connections is the speed mismatch between the optical and electrical components [4] and the performance/cost requirements. Optical fiber has a 30 THz bandwidth [5], much larger than the metallic bandwidth.

To simply replace metal connections with optical ones would result in many expensive under-utilized links.

1.2.1 Lightning Structure

LIGHTNING has a hierarchical structure in the form of a general m_i-ary tree. The leaves of the tree are processors and the interior nodes are routing elements. The system uses time-, space-, and wavelength-separation to increase the usable bandwidth in the system.

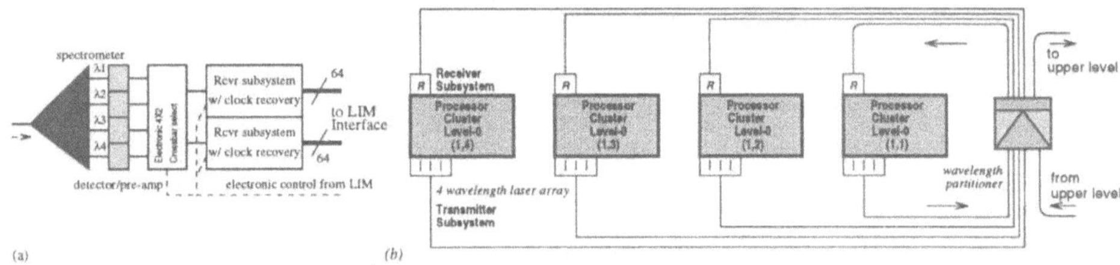

Figure 1.2 Lightning building blocks. (a) Receiver subsystem where each node can concurrently receive on any two of the wavelengths in this example; (b)A single level example showing that each node contains a receiver described in (a), a 4-wavelength laser array and are connected via the wavelength partitioning component. Note that the configuration is illustrated in a star configuration but a bus-type of configuration is possible depending on power budget considerations.

Internal nodes provide wavelength selective routing. The spatial and wavelength optical routing element is denoted as a Lambda Partitioner (Λ_x). The structure of a Lambda Partitioner is shown in Figure 1.1. The purpose of the Λ_x is to provide a wavelength routing capability where specific wavelengths could be routed to the upper level and other wavelengths could be retained in a lower level. This enables wavelength re-use at peer levels of the hierarchy and reduces the total number of required WDM channels.

A mixed radix system [6] is used to represent the node numbering. Let M (the total number of processors) be a decimal integer represented as a product of r factors:

$$M = \prod_{i=1}^{r} m_i. \tag{1.1}$$

The processor identifier P, $1 \leq P \leq M$, can be represented as an r-tuple

$$P = (p_r, p_{r-1}, \ldots, p_1), \tag{1.2}$$

where $1 \leq p_i \leq m_i$ for all $1 \leq i \leq r$. The term r denotes the number of hierarchical levels.

An example of a single level network is shown in Figure 1.2, and a 3-level example is shown in Figure 1.3. A bus based system is drawn for clarity in Fig. 1.3 but the actual implementation is a star-coupled based system.

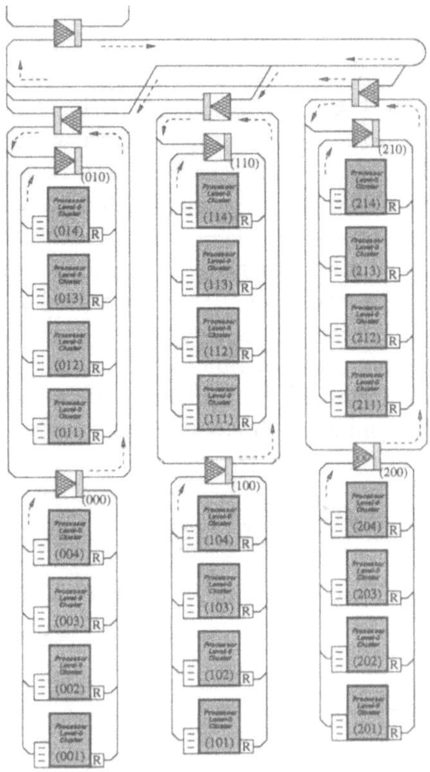

Figure 1.3 Example of a 3-level Lightning system. A bus based system is shown for clarity but implementation is closer to a star-couple configuration.

The function of the Λ_x is to serve as a 2×2 switch for each wavelength channel as illustrated in Fig. 1.1. Denote the number of wavelength channels as C and the ordered set of channels as $\mathbf{C} = \{\lambda_1, \lambda_2, \ldots, \lambda_C\}$. All the wavelengths below the i-level partition point are retained within the i-level cluster while all wavelengths above the partition point are routed to the $i+1$ level cluster. If the i-level partition point was set at X_i, then wavelengths $\{\lambda_1, \lambda_2, \ldots, \lambda_{X_i}\}$ are be retained within level i, and $\{\lambda_{X_i+1}, \lambda_{X_i+2}, \ldots, \lambda_C\}$ are routed to level $i+1$.

The set of partition points in an r-level system are denoted as $\{X_0, X_1, \ldots, X_r\}$. With the partition points staggered at each level such that $X_0 < X_1 < \cdots < X_{r-1} < X_r$, a separate set of wavelength channels are dedicated for communication at each level. Denote C_i as the set of channels allocated for i-level communication, $\mid C_i \mid = X_i - X_{i-1}$ where $X_0 = 0$ and $X_r = C$. Denote the collection of processors beneath an i-level node as an i-level cluster. If a processor wants to send a message to another processor in the same level 0 cluster, it uses a channel in C_0. If the second processor is not in the same level 0 cluster but is in the same level 1 cluster, a channel in C_1 is used.

The determination of which channel to transmit on is discussed in Section 1.2.2. Note that spatial re-use of the wavelength channels in different clusters occurs. A symmetric network would have all the i-level wavelength partitioners in the network set to the same partition point. Alternatively, Λ_xs in spatially separate clusters could partition the wavelengths differently, as longs as $X_i < X_{i+1}$. In either case, the partition points can be dynamically changed. This allows the network to adapt to changes in traffic patterns.

The total number of spatial/wavelength channels in the system, denoted as $\mathcal{C}$, can be calculated to determine the advantage of wavelength re-use. For example, consider a symmetric configuration where $M_j = \prod_{i=j+1}^{r} m_i$ denotes the number of j-level nodes in a hierarchy, and $\mathcal{C}_i$ as the number of channels allocated for i-level communication. Since there are a total of M_j j-level nodes, there is $\mathcal{C}_j M_j$ channels providing j-level communication.

Therefore the total number of channels in an r level system is

$$\mathcal{C} = \sum_{j=1}^{r} \mathcal{C}_j \prod_{k=j+1}^{r} m_k. \tag{1.3}$$

Consider a small system of $M = 4 \times 4$ with $C = 4$. In this example, the total number of effective channels increases to a maximum of $\mathcal{C} = 13$ through wavelength re-use.

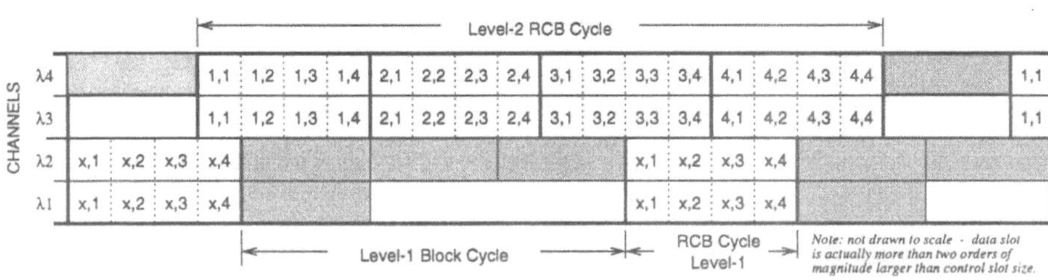

Figure 1.4 Assignment map for a two level media access protocol for $M = 4 \times 4$ and $C = (2, 2)$.

1.2.2 Media Access Protocol

The wavelength multiplexed environment requires a media access protocol to arbitrate access. With existing upper layer protocols in mind, the media access protocol has been designed to *exploit the bimodal pattern of network traffic.* There have been many studies of *typical* Internet traffic, with TCP/IP, and a common trait is the bimodal distribution of packet size: essentially all of the traffic consists of either big packets or small packets with little traffic in between [7, 8]. Batched applications, such as ftp, tend to use large data packets to get high throughput while interactive applications, like telnet, tend to use small packets for low latency.

This characteristic is also true with distributed shared memory computer systems. The traffic, generated by a memory coherence protocol needed to achieve DSM, has two major forms: memory *consistency control packets* (such as memory block requests, invalidations, acknowledgments) and *memory block packets* [9]. Memory consistency control packets are very small while the memory blocks can be up to 8 Kbytes. In a non-broadcast scheme there can be multiple memory consistency control packets for every memory block that needs to be transferred.

The media access protocol has been designed to exploit this characteristic. The media access protocol supports two different packet sizes. Small packets called *control packets* are used to reserve access to the network. These packets are only 64 bytes long, but they are also capable of carrying a small payload of 32 bytes. This can be used for transferring small packets from applications as well as the memory consistency control packets necessary to support DSM. Control packets reserve access for sending large packets referred to as *data packets*. These are 8 Kbytes long and can carry large packets from applications as well as memory block packets.

The protocol combines reservation access and fast tunable receivers. Typical reservation style WDMA protocols use one of the channels as a dedicated control channel [10]. However, the number of available WDM channels usually is small and there are usually many more nodes than number of channels. In such a situation, reserving one channel as the control channel is not very attractive. The media access protocol for LIGHTNING is a hybrid approach using reservation access but without a reserved control channel. Furthermore, it was designed specifically for the situation where each node had a fast tunable transmitter (based on a laser-array) and a fast tunable receiver.

The protocol reserves access on a channel through control packets. Transmission on each channel consists of two phases: the reservation control block (RCB) which is followed by the memory block transmission phase. Figure 1.4 shows the allocation map for a two level media access protocol where each level has two channels. The control packets are broadcast (on the subset of channels assigned to the level of the target node) while the data blocks can be sent on any channel available to that level, regardless of destination. A data packet causes the data cycle to be extended by a slot only when all channels in the previous slot are full or if there is a destination conflict. The only channel conflict that can occur during a data slot is when two or more nodes target the same destination. High utilization is maintained by having the second node transmit in the next data slot. Note that this enables a collisionless environment but significantly reduces the overall cycle time, when compared with TDMA, since data block slots are only assigned when they are needed. This significantly reduces the latency since Figure 1.4 is not drawn to scale and the data blocks are typically more that two orders of magnitude larger than a control slot.

The laser-array transmitter facilitates broadcast by simultaneously transmitting a control packet on all channels assigned to the target level of a system. This enables LIGHTNING to obtain the advantages of a reservation-based protocol without requiring a WDM channel to be a dedicated control channel. This is especially important in a multi-level architecture since one control channel would have to be assigned per level. A basic assumption and constraint on LIGHTNING is not to require a large number of WDM channels so LIGHTNING has no dedicated control channels and only uses the channels in that mode for a small fraction of the cycle duration.

1.3 DYNAMIC BANDWIDTH REALLOCATION

One of the unique features of the LIGHTNING architecture is the ability to change the number of channels assigned to each level of the hierarchy. This provides a utility for increasing the bandwidth to levels with higher traffic loads thereby exploiting temporal locality. Increasing the bandwidth would lead to a reduction in overall system latency.

This is accomplished by monitoring the traffic patterns of the system, in a decentralized fashion, with dynamic reallocation of system bandwidth when an imbalance occurs. Monitoring occurs at the media access level - *operating system intervention is not required.* The monitoring routines measure the traffic intensity of each level and detect imbalances. When an imbalance is detected, the routines decide on the best course of action and initiate the system resource reallocation. Close interaction between the reconfiguration routines and the media access protocol is required to ensure proper operation of the system and to place no additional burden on the system.

The system architecture is based on Λ_x partitioners. In the current implementation, the role of the Λ_x processor falls on one of the leaf node processors within its sub-cluster. The LIGHTNING Interface Module (LIM) that each node possesses performs the protocol, synchronization, and reconfiguration functions. Furthermore, each LIM can also interface into a Λ_x and can be configured to perform the role of the individual processor and also the processing responsibilities of the Λ_x. Such a configuration is referred to as a Λ_x node.

As defined in Section 1.2, an i–level Λ_x is configured to a *crossover channel* denoted as X_i, $\lambda_{X_i} \in \mathbf{C}$, where channels $\{\lambda_{X_i+1}, \ldots, \lambda_C\}$ are passed to level $i+1$ and channels $\{\lambda_1, \ldots, \lambda_{X_i}\}$ are contained locally for local intra–cluster communication.

Symmetric Configuration: Connection patterns of channels to nodes at the cluster are governed by the allocation strategy of the access protocol. Essentially, this mapping is used to retain the local i–level traffic $\{\lambda_1, \ldots, \lambda_{X_i}\}$, and forward the global traffic $\{\lambda_{X_i+1}, \ldots, \lambda_C\}$ to the $(i+1)$–level Λ_x switch. This limits the partition point of the $(i-1)$–level Λ_x to be within $\{\lambda_1, \ldots, \lambda_{X_i-1}\}$. All of the $(i-1)$–level Λ_x partitioners can then be configured to the same crossover channel X_{i-1}. This gives rise to a **symmetric configuration**, one example of which is shown in Figure 1.5(a).

Asymmetric Configuration: However, it is not necessary for each $(i-1)$–level Λ_x to have the same partition point. Each $(i-1)$–level Λ_x can be individually set to its own partition point, independent of the others on its

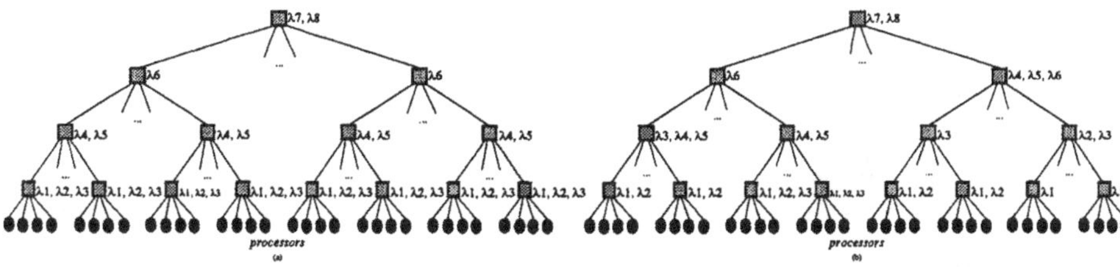

Figure 1.5 Example of some possible configurations of Lightning. (a) A *symmetric* configuration. (b) An *asymmetric* configuration.

level, assuming it stays within the limits set by the i–level Λ_x. This gives rise to a great deal of flexibility in reallocating bandwidth. If an imbalance is localized to a section of the system that section can be reconfigured without impacting the rest of the system. This results in an **asymmetric configuration**, an example of which is shown in Figure 1.5(b).

For the remainder of this discussion only the asymmetric configuration will be considered. The symmetric configuration can be treated as a special case of the asymmetric configuration. All of the algorithms that will be developed for the asymmetric case are also applicable to the symmetric case. The only necessary modification is that all peer Λ_x nodes must have the same partition point. This maintains the symmetric nature of the system while still allowing it some freedom to reconfigure to take advantage of spatial and temporal locality.

1.3.1 Detecting An Imbalance

An integral part of reconfiguring the network is deciding when reconfiguration is necessary. Since reconfiguration will cause some disruption of normal network operation, it is desirable to avoid unnecessary reconfigurations. The decision to reconfigure should be based on the traffic patterns in the network. Since the network is hierarchical it is not possible for a node to monitor traffic in the entire network. This makes it necessary for nodes to distribute traffic information to different parts of the network.

Although each node records the traffic intensity for every channel in the system it can receive, this information does not give a complete picture of the state of the system. Due to the spatial reuse of wavelengths a node is only capable of tracking the load within its clusters. While some load balancing algorithms probe the system to determine the loads in other parts of the system [11], this would greatly increase the time required to reconfigure the system. This would make reactions to temporal locality slow, possibly even causing the system to reconfigure too late to take advantage of the locality. Another possibility would be to have the nodes periodically distribute their load information [12]. While this requires more state information and increases the processing burden at the interface card, it allows for faster reaction times since the current state of the network is known.

1.3.1.1 Symmetric Channel Allocation. It is possible to determine the optimal channel allocation (OCA) for a system given the total traffic intensity, expressed in number of data packets per data cycle, of each level of the system. The OCA achieves a uniform saturation load at each level of the system. This allocation was derived in [1] and is given by

$$C_i = \frac{I_i}{I_T} C, \tag{1.4}$$

where I_i is the traffic intensity for level i, and I_T is expressed as

$$I_T = \sum_{j=1}^{r} I_j. \tag{1.5}$$

This expression was derived for a symmetric system where the probability of a node transmitting on the ith level was the same for all nodes.

This allocation scheme would require that each node have complete information of the traffic intensities for the entire system, which would require a great deal of communication as well as a great deal of state information kept at each node. This would significantly increase the processing burden of the interface card. It also requires that the system maintain a symmetric configuration, which may in fact not be optimal under imbalanced traffic loads.

1.3.1.2 Asymmetric Channel Allocation. It is possible to derive an equivalent expression for the optimal number of channels for level i that considers only the traffic patterns below level i. The partition point X_i is the number of channels allocated for communication within the level i cluster. This can be expressed as

$$X_i = \sum_{j=1}^{i} C_j = \sum_{j=1}^{i} \frac{I_j}{I_T} C. \tag{1.6}$$

Solving for C in Equation 1.6 gives

$$C = \frac{X_i \times I_T}{\sum_{j=1}^{i} I_j}. \tag{1.7}$$

Substituting Equation 1.7 into Equation 1.4 gives

$$C_i = \frac{I_i}{I_T} \times \frac{X_i \times I_T}{\sum_{j=1}^{i} I_j} = \frac{I_i}{\sum_{j=1}^{i} I_j} X_i. \tag{1.8}$$

Since Equation 1.8 is targeted to an asymmetric channel allocation it is necessary to redefine I_i. Previously it was assumed that the traffic intensity would be uniform throughout a level. This is not a valid assumption with an asymmetric channel allocation since the underlying cause of an asymmetric channel allocation would be an asymmetric traffic load. The goal of reconfiguring the network is to minimize the system latency. Since latency is inversely proportional to the bandwidth available the only way to minimize the latency is to maximize the bandwidth. Therefore, in an asymmetric channel allocation the traffic intensity I_i is the maximum traffic intensity on level i.

Equation 1.8 makes it possible to determine the optimal number of channels without keeping state information for the whole network. It also allows different channel allocations for peer level clusters, which allows asymmetric configurations. However, some communication and state information is still necessary. The amount of communication and state information can be kept to a minimum by only allowing Λ_x nodes to control reconfiguration. This does not significantly affect fault-tolerance because it is not possible to reconfigure the system if a Λ_x node is not functioning even if another node is controlling the reconfiguration.

The traffic intensities will be tracked by the Λ_x nodes. The level 1 Λ_x node will send the maximum intensity for its level 1 cluster to its level 2 Λ_x node. The level 2 Λ_x node will determine the maximum value for level 1 as well as tracking the intensity on level 2. It will then send both values to the level 3 Λ_x node, where the process will be repeated. The amount of state information kept at each node will be proportional to $(i-1)m_i$. The communication can be done periodically using control packets. This will eliminate the need for any explicit probing of the system when attempting to detect an imbalance. Each Λ_x node is then responsible for reconfiguring the channel allocation for its level.

Once the OCA for a particular level is computed it is compared against the actual number of channels allocated to the level. A level is considered overloaded if it has fewer channels than its OCA. Similarly, a level is considered underloaded if it has more channels than its OCA.

1.3.2 Reconfiguration Algorithm

Once a node determines that a level is over- or underloaded it initiates the reconfiguration algorithm. This is a distributed algorithm which will be used at each node within the cluster affected by the reconfiguration. At the node which initiates reconfiguration, called the *controlling node*, the reconfiguration algorithm determines the appropriate response to an imbalance and triggers the actual mechanism of reconfiguration. This involves informing the other nodes in the cluster of its decision. At the remaining nodes the algorithm determines what must be done at that node to carry out the reconfiguration. This involves setting the correct partition point at all Λ_x nodes within the system. This distributed approach takes some of the processing burden off of the controlling node.

1.3.2.1 Overloaded Level. The only way to alleviate an overloaded level is to allocate another channel to that level. It is also necessary that the algorithm determine where the additional channel will be taken from.

As long as there is a level somewhere below the overloaded level with more than one channel it is possible for the algorithm to reallocate the channel. It is not necessary for the controlling node to know which level will give up the channel. In fact, it is not even possible for the controlling node to know the partition point of every Λ_x node in all of the clusters below it. The new configuration is determined in a distributed fashion. The controlling node simply indicates that it intends to shift the partition point for one level. The partition point that is changed is X_{i-1}. The other Λ_x nodes then determine whether this will require a shift in their partition points. This is possible at the lower nodes because they do know the partition points of every Λ_x node above them in the hierarchy. For example, consider a three level system with $m_i = 4, \forall i$, $C = 4$ and the allocation shown in Figure 1.6(a). Assume that the top level is determined to be in need of the extra channel in the system. There are two possible configurations of the lower levels. The extra channel is either on the second level or the bottom level, as indicated in Figure 1.6(b). The Λ_x node for the top level shifts its partition point by one channel, setting $X_2 = 2$. In Cluster A, this is all that is necessary. However, in Cluster B this would leave the second level without any channels. Therefore the Λ_x nodes on the bottom level must realize this and shift their partition points by one channel as well, setting $X_1 = 1$. The final configuration is shown in Figure 1.6(c).

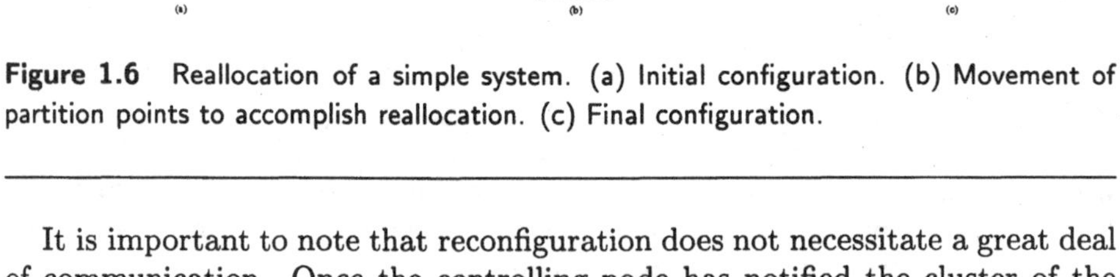

Figure 1.6 Reallocation of a simple system. (a) Initial configuration. (b) Movement of partition points to accomplish reallocation. (c) Final configuration.

It is important to note that reconfiguration does not necessitate a great deal of communication. Once the controlling node has notified the cluster of the new configuration, all further changes can be made by the nodes with no other communication. All the information needed is already present at the node because it is needed for normal operation of the media access protocol. The only other necessary communication is to acknowledge the new configuration and synchronize the system, which will be explained in Section 1.3.3.

1.3.2.2 Underloaded Level. Once a controlling node has decided to reallocate a channel from an underloaded level the algorithm is very similar to that for an overloaded level. The controlling node indicates that it is shifting the partition point for the underloaded level. The Λ_x nodes for the level below this then determine whether to keep the channel at this level or push it down again. It is only kept at a level if the level is considered overloaded. The Λ_x nodes on lower levels are able to determine whether the Λ_x nodes above them will keep the channel because they have the same traffic indices for that level. This allows all Λ_x nodes to determine the new configuration in a distributed fashion with no further communication other than the acknowledgment and synchronization, which will be discussed in Section 1.3.3.

1.3.3 Reconfiguration Mechanism

Implementing a new configuration does require a certain amount of communication among the nodes. This includes acknowledging the new configuration and synchronizing the change from the old configuration to the new. Synchronization is necessary to keep nodes in different clusters from inadvertently transmitting on the same channel before reconfiguration is complete and causing collisions. It is also necessary to prevent more than one node from attempting to reconfigure the same portion of the system simultaneously. This could result in an invalid configuration, such as a level using the same channel as its parent, or a level not having any channels at all. Such a situation would be very difficult to correct without re-synchronizing the entire system.

A controlling node initiates a reconfiguration by continuously broadcasting a configuration packet in its control slot on the affected level, as illustrated in Figure 1.7(a). This packet contains the reallocation the controlling node has determined appropriate. It also contains a time stamp to allow for ordering of reconfiguration requests. In the case of two requests arriving in the same control cycle or with time stamps that differ by less than a control cycle length, precedence is given to the node with the earlier control slot. If nodes on two different levels attempt to reconfigure at the same time precedence is given to the higher level. If an upper level attempts to reconfigure while one of its sub-clusters is reconfiguring, the nodes in the sub-cluster will ignore the reconfiguration packet and the upper level controlling node will timeout and abort.

The configuration packet will identify the level and what action is to be taken, either allocate a new channel to this level or deallocate one channel from this level. This triggers the distributed reconfiguration algorithm in all nodes from the indicated level down to the bottom of the system. Once a node has applied the reconfiguration algorithm and arrived at a new configuration it will send an acknowledgment packet containing this new configuration to its Level 1 Λ_x node as shown in Figure 1.7(b). The Λ_x node will compare the configurations to make sure that all nodes arrive at the same configuration. Conflicts are resolved by using a simple voting scheme, counting the Λ_x node twice if there is an even number of nodes. This is necessary for fault-tolerance, to ensure that all nodes will have the same configuration once reconfiguration is completed.

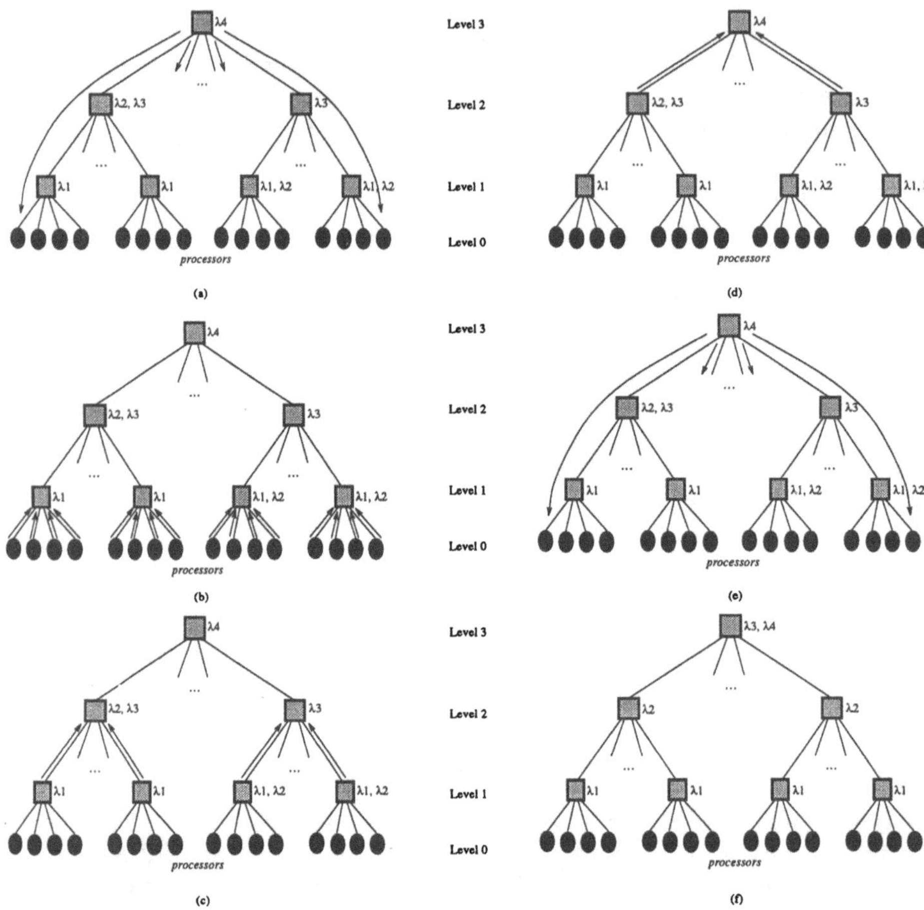

Figure 1.7 The communication required for reconfiguration in a Level 3 cluster. (a) The initial configuration packet is broadcast to all nodes in the Level 3 cluster. (b) All nodes send acknowledgments to their Level 1 Λ_x node. (c) Level 1 Λ_x nodes send acknowledgments to their Level 2 Λ_x nodes. (d) Level 2 Λ_x nodes send acknowledgments to the Level 3 Λ_x node. (e) The synchronizing packet is broadcast to all nodes in the Level 3 cluster. (f) The system restarts with the new configuration.

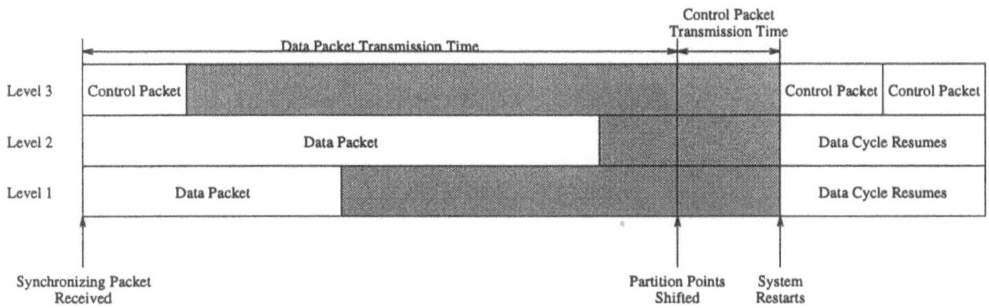

Figure 1.8 Restarting after reception of the synchronizing packet. (NOTE: Data packets are not drawn to scale.)

The Level 1 Λ_x node then sends an acknowledgment packet to its Level 2 Λ_x node and the process is repeated as shown in Figure 1.7(c)-(d). Nodes continue to broadcast their acknowledgments in their control slots until they receive the acknowledgment of the Level 1 Λ_x node to the Level 2 Λ_x node. This assures them that their acknowledgment was received. This is also repeated for each level. Acknowledgments continue up the hierarchy until they arrive at the controlling node. The controlling node then stops broadcasting the configuration packet and begins broadcasting a synchronizing packet. It continues to broadcast this synchronizing packet in its control slot until the system has accomplished the reconfiguration as shown in Figure 1.7(e)-(f).

The synchronizing packet causes all nodes to stop transmitting on the cluster being reconfigured. The remaining portion of the system continues as normal. If a node is transmitting when it receives the synchronization packet it completes transmission of the packet. All nodes will have stopped transmitting one data packet transmission time after receiving the synchronization packet. All Λ_x nodes then adjust their partition points to the new configuration. The nodes pause for one additional control packet transmission time to allow the Λ_x nodes time to change their configuration. (This also relaxes the switching time constraints on the optical devices that comprise the Λ_x) The nodes within that cluster then resume transmitting. Any reservations made for a channel which is no longer allocated to a level will be rescheduled for the end of the data cycle. The nodes resume operations at the same point transmissions were interrupted. This is illustrated in Figure 1.8.

1.4 PERFORMANCE ANALYSIS

This section illustrates the performance of the reconfiguration strategy through simulation analysis. The simulation models the media access protocol with the addition of the reconfiguration mechanism. The simulation is process oriented and uses the YACSIM [13] package to take care of the process switching and

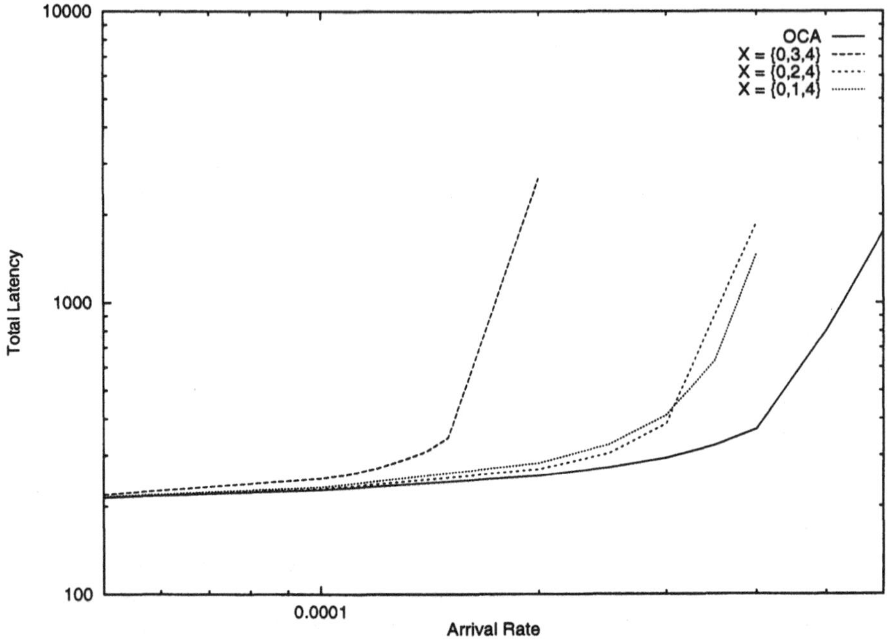

Figure 1.9 System latency for a symmetric traffic load and a symmetric channel allocation in a two level system with $N = 64$ and $C = 4$.

---

statistics gathering. Also, the statistics collection mechanisms in YACSIM collected the data used in the reconfiguration mechanism.

This simulation essentially ignores control packets, except those used for data cycle reservation and reconfiguration messages. The primary emphasis is on data packets and how reconfiguration affects their transmission. It also does not consider propagation delay and device latencies. The only delay through the network is transmission time. Despite these simplifications, the model captures enough of the details of the media access protocol and underlying network needed to see how a reconfiguration strategy affects it.

The traffic load placed on this simulation is a series of streams of Poisson distributed traffic. Each node in the network generates its own stream which enters a queue before transmission. Each packet a node generates in the stream randomly targets a level of the system and then uniformly targets a node on that level. A node can not target itself or any node within the same cluster at any level lower than the one targeted. The division of traffic between the levels is randomly and uniformly generated and changed at random intervals with an exponential distribution. This provides an ever-changing traffic load to test the system over a wide range of traffic patterns.

A symmetric traffic load occurs when the traffic load placed on the system is constrained such that all of the nodes on level i share the same traffic intensity.

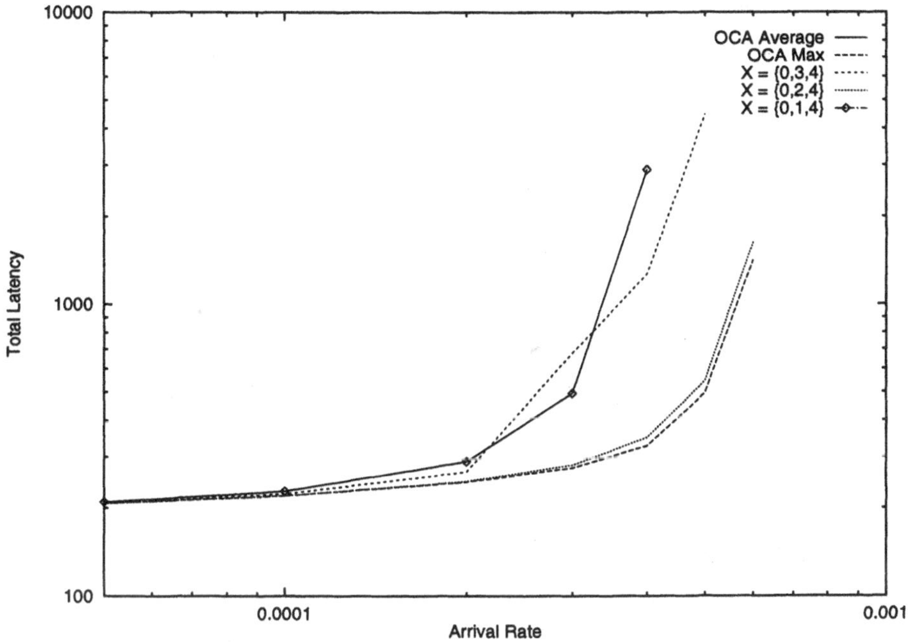

Figure 1.10 System latency for an asymmetric traffic load and an asymmetric channel allocation in a two level system with $N = 64$ and $C = 4$.

This traffic load can easily be accommodated by a symmetric channel allocation. The system latency for a two level system using a symmetric configuration is shown in Figure 1.9. The figure clearly shows that the reconfigurable system using the OCA equation to determine the channel allocation performs much better than any of the possible static configurations. The system is able to accommodate a much larger load using the reconfiguration algorithm.

While a two level system is only capable of generating a symmetric channel allocation it is still possible to have an asymmetric traffic load. This occurs when different clusters on level i have different traffic intensities. Such a situation would be very common in a system running multiple jobs having different communication patterns. Figure 1.10 shows the system latency for an asymmetric traffic load. This figure shows that using the original OCA equations and averaging the various traffic intensities the reconfigurable system performs only as well as a static system with a channel allocation of $X = 0, 1, 4$. However, by using the maximum traffic intensity from each level the reconfigurable system performs better than any static configuration.

Since a two level system can only have a symmetric channel allocation it does not benefit from the new OCA given in Equation 1.8. Figure 1.11 shows the system latency in a three level system when an asymmetric traffic load is applied to the system. The figure shows that, even using the new OCA equations,

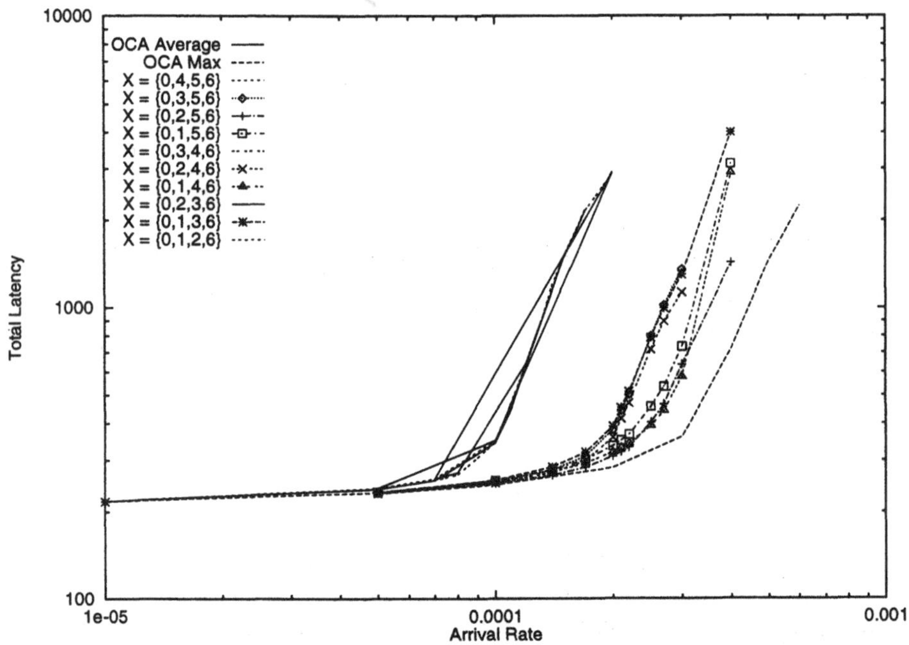

Figure 1.11 System latency for an asymmetric traffic load and an asymmetric channel allocation in a three level system with $N = 256$ and $C = 6$.

averaging the traffic load causes the reconfigurable system to perform worse than many of the static systems. However, by using the new OCA equation and the maximum traffic intensity for each level the reconfigurable system performs much better than any possible static configuration. This allows the system to carry a larger load before saturating.

1.5 CONCLUSIONS

This chapter described a testbed for high-performance clustering based on a WDM hierarchical network. It creates a shared memory environment for the supercomputers interconnected with the high-speed network, creating an extremely powerful computing platform. The system is dynamically adaptable where the traffic characteristics are sensed and the system is dynamically reconfigured to improve performance. The users of the system do not have to explicitly configure the system to suit their applications - the system does this automatically. However, we have built in support to allow users to indicate to the operating system a shift in communication patterns. This allows the operating system to get ready for the reconfiguration thereby overlapping the latencies. This facility is useful to certain users/programmers who have extremely high-performance requirements and who do not mind incorporating system specific code in their applications. The architecture of LIGHTNING has been generalized to include both metal and optical interconnects so that it can support both work-station and and supercomputer interconnection with the appropriate price/performance characteristics.

A three node demonstration LIGHTNING system has been constructed to verify the feasibility of the media access protocol and the optical components. The network interface consisted of several FPGAs to implement the media access protocol and a commercially available chipset to implement the physical interface to the optics. A second system is currently under construction that will consist of eight nodes in a two level hierarchy. Again FPGAs are being used to implement the media access protocol while commercially available chipsets are being used to provide the physical interface to the optical components.

Acknowledgment

This work was supported by the U.S. Department of Defense, Laboratory for Physical Sciences, College Park, Maryland. Preprints are available via WWW at http://tebbit.eng.umd.edu/.

References

[1] P. W. Dowd, K. Bogineni, K. A. Aly, and J. Perreault, "Hierarchical scalable photonic architectures for high-performance processor interconnection," *IEEE Transactions on Computers*, vol. 42, pp. 1105–1120, Sept. 1993.

[2] C. E. Leiserson, "Fat-trees: Universal networks for hardware-efficient supercomputing," *IEEE Transactions on Computers*, vol. c-34, pp. 892–901, Oct. 1985.

[3] L. Hutcheson, "High speed optical interconnects for computing applications," in *Proc. SPIE*, vol. 862, pp. 2–10, 1987.

[4] P. W. Dowd, "Optical bus and star-coupled parallel interconnection," in *Proc. 4th International Parallel Processing Symposium*, (Los Angeles, CA), pp. 824–838, Apr. 1990.

[5] C. A. Brackett, "Dense wavelength division multiplexing networks: Principles and applications," *IEEE Journal on Selected Areas of Communications*, vol. 8, pp. 948–964, Aug. 1990.

[6] L. Bhuyan and D. P. Agrawal, "Generalized hypercube and hyperbus structures for a computer network," *IEEE Transactions on Computers*, vol. c-33, pp. 323–333, Apr. 1984.

[7] J. Mogul, "Observing TCP dynamics in real networks," in *Proceedings of ACM SIGCOMM*, pp. 305–317, Oct. 1992.

[8] K. Claffy, G. Polyzos, and H. Braun, "Traffic characteristics of the T1 NSFNET backbone," in *IEEE INFOCOM'93*, pp. 885–892, Mar. 1993.

[9] P. W. Dowd and J. Chu, "Photonic architectures for distributed shared memory multiprocessor systems," in *IPPS'94 Workshop on Massively Parallel Processing Through Optical Interconnects*, (Cancun, Mexico), IEEE, 1994.

[10] I. M. I. Habbab, M. Kavehrad, and C. E. W. Sundberg, "Protocols for very high-speed optical fiber local area networks using a passive star topology," *IEEE Journal on Lightwave Technology*, vol. LT-5, pp. 1782–1793, Dec. 1987.

[11] R. Mirchandaney, D. Towsley, and J. Stankovic, "Adaptive load sharing in heterogeneous distributed systems," *Journal of Parallel and Distributed Computing*, vol. 9, pp. 331–346, 1990.

[12] A. P. Reeves and M. H. Willebeek-LeMair, "Strategies for dynamic load balancing on highly parallel computers," *IEEE Transactions on Parallel and Distributed Systems*, vol. 4, pp. 979–993, Sept. 1993.

[13] J. R. Jump, *YACSIM Reference Manual.* Rice University, Department of Electrical and Computer Engineering, 1.2 ed., August 1992.

[14] M. Lin, R. Tsang, D. Du, A. Klietz, and S. Saroff, "Performance evaluation of the CM-5 interconnection network," Tech. Rep. AHPCRC Preprint 92-111, University of Minnesota, Army High Performance Computing Research Center, Oct. 1992.

[15] T. Bujewski, J. Bannister, and M. Gerla, "ISONet: A dynamically reconfigurable integrated services optical network," in *7th IEEE LAN/MAN Workshop*, pp. 338–341, Mar. 1995.

[16] J. Feehrer, J. R. Sauer, and L. Ramfelt, "Design and implementation of a prototype optical deflection network," in *Proc. ACM SIGCOMM'94*, Aug. 1994.

[17] S. Scott, J. Goodman, and M.K.Vernon, "Performance of SCI Ring," in *Proc.* 19th *International Symposium Computer Architecture*, pp. 403–414, May 1992.

[18] P. Dowd and K. M. Sivalingam, "A multi-level WDM access protocol for an optically interconnected parallel computer," in *Proc. IEEE INFOCOM'94*, (Toronto, Ontario Canada), pp. 400–408, June 1994.

[19] R. G. Minnich and D. V. Pryor, "Mether: Supporting the shared memory model on computing clusters," in 38th *Annual IEEE Computer Society International Computer Conference*, pp. 558–567, February 1993.

[20] R. Minnich, "Mether-NFS: A Modified NFS Which Supports Virtual Shared Memory," in *Proc SEDMS'93*, 1993.

[21] L. D. Wittie, G. Hermannsson, and A. Li, "Eager sharing for efficient massive parallelism," in *International Conference on Parallel Processing*, (St, Charles, IL), pp. 251–255, August 1992.

[22] F. Hady, R. Minnich, and D. Burns, "The memory integrated network interface," in *ACM Hot Interconnects Symposium*, (San Jose, CA), 1994.

[23] W. G. Schmidt, "The Application of TDMA to the INTELSAT IV Satellite Series," *COMSAT Tech. Rev.*, vol. 3, pp. 257–276, Fall 1973.

[24] C. R. Carter, "Survey of Synchronization Techniques for a TDMA Satellite Switched System," *IEEE Transactions on Communications*, vol. COM-28, pp. 1291–1301, Aug. 1980.

[25] K. Aly and P. Dowd, "A class of scalable optical interconnection networks through discrete broadcast-select multi-domain wdm," in *Proc. IEEE INFOCOM'94*, (Toronto, Ontario Canada), pp. 392–399, June 1994.

[26] Y. Oie, T. Suda, M. Murata, D. Kolson, and H. Miyahara, "Survey of switching techniques in high-speed networks and their performance," *IEEE INFOCOM '90*, June 1990.

2 PARALLEL COMPUTING WITH "INTELLIGENT OPTICAL NETWORKS"

Ted H. Szymanski

Department of Electrical Engineering
McGill University
Montreal, Quebec H3A 2A7
Canada

teds@macs.ece.mcgill.ca

Abstract: Parallel computing on a class of networks called *"Intelligent Optical Networks"* is examined. Traditional All-Optical and Passive-Optical networks have very limited abilities to process data in the all-optical layer. In contrast, we propose the concept of an "Intelligent Optical Network", which has the ability to process and buffer vast amounts of data within the optoelectronic layer. Intelligent Optical Networks are can be realized with CMOS substrates with optical I/O which have been flip-chip bonded over the surface. This technology has the potential for thousands of optical I/O per substrate, and supports rapid and transparent conversion of parallel bits between the electrical and optical domains. The design of an Intelligent Optical Network for a Network-of-Workstations (NOW) is described. Each workstation has a parallel fiber ribbon optical datalink to a centralized switching core, implemented on a single optoelectronic CMOS integrated circuit. A centralized integrated circuit interconnecting 32 fiber ribbons, each with 32 fibers clocked at 1 GHz, would support an aggregate bandwidth of 1 Tb/s. The switching is based on a conventional *"Broadcast-and-Select"* architecture. We consider the impact of an Intelligent Optical Network on the parallel computing of Butterfly-based algorithms. It is shown that a multi-dimensional NOW with DSP coprocessors can sustain performances in the 100 TeraFLOP per second range for Butterfly-based scientific algorithms. This class of interconnects thus represents one viable approach to achieve the US Accelerated Strategic Computing Initiative computing targets of 100 TeraFLOP machines by the year 2004.

2.1 INTRODUCTION

The performance of microprocessors and communication networks has been growing *exponentially* over the last decade, and these trends are expected to continue well into the future [7, 20]. The U.S. *Accelerated Strategic Computing Initiative* (ASCI) has undertaken a program to further accelerate the growth in computing performance, aiming to triple computing power every 18 months. The program aims to deliver 10 TeraFLOP machines in the year 2001 and 100 TeraFLOP machines in the year 2004 [3, 7]. IBM estimates that a 100 TeraFLOP machine will require between 128 - 256 cabinets of electronics occupying 20×40 meters of floor space [3]. Such a machine will require an interconnection network with approximately 300 Terabits of bandwidth per second, which will challenge existing approaches to networking, and which will almost certainly require optical technologies .

Several classes of high capacity optical networks have been proposed. For example, *"All-Optical"* networks using very high speed Time Division Multiplexing (TDM) over fiber have been proposed in [12, 21]. *"Passive Optical Networks"* using Wavelength Division Multiplexing (WDM) over fiber have been proposed in [4, 5]. These approaches have very high bandwidths, but they have a very limited ability to process data in the optical domain. Passive-Optical networks do not perform any processing on the optical data; they are completely passive as their name implies. All-Optical networks can perform very limited processing of data, usually simple address bit recognition along with optical Add/Drop multiplexing functions, which can add an optical packet to a stream, or extract an optical packet from a stream. These networks are not well suited to perform processing or buffering on vast amounts of optical data, since it is very difficult to process data in the all-optical domain with these technologies.

A class of *"Intelligent Optical Backplanes"* was proposed in [24, 25]. This class of interconnects exploits new optoelectronic technology, which merges traditional CMOS substrates with optical I/O, and exploits the processing power made available by the CMOS logic [16, 24]. This technology allows for the rapid processing of vast amounts of data, a feature which is not possible with All-Optical and Passive-Optical-Networks. An Intelligent Optical Backplane is under development in Canada, as part of a multi- institutional national research program, funded in part by the Canadian Institute for Telecommunications Research (CITR). The program is expected to demonstrate a shared memory multi- processor executing over a free-space optical backplane with a fraction of a Terabit per second bandwidth before the year 2000.

In this chapter, we extend the concept of "Intelligent Optical Backplanes" to a more general class called "Intelligent Optical Networks". This class consists of collections of optoelectronic crossbar-like switches interconnected with parallel fiber ribbons, as proposed in [26, 27]. This class of optical networks has the ability to process and buffer vast amounts of data within the optoelectronic layer, which is in contrast to the philosophy of all-optical networks and passive-optical networks. These networks can be realized with CMOS

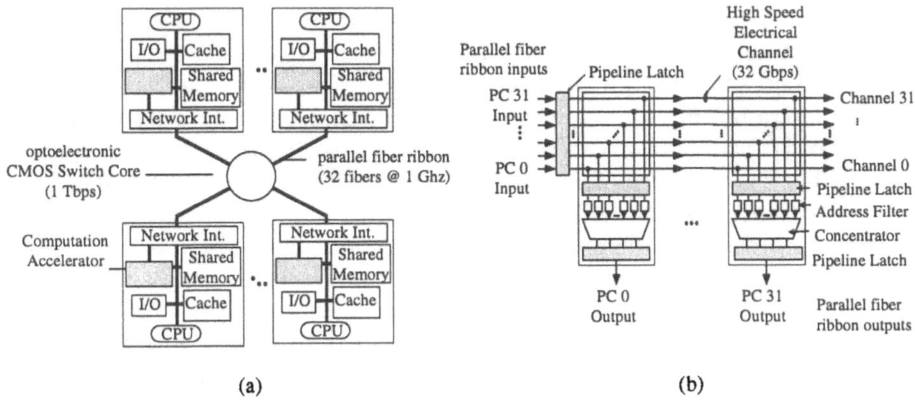

Figure 2.1 (a) An Intelligent Optical Network for a Network-of-Workstations or Personal Computers (PCs). (b) A Broadcast-and-Select switch core interconnecting several PCs.

processing substrates with optical I/O which have been flip-chip bonded over the surface. Optoelectronic CMOS integrated circuits have the potential to support thousands of optical I/O with potentially millions of logic gates. This technology supports rapid and essentially transparent conversion of parallel bits between the electrical and optical domains, a feature which Intelligent Optical Networks exploit.

An Intelligent Optical Network can exploit several attractive system concepts, as described in [26, 27]. The fiber ribbons and crossbar-like switches can be multiplexed to support multiple "Virtual Circuits" (VCs) per fiber ribbon. The use of multiple VCs per datalink will lower the blocking probability, increase the reliability, and decrease variance in packet delivery time. The crossbar-like switches with multiple VCs per I/O port can be designed to deliver most packets in their first transmission attempt, minimizing the need for buffering of packets within the optical network, i.e., most buffering can be performed within the workstations themselves, where there is plenty of memory already. The parallel fiber ribbons can also allocate a few fibers for clocking and timing information, thus simplifying the task of synchronization and timing in large distributed systems. Other attractive system concepts are described in [26, 27].

In this chapter, we outline the design of a single chip self-routing optoelectronic switch core for the system shown in Fig. 2.1. The self-routing single chip switch core will interconnect 32 parallel fiber ribbons, each with 32 fibers clocked at 1 GHz. We point out that prior single chip optoelectronic switches using CMOS technology have also been described in [14, 16, 18], however these switches do not have any "intelligence", i.e., they do not include any logic for processing packet headers or for self-routing and configuring the switch, which represents one of the most challenging aspects of our switch design, and which would be essential in any realistic multiprocessor system.

The design of the self-routing logic to process and switch vast amounts of data on a single optoelectronic CMOS substrate represents a significant design challenge, far beyond the requirements for conventional "all-electrical" switches. Traditional all-electrical CMOS switches typically support a small number of I/O ports, each with 16 - 32 bits, at electrical clock rates of about 100 - 200 MHz, as described in [6]. According to [6], the design of all-electrical self-routing switches is a challenging process with no clear guidelines. Blocking in the switch can degrade performance. The blocking can be lowered by multiplexing of multiple "Virtual Lanes" or "Virtual Circuits" onto each electrical data link, but this approach increases the logic gate complexity of the switch, thereby slowing down the clock rate and actually decreasing the capacity of the switch [6]. According to [6], it is not feasible to have the low blocking associated with multiple VCs along with a very fast clock rate (for electrical technology). The problem is compounded by the bandwidth and clock rate increases due to optical technology.

The logic to perform self-routing and switching of large amounts of data on a single substrate has not previously been demonstrated, and will place very stringent design requirements on the CMOS substrate. Furthermore, in order to compete with other optical technologies, we anticipate that within a decade the single chip optoelectronic switch in Fig. 2.1 should scale to a 10 times increase in the number of bits per port and a 10 times increase in clock rate. Within a decade, a single chip optoelectronic switch should thus support a 100 times increase in bandwidth over today's design. Given the complexity of the design tradeoffs for the conventional all-electrical switches described in [6], it is not even clear whether the design of such a self-routing single chip optoelectronic switch is feasible. We argue that the designs for self-routing switch cores supporting Terabits of bandwidth are feasible, and refer the reader to [26, 27] for detailed supporting arguments. If such single chip switches are feasible, then it is possible to interconnect such switches in various topologies, such as 3 stage Clos networks, to achieve low latency optical networks with hundreds of Terabits of bandwidths [27]. Such networks can potentially be used to meet the US ASCI performance targets for the year 2004. In the second part of this chapter, we will consider parallel computing on a Network-of-Workstations (NOW) [1] interconnected with an Intelligent Optical Network. Consider a NOW computing system where each workstation has a parallel fiber ribbon optical datalink, with 32 parallel fibers clocked at 1 GHz, to a centralized switch core. The centralized core has an aggregate bandwidth of 1 Terabit per second, which is significantly greater than the bandwidth of existing commercial NOWs. Furthermore, we assume that each Workstation is equipped with a Computational Accelerator card, containing several Digital Signal Processor (DSP) coprocessors interconnected with shared memory, as shown in Fig. 2.1.

The Butterfly class of algorithms spans a range of scientific applications, including parallel sorting, parallel merging, parallel prefix, the Fast Fourier Transform, the Discrete Cosine Transform, and matrix-vector and matrix-

matrix algorithms, including matrix multiply and Gaussian elimination, to name a few. The Butterfly algorithms have very demanding bandwidth requirements, and there are several other classes of scientific algorithms which are easier to map onto a multiprocessor system, i.e., particle systems based on meshes. By considering a bandwidth-intensive Butterfly-based algorithm such as parallel sorting, it follows that other scientific algorithms with less demanding bandwidth requirements can also be implemented in the proposed system.

Parallel sorting is often used as a metric when comparing different architectures. We present a performance analysis of parallel sorting, based upon a Berkeley methodology used to evaluate the performance of NOWs [1]. In the analysis, we consider all significant components of delay. We assume the use of hypothetical DSP coprocessors, whose performance is limited by the time needed to *issue instructions*. This methodology can be easily adapted to evaluate other algorithms with other specific types of DSP or RISC coprocessors.

The analysis yields an *accurate relationship* between the system performance measured in TeraFLOPs per second, and the network bandwidth needed to sustain the computation, measured in Terabits per second. It is the relationship between these two quantities, rather than their absolute values, which is of interest in this chapter. The system performance and optical bisection bandwidth can be scaled accordingly, given the use of other types of DSP coprocessors. Thus, we present an accurate methodology for determining the bisection bandwidth needed to sustain computations at a desired rate, which can be applied to all Butterfly algorithms and to all types of DSP or RISC coprocessors. Our analysis confirms IBM's preliminary estimate [3], that a 100 TeraFLOP computing system will require an interconnection network with a bisection bandwidth of several hundred Terabits per second.

Using 1998 technologies, a multidimensional NOW with $1K$ PCs, each with 16 DSP coprocessors, can attain a performance of approx. 3 TeraFLOPs for parallel sorting. Using PC, RAM and disk technology expected to be available in the year 2004, a NOW with $1K$ PCs can attain a performance of approx. 30 TeraFLOPs for parallel sorting. The multidimensional topology to interconnect $1K$ PCs is described in Section 2.6. Once again, we point out that Butterfly algorithms are among the most bandwidth-intensive algorithms, so other algorithms with localized communication will have substantially higher performances. However, two *supporting technologies* are critical to achieve these figures: the availability of an *optical interconnect* with hundreds of terabits per second of bandwidth, and sufficient *distributed shared memory* to avoid disk access. Other Butterfly-based algorithms will have similar performances, which can easily be computed using our methodology.

This chapter is organized as follows. Section 2.2 includes a summary of Intelligent Optical Networks, and the proposed NOW system. Section 2.3 presents a description of the Butterfly class of parallel algorithms, and summarizes optimal transformations and multiway partitions which minimize the bandwidth required to implement Butterflies. Section 2.4 describes the embedding of a parallel sorting Butterfly algorithm into the proposed system. Section 2.5 presents the performance rating of the system and Section 2.6 discusses applications to the US ASCI program. Section 2.7 includes a summary.

2.2 INTELLIGENT OPTICAL NETWORKS

Intelligent Optical Networks include *intelligence* directly in the optoelectronic integrated circuits, and can process vast amounts of optical data as it moves through the optoelectronic layer. The intelligence can include the functions associated with conventional systems, i.e., packet address processing and routing, acknowledgment of successfully transmitted packets, and negative acknowledgment of unsuccessfully transmitted packets [26, 27]. This class interconnects thus represents a departure from the traditional view of "all-optical" networks, where very limited functionality is performed on optical data in the all- optical layer.

The design issues for an optoelectronic CMOS switch core supporting hundreds of Gigabits or Terabits of data will represent a significant challenge, and will be several times more complex than the design for a conventional all-electrical switch. Existing optical I/O support clock rates in the 500 MHz - 1 GHz range. In a decade, Vertical Cavity Surface Emitting Laser (VCSEL) technology [19] should support clock rates of 10 -20 Gb/s range. The very fast clock rates associated with optics, and the large scale bit parallelism offered by optics, will place very stringent demands on the design of an optoelectronic IC. At these clock rates, packet address recognition and packet extraction functions will have to be performed within a few nanoseconds. To the best of our knowledge, there are no prior CMOS integrated circuits which perform switching functions at anywhere near these clock rates. We argue that such switches are feasible, and refer the reader to [26, 27]. The detailed design of optical networks with up to 700 Terabits of bandwidth are described in [27], and the implementation of a prototype optical network we are developing is described in [26].

Each workstation has a dedicated parallel fiber ribbon connection [9] to a centralized optoelectronic CMOS-based switch core. Parallel fiber ribbons, where each ribbon supports 32 fibers at 1 Gb/s each, are feasible and should be available soon. Each ribbon provides typically 32 Gb/s per link. A centralized optoelectronic switch core interconnecting 32 workstations would support a bandwidth of 1,024 Gb/s or 1 Tb/s.

The optical network supports the "broadcast and select" connectivity model shown in Fig. 2.1, with multiple-broadcast busses within the CMOS substrate. Each workstation can transmit up to 32 Gb/s over its parallel fiber datalink to the switch core, where the data is broadcasted on its own reserved broadcast channel within the CMOS substrate. In absence of collisions at a destination, all transmissions proceed in parallel. To minimize destination conflicts, the application algorithms should schedule transmissions through the switch core, as in [2]. (This model is similar to the early Cray-style supercomputers, which scheduled memory accesses to avoid conflicts in parallel memory banks).

One of the most challenging aspects is the design of the arbiters or concentrators. Suppose each parallel fiber ribbon with 32 bits is subdivided into two Virtual Channels, each 16 bits wide. To implement the switching, each output port typically will require a 64-to-2 concentrator. Each concentrator acts as a *arbitrator* for one output port of the switch, controlling access to two outgoing VCs from up to 64 contending incoming VCs. The design of efficient concentrators is a classic and well studied problem in the field of telecommunications. However, even in conventional all-electrical switches, concentrators are usually limited to clock rates of about 100 - 200 MHz [6]. To be useful in optoelectronic switches, concentrators should support clock rates of 500 MHz - 10 GHz, which will clearly require some significant innovation when compared to existing approaches. Designs for some efficient concentrators are described in [23, 27].

First consider an unbuffered switch core with no internal packet buffers operating in a synchronous manner. Workstations first wait for a signal denoting the start of a new time slot from the central switch; workstations with a packet will transmit that packet during the time slot; if the packet is not blocked at the core it will be delivered to the intended destination by the core, and the sender will receive an acknowledgment from the core. If the packet is blocked, the sender is notified and will attempt another transmission in the next time slot. This scheme eliminates buffers from the core, a significant departure from traditional view of all-electrical switches, which tends to favor extensive buffering. Our scheme can be generalized to support limited buffering, by adding a packet buffer(s) at the input side, and packet buffer(s) at the output side. The use of buffers will *pipeline* the movement of packets through the switch, i.e., in the first time slot a packet moves from the sending workstation to the input side of the core, in the second time slot it moves to the output side of the core, and in the third time slot it moves from the core to the destination workstation(s). This pipelining will result in high aggregate throughputs [26, 27].

Within the switch core, any packet can broadcasted to one or more destinations, by setting the appropriate request bits in the packet header. In a 32×32 switch, every packet header has a vector of 32 Request bits, where each bit specifies one workstation as a destination. Each packet can be destined to any subset of 32 workstations, by setting packet header appropriately. The detailed logic to implement such self-routing and switching is for example described in [23, 27].

The proposed single chip switch core will perform self-routing functions including arbitration, and can provide a sender with positive acknowledgment of successfully transmitted packets, and negative acknowledgment of unsuccessfully transmitted packets, at the end of each time slot. This acknowledgment feature will require nontrivial processing or *"intelligence"* within the switch core, and represents a significant departure from previous optoelectronic CMOS based switches which do not perform such functions [14, 15, 16, 17]. It also represents a significant departure from prior all-optical and passive-optical networks, which cannot perform such processing. Acknowledgment is required to

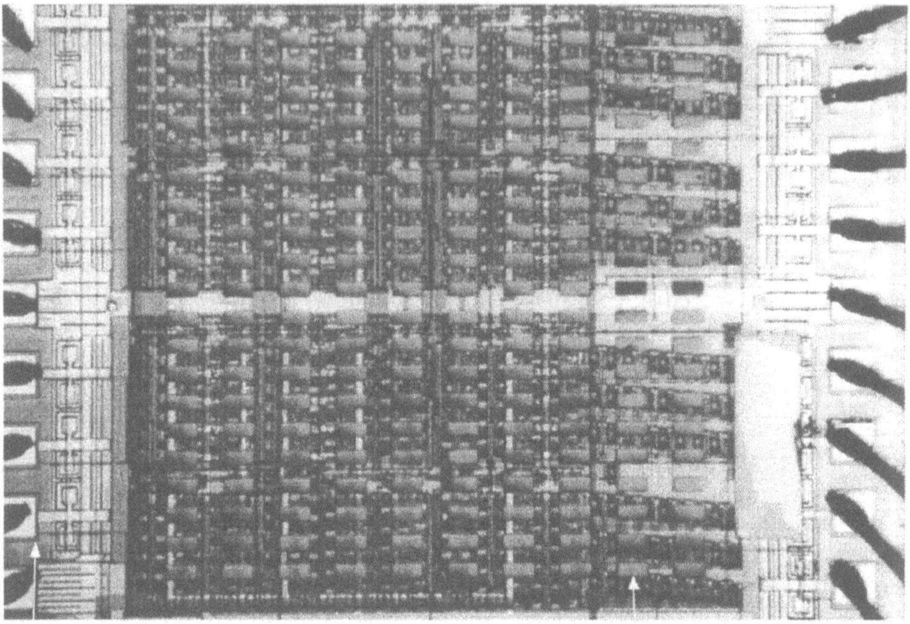

Figure 2.2 An optoelectronic CMOS integrated circuit with 200 optical I/O and 44 electrical I/O on a 2×2 millimeter die. [22, 27].

ensure reliable communications. These other systems could perform acknowledgments by having the communicating workstations send special acknowledgment packets among themselves, which will not require any intelligence within the network. This approach introduces complexity to the protocol (i.e., should the acknowledgment packets also be acknowledged?), and will introduce significant delays before a packet is acknowledged, as the packet must be processed by the workstation.

Intelligent Optical Networks will have very high bandwidths and low latency. The propagation delay of light in fiber is about $2 \cdot 10^8$ meters per second, or about 5 ns per meter. Assuming fiber ribbons are about 20 meters in length, the propagation delay over the fiber ribbon is approx. 100 ns. The round-trip delay from a workstation to the core and back is about 200 ns, plus the latency through the switch core. Assuming a delay of 100 ns through the core, the total end-to-end delay of an unbuffered system is approx. 300 ns. End-to-end Error and flow control protocols are implemented within the Network Interface module in the workstation as shown in Fig. 2.1 [26].

A photograph of a prototype optoelectronic switch core we have developed is shown in Fig. 2.2. This device was designed and fabricated in the 1995/96

ARPA/Lucent/COOP (Consortium on Optoelectronic Technologies) workshop. The 2 × 2 millimeter device contains 200 optical I/O over the surface, and approx. 44 electrical I/O along the perimeter. The device demonstrates the feasibility of dense CMOS logic with optical I/O, and supports 500 logic gates per optical I/O bit. This device is described in more detail in [22, 26]. A larger color photograph is available in [26].

2.3 PARALLEL COMPUTING WITH INTELLIGENT OPTICAL NETWORKS

Consider a Network-of-Workstations with 32 PCs or workstations as shown in Fig. 2.1. Each workstation requires (i) a local Distributed Shared Memory module based upon DRAM supporting virtual addressing, (ii) a local cache module, (iii) local disk I/O, and (iv) a *"Computation-Accelerator"* card.

Each Accelerator card has (i) an array of Digital Signal Processor (DSP) coprocessors, and (ii) a local shared memory accessible to all DSPs based on high speed SRAM which supports physical addresses, and (iii) a high-speed CPU interface. Each card also has (i) a "Network Interface" module to interface to the very high bandwidth Intelligent Optical Network, and (ii) pre-processor and post-processor circuits to implement application-oriented packet processing in hardware (these are not shown in Fig. 2.1). The Network Interface modules implement light- weight error and flow protocols, in programmable hardware [26]. All these modules interact with the workstation CPU as shown in Fig. 2.1, thereby gaining access to the conventional computing infrastructure.

The Accelerators communicate with the host CPU by raising interrupts or by responding to polling queries from the CPU. In this manner, the host CPU can load the memory of the Accelerator, initialize the DSPs to implement the desired task by downloading of the appropriate control bits, and start the accelerators. When the accelerators are finished, they inform the hosts through interrupts.

A DSP testbed consisting of an array of parallel DSPs along with a reconfigurable electrical network was recently proposed in [30]. Our model also explores the use of parallel DSPs are coprocessors, however there are some significant differences. The system in [30] is not tightly coupled with RISC coprocessors, and thus the DSPs do not have easy access to the conventional computing infrastructure. The system in [30] uses a relatively low bandwidth electronic network to connect clusters of parallel DSPs, and will not scale to the levels needed to achieve performances mandated by the ASCI program.

2.3.1 *Programming the Computing System*

For pipelinable computations, the proposed system operates in a *"pseudo-synchronous"* manner; the PC microprocessors act as overall controllers, which supervise and coordinate the computations of the Accelerators. Within a time step, all Accelerators execute their course grained partitions in parallel, and then synchronize using a barrier synchronization primitive. After synchronization, all Accelerators move forward to the next time step. This model is similar to Valiant's "Bulk Synchronous Parallel" (BSP) model [11].

Programming the computing system consists of a series of steps including: (1) *Data Flow Graph formulation*, (2) *global transformation*, *partitioning* and *mapping* to Accelerator cards within the NOW, (3) *local transformation*, *partitioning* and *mapping* to DSPs within an Accelerator card, (4) *synthesis* onto the DSP coprocessors, and finally; (5) *global communication scheduling* between the Accelerator cards, and (6) *local communication scheduling* between the DSP coprocessors.

In more detail, the task to be performed is first expressed as a very large data flow graph, typically in the C language. Secondly, a global topological transformation is performed yielding an optimal multi-way partition. The partitioned graph will be mapped onto the collection of 32 Accelerator cards in the NOW. This global partitioning minimizes the network bisection bandwidth required in later steps, and maximizes the computation- to-communication ratio.

In the local transformation process, the flow graphs assigned to each Computational Accelerator card must again be partitioned and assigned to the array of DSPs within the card. In the 5th step, the global packet exchanges between Accelerator cards must be scheduled over the *Optical LAN* to avoid destination conflicts. In the 6th step, the local data exchanges between DSPs must be scheduled to avoid memory conflicts. At this point the system is ready to execute the original task, in a deterministic manner with synchronization occurring after each course grained time step.

2.4 ANALYSIS OF BUTTERFLY ALGORITHMS (PARALLEL SORTING)

Parallel sorting is a classic problem which has been studied for decades [15]. In practice one of the fastest known algorithms is based on Batcher's bitonic sort. Stone has presented a variation of Batcher's $N \times N$ Bitonic sort which consists of $O(\log N)$ Omega networks in series, as shown in Fig. 2.3a [15]. Each node represents a binary *Compare-Exchange* (CE) element, which accepts two numbers, compares them and outputs them in sorted order.

Parallel sorting is a commonly used metric for comparing different architectures. We present a detailed methodology for analyzing the performance of Stone's algorithm on the proposed system. However, the analysis and methodology is general and applies to any Butterfly algorithm, since every Butterfly algorithm undergoes the same global partition steps, the same local partition steps, and realizes the same global and local communication scheduling. Other Butterfly algorithms such as the Fast Fourier Transform (FFT), the Discrete Fourier Transform (DFT), parallel merging or parallel prefix, will only change the nature of the task to be synthesized within each DSP.

Each floating point number is represented in the standard 32 bit format, with a 24 bit mantissa and 8 bit exponent field. Numbers are entered into the sorting network. Each CE element sorts two numbers and propagates them to the next stage. After 400 stages, all $1M$ numbers emerge in sorted order.

A $1M$ point DFG based on Stone's sorting network consists of 20 Omega networks of size $1M \times 1M$ in series; see Fig. 2.3a, which illustrates a 16×16 sorting network. For $N = 1M$, this flow graph has tremendous bandwidth requirements. Each Omega network transfers 320 million bits over the bisector. Hence, Stone's flow graph with 20 Omega networks transfers 6400 million bits over the bisector.

To minimize bandwidth requirements, we can perform a global transformation where each $1M \times 1M$ Omega network can be transformed into a *"Recursive Omega"* network, which minimizes the bisection width. See [28] for a description of Recursive-Omega networks. Hence, the original $1M \times 1M$ sorting flow graph can be transformed from a series of 20 Omega Networks into a series of 20 Recursive- Omega networks. Such a transformation is shown in Fig. 2.3b, for a 16×16 sorting network. The equivalence between these two topologies is established in [28]. Each $1M \times 1M$ Recursive-Omega network will consist of 2 stages of "global factors". Each stage will have $1K$ global factors, where each global factor is a smaller $1K \times 1K$ Omega network. Each stage is preceded by a radix-$1K$ shuffle of $1M$ numbers [28], which we call the global communication pattern.

The advantages of performing the global topological transformation are now apparent. In Stone's original topology, global communications will occur 400 times, and the topology moves 6,400 million bits over the bisector, as illustrated in Fig. 2.3a for $N = 16$. In the transformed recursive topology, the global communications will occur 40 times, and the topology moves 640 million bits over the bisector, as illustrated in Fig. 2.3b for $N = 16$. For $N = 1M$, the data communication time is reduced by an order of magnitude by the transformation, and the ratio of computation-to-communication time is maximized. The transformation also yields a straightforward partitioning of the task onto the Accelerator cards. The topology can be partitioned into several *horizontal slices*, as shown in Fig. 2.3b. Each Accelerator card will simulate 32 $1K \times 1K$ Omega networks from both stages of the Recursive Omega network.

In summary, at the global level $1M$ numbers can be sorted in these main global steps. Steps (1)..(4) are repeated twenty times, once for each Recursive Omega network in the series, at which point the numbers emerge in sorted order. (1) All Accelerators broadcast their data to implement the global communication pattern. It can be proven that this pattern can be conflict-free, although the proof will be presented elsewhere. (2) Each Accelerator card computes 32 $1K \times 1K$ Omega networks in its partition. (3) All Accelerators broadcast their data to implement the global permutation. (4) Each Accelerator card computes 32 $1K \times 1K$ Omega networks in its partition. Communication steps (1) and (3) are identical, and computations steps (2) and (4) are largely identical, so that programming the DSP coprocessors is highly regular.

2.4.1 Partitioning Horizontal Slices onto DSPs

In the local transformation step, another topological transformation is required to minimize the communication bandwidth between the DSPs within an Accelerator card. In the local transformation, each $1K \times 1K$ Omega global factor is once again transformed to a Recursive Omega network with two stages of *"local factors"*. Each stage will have 32 local factors, where each local factor is a smaller 32×32 Omega network (or Butterfly). Hence, each $1K \times 1K$ Omega network is replaced by 2 stages of smaller Omega networks (of size 32×32) with the appropriate interconnections between the stages. These interconnections are the local permutations consisting of radix- 32 shuffles of $1K$ numbers [28].

In summary, within each Accelerator card each $1K \times 1K$ Omega network can be simulated by 16 DSPs in parallel with these 4 local steps. (1) All DSPs permute their data to implement the required local permutation. (2) Each DSP simulates 2 32×32 Omega networks. (3) All DSPs permute their data to implement the local permutation. (4) Each DSP simulates 2 32×32 Omega networks. We observe that local communication steps (1) and (3) are identical, and local computation steps (2) and (4) are largely identical, so that programming the DSP coprocessors is highly regular.

2.5 PERFORMANCE AND BANDWIDTH RELATIONSHIP

Berkeley researchers developed a methodology to compare NOWs with conventional Cray supercomputers. The methodology using a well known application, parallel sorting, and considered all significant delays, including computation time, communication time, and disk I/O times. We use the same methodology to evaluate the performance of the proposed NOW with DSP based Accelerator cards.

Consider the time it takes for an Accelerator card to compute one $1M \times 1M$ Recursive Omega network. Each Accelerator card must simulate the first stage of 32 $1K \times 1K$ Omega networks, synchronize, then simulate the second stage of 32 $1K \times 1K$ Omega networks and synchronize. (In the following, we will often refer to Omega networks as Butterflies, since they are topologically equivalent).

For maximum performance, *computation must be pipelined with communication.* To achieve maximum performance, the following mode of operation is employed. At the global level, communication step (1) and computation step (2) overlap in time, and similarly for steps (3) and (4). To start the computation in step (2) as soon as possible, the global communication in step (1) is partitioned into several "waves". In the first wave, a $1K \times 1K$ Butterfly in every Accelerator is initialized in parallel. At this point, all Accelerators can start executing computation step (2). When the computation of the i-th Butterfly is completed, the data for the $(i + 1)$-th Butterfly will have arrived, so that its computation can begin. To support this mode of operation, the optical network must be able to support the traffic requirements of any wave within the time required to compute the Butterfly. The total execution time of steps (1) and (2) is therefore equivalent to (33/32) times the execution time of one $1M \times 1M$ Omega network (the extra fraction of 1/32 is due to the overhead of the first wave). A proof that this technique allows *computation to largely overlap with communication* will be presented elsewhere.

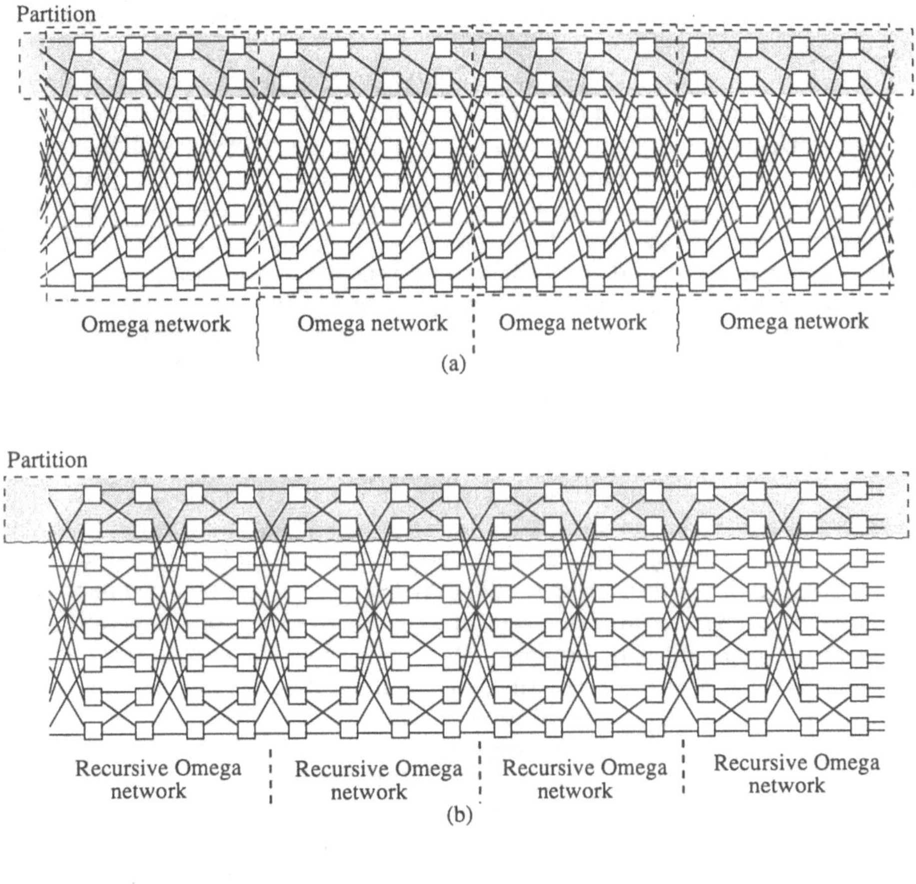

Figure 2.3 (a) Sorting $N = 16$ numbers using a sequence of Omega networks corresponding to Stone's algorithm. (b) Sorting $N = 16$ numbers using a sequence of Recursive Omega networks.

To support this mode of operation, within an Accelerator card the pre and post processors must be able to access data into the DSP's shared memory, simultaneously while DSPs are executing from the shared memory. It will be shown later that the DSPs spend $> 60\%$ of their time computing, and $< 40\%$ of their time accessing the shared memory. If we assume the shared memory consists of 4 banks, the pre and post-processors could be accessing banks 1 and 2, while the DSPs read and write with banks 3 and 4. Hence, this mode of operation is achievable, provided that the DSP's shared memory be organized into independent banks.

In the local transformation process, each $1K \times 1K$ Omega network is replaced by 64 32×32 Omega networks. It follows that for each $1K \times 1K$ network, each Accelerator card must simulate 2,048 32×32 Omega networks, complete the first wave of a global communication and synchronize, and then repeat the process. Since there are 16 DSPs per Accelerator card, it follows that each DSP must simulate 128 32×32 Omega networks, complete the first wave of a global communication and synchronize, and repeat the process..

We now determine the clock cycles required for one DSP to simulate a 32×32 Butterfly. Assume a hypothetical but realistic DSP or RISC coprocessor, with up to 8 instruction issues per clock cycle and a 300 MHz clock. Assume that up to 2 Loads or Stores, 2 Floating point operations, or 4 conditional moves can be issued per clock cycle. This assumption may be aggressive for 1998 technology, but it is straightforward to model any specific DSP coprocessor using this methodology. The superscalar assumption will allow the instructions for a 2×2 Compare-Exchange node to be issued in one clock cycle, given that the data has already been loaded. To implement a single 2×2 Compare-Exchange node in Fig. 2.3, the instructions will consist of a floating point compare with 2 source operands, and 3 conditional moves of the source operands after the result of the compare is determined. We will assume the coprocessors schedule the execution of these instructions internally, using for example Tomasulo's algorithm. Hence, the key limitation to performance is issuing the instructions. The instructions will finish execution a few clock cycles after they issue, depending upon the internal organization of the DSP.

Fig. 2.4 illustrates the computation of a 32×32 Butterfly by each DSP. Initially, 16 clock cycles are needed to issue the Loads of 32 words from shared memory. Each of the 5 stages of the Butterfly has 16 CE nodes, and each CE node requires one clock cycle to issue. Hence, it requires 16 clock cycles just to issue the instructions for each stage of the Butterfly. Finally, 16 clock cycles are needed to issue the Stores of 32 words to shared memory. The Loads will be limited to the speed of the shared SRAM, which we assume can support a 100 MHz clock rate. Hence, each pair of Loads will require approx. 3 clock cycles of a 300 MHz clock on average. The first stage of Compare-Exchange instructions can be scheduled by the compiler to issue with the Loads (this may add a few extra clock cycles, depending upon the internal DSP architecture). The last 4 stages of Compare-Exchange nodes will require 16 clock cycles each to issue. The Stores can be scheduled to issue with the last set of Compare-Exchanges (this

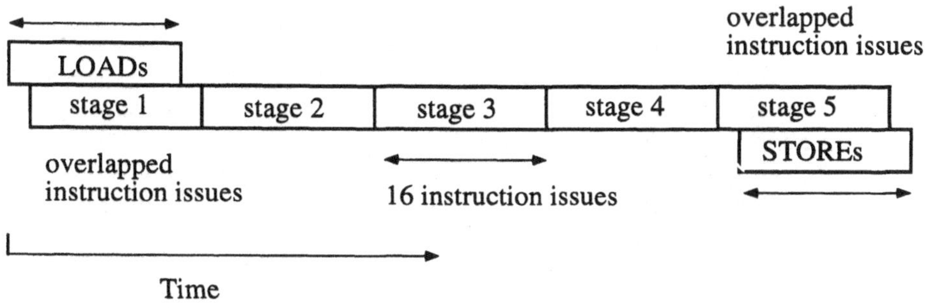

Figure 2.4 Execution time of one 32×32 Butterfly determined by instruction issues.

may add a few extra clock cycles, depending upon the internal DSP architecture). Hence, the total number of clock cycles for a DSP coprocessor to simulate a 32×32 Butterfly is $16 \times (3$ clock cycles per Load$) + 16 \times 4 + 2 = 40 + 64 + 2 = 118$ clock cycles.

Each DSP must simulate 128 Butterflies, complete the first wave of a global communication and synchronize, and repeat the process. The local permutations are accomplished implicitly, by having the DSPs Load their data from the appropriate addresses in shared memory. We assume that the DSPs operate in tight synchronization, and that these Loads can be accomplished conflict-free. Hence, the local permutations do not require additional time. The number of clock cycles required for computation in global steps (2) or (4) is thus $(33/32) \times 128 \times (118 + 118) = 31,152$ clock cycles.

The specification of the superscalar DSP will largely determine the execution time for each Butterfly. In our example, *the computation time is largely determined by the time needed to issue the instructions.* The final results may take a few extra clock cycles to appear, but these extra clock cycles are insignificant compared to the time needed to issue the instructions in the first place. An off-the-shelf DSP using 1998 technology may require on average 2 clock cycles to issue the instructions for each 2×2 Compare-Exchange node. In this case, the execution time is essentially doubled. For computationally intensive Butterflies, the same methodology of determining the clock cycles to issue the computations, loads and stores can be used to estimate performance.

Assume that the $1M$ input and output numbers are stored in shared memory within the Accelerator cards, so that access to a disk is not necessary. The total time to sort the $1M$ numbers on the NOW system is therefore $20 \times (31,152$ clock cycles$) \times (3.33$ ns per clock cycle$) = 2,074\mu s$. Assume that synchronization will add 2.5% to the execution time, and retransmissions of occasionally erroneous packets [26] will be add 2.5% to the execution time, yielding approx. 2,178 microseconds. (Retransmissions of occasional packets will not cause any blocking, as the switch core can buffer a small number of packets per input port, and it can support multiple Virtual Circuits per output port). The number of operations performed to sort $1M$ numbers using this algorithm is approximately 200 MFLOPs and 400 MOps. (We count each CE node as one floating point compare, and 2 integer assignments. In practice, some of the CE nodes simple pass their data through unchanged, and this relatively minor effect is not considered.) The sustained computation rate is $(10^6/2,178) \times 200MFLOPs =$ 92 GigaFLOPs (for the system with 32 PCs and 512 DSPs). This is equivalent to a sustained rate of 180 MegaFLOPs per DSP.

The throughputs are significantly lower if the data must be fetched from the disk, and stored to the disk. In this case, each PC must read 32K numbers or 128K bytes from disk; Assuming a modest 10MB/s disk I/O rate per PC, the data input time is 12.8 ms. Hence, the total time to sort the $1M$ numbers on the cluster-based system is: $12.8ms + 2.2ms + 12.8ms = 27.8ms$. The sustained computation rate is $(103/27.8) \times 200MFLOPs = 7.2GFLOPs$. When data must be read from and written to the disk, the sustained performance is approx. an order of magnitude lower than when the data is available in memory.

The previous analysis highlights a key concept. The concept of caching pages of data in network-wide Shared Memory or shared Disk Cache is essential to realizing maximum potential. The disk- cache concept was proposed by the Berkeley NOW group, to cache VM pages in memory to avoid repeated disk accesses [1]. The Accelerator's high speed shared memory based on SRAM is reserved for tasks under computation. Temporary data sets should be stored in the microprocessor's larger shared memory based upon DRAM, to avoid disk accesses. The performance of caching will be critical to realize the full potential of the DSP coprocessors.

2.5.1 System Bandwidth Requirements

Consider the NOW computing system with 32 Personal Computers and 16 DSP coprocessors per Accelerator Card. When the Accelerators perform the "All-to-All" global communications in global steps (1) and (3), the optical LAN must support the total bandwidth of this traffic pattern. As shown in Fig. 2.2, each accelerator card reserves its own a contention-free broadcast bus for its own broadcasts [19].

From the previous section, it requires 31,152 clock cycles to compute the $1M \times 1M$ Recursive Omega networks in each Accelerator card. By earlier assumption, the communication is overlapped with computation. Therefore, all the input data for each $1M \times 1M$ Butterfly must be received by the Accelerator card from the optical LAN in approx. 31,152 clock cycles, and all the output data for each $1M \times 1M$ Butterfly must be transmitted by each Accelerator card. However, twice this amount of data must be transmitted in this interval, since each $1M \times 1M$ Recursive Omega network requires 2 global communications. We assume the pre and post processors handle the packet preparation, transmission and reception, in a manner transparent to the DSPs. This assumption is valid if the pre and post processors can execute out of separate banks of shared memory, without interfering with the DSPs.

Hence, each Accelerator card generates traffic for the Optical LAN at the rate of two sets of 32K numbers every 31,152 clock cycles, or equivalently 2.097 Mbits every 104 microseconds, or approx. 20 Gb/s. Assume that overhead per packet is 25%. Hence each Accelerator card generates data at the rate 25 Gb/s. In order to avoid stalling the pipeline, each Accelerator must be able to transmit and receive data over the Optical LAN at the rate of 25 Gb/s. The Intelligent Optical Network must support the simultaneous transmissions of 32 Accelerator cards, for a total bandwidth of 800 Gigabits per second.

These bandwidths are far beyond the capability of conventional shared medium LANs. The proposed Intelligent Optical Network can support such bandwidths with careful design. Using fiber ribbons with 32 fibers clocked at 1 Gb/s, a datalink supports 32 Gb/s, which is sufficient for the system.

2.5.2 Other Butterfly Applications - FFT, DCT, Matrix Multiply

The same partitioning strategy applies to all Butterfly parallel algorithms. For example, a large parallel FFT can be expressed as a single Butterfly network. The *only significant difference* between the parallel sorting and parallel FFT application or any other Butterfly-based algorithm will be in the programming of the 32×32 Butterflies onto the DSPs. The methodology presented earlier can be used to determine the performance of specific types of coprocessors with specific types of Butterfly algorithms.

2.6 SCALABILITY TO U.S. ACCELERATED COMPUTING INITIATIVE TARGETS

The U.S. Accelerated Strategic Computing Initiative (ASCI) has undertaken a program to develop 10 TeraFLOP machines in the year 2001 and 100 TeraFLOP machines in the year 2004 [7]. A 100 TeraFLOP computing system will occupy rows of electronic cabinets, and will require a complex high performance interprocessor communication network, the scale of which has not previously been considered. IBM has described a tentative design based upon conventional IBM workstation, microprocessor, disk and memory technologies [3]. IBM estimates indicate that the system will require $8K$ RISC processors and between 128 and 256 cabinets of electronics, occupying 20 by 40 meters of floor space. The interconnect must provide in the neighborhood of 300 Terabits per second of usable bandwidth. These parameters indicate the magnitude of the problem that computer manufacturers will face to meet the ASCI targets.

Table 2.1 illustrate several ranges of the proposed NOW system's performance based upon current 1998 technology; including a single PC, a 1 dimensional NOW with 32 PCs, and a 2 dimensional NOW of 1,024 PCs. Using 1998 technology, a 2 dimensional NOW could realistically support 1,024 PCs, 2 TeraBytes of Main Memory, and 16 TeraBytes of disk. (Using commercially available software such as RAM-Doubler, the amount of DRAM can be doubled.)

Table 2.1 Projected performance of a single personal computer/workstation, a 1 dimensional NOW, and a 2 dimensional NOW in a Hypermesh topology, using 1998 technology.

Component	*Single System*	*1D 32×1*	*2D 32×32*
CPU	100 MFLOP	3.2 GFLOP	100 GFLOP
DRAM	2 GB	64 GB	2 TB
Disk	16 GB	512 GB	16 TB
Accelerator Card	2.88 GFLOP	92 GFLOP	2.94 TFLOP
Optical Interconnect	N/A	800 Gb/s	25.6 Tb/s

The 2D NOW consists of a 2D array of 32 × 32 PCs, with an Intelligent Optical Network interconnecting the PCs in each row and each column. This topology is called a Hypermesh [28], and it realizes a superset of the Recursive Omega and binary Omega networks. Hence, it will realize all Butterfly-based algorithms without conflict.

Using 1998 technology, the performance of the 2D system for Butterfly-based parallel sorting algorithms would be about 3 TeraFLOPs per second, provided that the data was found in the shared memory or disk cache. This performance requires an interconnect bisection bandwidth of about 25 Terabits per second. To ensure maximum performance, the size of the DRAM and Disk Cache should be maximized, by including as much DRAM as possible.

These parameters yield some interesting insights. There have been reports of specialized 1 TeraFLOP machines in the literature, however to the best of this author's knowledge no existing system supports a bisection bandwidth of 25 Terabits per second. Our detailed analysis indicates that a bisection bandwidth of about 25 Terabits per second is an essential pre-condition to realize a performance of 3 TeraFLOPs for Butterfly-based algorithms such as parallel sorting. The bandwidth requirement is essentially a lower bound based upon the number of bits that must be moved to realize the computation, and is independent of the type of processor used in the computation. The bandwidth requirements change for different task sizes N and different algorithms such as FFT or matrix multiply, but these variations only have a moderate affect on the bandwidth, roughly within an order of magnitude.

Table 2.2 illustrates several ranges of system performance based upon projected technology available in the year 2004; including a single PC, a 1 dimensional NOW of 32 PCs, and a 2 dimensional NOW of 1,024 PCs. A 2 dimensional NOW could realistically support 1,024 PCs, 20 TeraBytes of Main Memory, and 160 TeraBytes of disk. The performance for Butterfly-based parallel sorting algorithms would be about 23 TeraFLOPs per second, if the data was found in the shared memory or disk cache. This computation requires a bisection bandwidth of about 200 Terabits per second.

Table 2.2 Projected performance of a single personal computer/workstation, a 1 dimensional NOW, and a 2 dimensional NOW in a Hypermesh topology, using 2004 technology (following the growth trends of last 2 decades, with performance doubling every 18 months, equivalent to an 8× increase in approx. 6 years).

Component	*Single System*	*1D 32×1*	*2D 32×32*
CPU	800 MFLOP	25.6 GFLOP	800 GFLOP
DRAM	16 GB	512 GB	16 TB
Disk	128 GB	4 TB	128 TB
Accelerator Card	23 GFLOP	736 GFLOP	24 TFLOP
Optical Interconnect	N/A	6.4 Tb/s	204 Tb/s

To reach 100 TeraFLOP performance for Butterfly algorithms, a 3 dimensional $16 \times 16 \times 16$ Hypermesh system could be used, where a 16×16 Intelligent Optical Network interconnects all the PCs aligned along a each row, column, or dimension [28]. This system would require smaller switch cores (16×16), and would interconnect $4K$ PCs. Using 2004 technology, the performance for Butterflies would be about 100 TeraFLOPs.

Based upon the analysis, we can make an interesting observation. To achieve a sustained performance of 100 TeraFLOPs, the aggregate network bandwidth must be in the neighborhood of several hundred Terabits per second; this fact is inescapable, and it simply reflects the amount of bandwidth than must be moved in the Butterfly-class of parallel algorithms.

2.7 CONCLUSIONS

We have considered the impact of Intelligent Optical Networks on parallel computing machines. A computing system merging three powerful technologies, high performance Personal Computers (PCs), DSP coprocessors and Intelligent Optical Networks, was proposed and analyzed. An analysis of the relationship between system performance (in TeraFLOPs per second) and system bandwidth (in Terabits per second) was presented. The analysis applies to all Butterfly-based algorithms, including parallel sorting, parallel merging, parallel prefix, the Fast Fourier Transform, the Discrete Cosine Transform, and a large collection of matrix-matrix and matrix-vector algorithms, including matrix multiply and Gaussian elimination. The NOW with $1K$ PCs, each with 16 DSP coprocessors, can potentially attain performances in the neighborhood of 23 TeraFLOPs for Butterfly-based scientific algorithms, using year 2004 technology. However, to achieve such performances two supporting technologies are critical. The optical interconnect must scale to support bandwidths of 100s of Terabits per second, and the system must have sufficient DRAM to support a very large network-wide shared memory or disk cache, to avoid data fetches from the disks.

Acknowledgments

This research was supported by NSERC Canada Grant OPGO-0121601. The use of CAD software and silicon CMOS VLSI foundry services from the Canadian Microelectronics Corporation (CMC) is acknowledged. Thanks to the ARPA program for funding the fabrication of the optoelectronic IC in Fig. 2.2. Thanks to Sherif Sherif for contributing to the VLSI layout of the optoelectronic IC.

References

[1] T.E. Anderson, D.E. Culler, and D.A. Patterson, "A Case for NOW (Networks of Workstations)", *IEEE Micro*, vol. 16, no. 1, Feb. 1995, pp. 54-64.

[2] T.E. Anderson, S.S. Owicki, J.B. Saxe, and C.P.Thacker, "High- Speed Switch Scheduling for Local-Area networks", *ACM Transactions on Computer Systems*, vol. 11, no. 4, Nov. 1993, pp. 319- 352.

[3] A. Benner, L. Rudolph, E. Schenfeld, T. Sterling, T. Szymanski, "Panel Discussion: Design Options for Interconnecting 100+ TFLOP/sec Parallel Supercomputer in 2004", *IEEE 4th Int. Conf. Massively Parallel Processing using Optical Interconnects (MPPOI)*, June 1998.

[4] H. Bourdin, A. Ferreire and K. Marcus, "A Comparative Study of One-to-Many WDM Lightwave Interconnection Networks for Multiprocessors", *2nd Int. Conf. Massively Parallel Processing using Optical Interconnects (MPPOI)*, Oct. 1995, pp. 257-264.

[5] D. Chiarulli, S. Levitan, R. Melhem, J. Teza, and G. Gravenstreter, "Partitioned Optical Passive Star (POPS) Topologies Multiprocessor Interconnection Networks with Distributed Control", *IEEE Journal of Lightwave Technology*, vol. 14, no. 7, 1996, pp. 1601-1612.

[6] A. Chien, "A Cost and Speed Model for k-ary n-Cube Wormhole Routers", *IEEE Transactions on Parallel and Distributed Systems*, vol. 9, no. 2, Feb. 1998.

[7] D. Clark, "Breaking the Teraflops Barrier", *IEEE Computer*, vol. 30, Feb. 1997, pp. 12-14.

[8] W.A. Crossland and T.D. Wilkinson, "Optically Transparent Switching in Telecommunications Using Ferroelectric Liquid Crystals Over Silicon VLSI Circuits", *Proceedings of the 1996 IEEE/LEOS Summer Topical Meeting*, 1996, pp. 22-23.

[9] D.R. Engebretsen, D. M. Kuchta, R. C. Booth , J.D. Crow, and W.G. Nation, "Parallel Fiber-Optic SCI Links," *IEEE Micro*, vol. 16, no. 1, Feb. 1996, pp. 20-26.

[10] M. Galles, "Spider: A High-Speed Network Interconnect", *IEEE Micro*, vol. 17, no. 1, Jan/Feb. 1997, pp. 34-39.

[11] A.V. Gerbessiotis and L. Valiant, "Direct Bulk-synchronous parallel algorithm", *Journal of Parallel and Distributed Computing*, vol. 22, 1994, pp. 251-257.

[12] Glesk, I. Kang, K.I. Prucnal, P. R., "Demonstration of ultrafast all-optical packet routing", *Electronics Letters*, vol. 33, no. 9, Apr 1997, pp. 794-795.

[13] J.L. Hennessy and D.A. Patterson, *Computer Architecture: A Quantitative Approach*, Second Edition, Morgan Kauffman, 1995.

[14] S.J. Hinterlong, A.L. Lentine, D.J. Reiley, J.M. Sasian, R.L. Morrison, R.A. Novotny, M.G. Beckman, D.B. Buchholz, T.J. Cloonan, and G.W. Richards, "An ATM Switching System Demonstration Using a 40 Gb/s Throughput Smart Pixel Opto- electronic VLSI Chip", *Proc. 1996 IEEE/LEOS Summer Topical Meeting*, 1996, pp. 47-48.

[15] D. Knuth, *The Art of Computer Programming, Vol. 3, Sorting and Searching*, Second Edition, Addison-Wesley, 1973.

[16] A.V. Krishnamoorthy, and D.A.B. Miller, "Scaling Optoelectronic-VLSI Circuits into the 21st Century: A Technology Roadmap", *IEEE Journal of Selected Topics in Quantum Electronics*, vol. 2, no. 1, Apr. 1996, pp. 55-76.

[17] A.V. Krishnamoorthy, J.E. Ford, K.W. Goossen, J.A. Walker, B. Tseng, S.P. Hui, J.E. Cunningham, W. Y. Jan, T. K. Woodward, M.C. Nuss, R. G. Rozier, F. E. Kiamilev, and D. A. B. Miller, "The AMOEBA Chip: An Optoelectronic Switch for Multiprocessor Networking Using Dense-WDM", *Proceedings of the Third Int. Conf. on Massively Parallel Processing Using Optical Interconnections (MPPOI'96)*, 1996, pp. 94-100,

[18] A.L. Lentine, K.W.Goossen, J.A. Walker, L.M.F. Chirovsky, L.A. DAsaro, S.P. Hui, B.T. Tseng, R.E. Leibenguth, J.E. Cunningham, W.Y. Jan, J.M. Kuo, D. Dahringer, D. Kossives, D.D. Bacon, G. Livescu, R.L. Morrison, R.A. Novotny, and D.B. Buchholz, "Optoelectronic VLSI Switching Chip with Greater Than 4,000 Optical Flip-chip-bonded GaAs/AlGaAS MQW Modulators and Detectors on Silicon CMOS Circuitry", *Conf. Lasers and Electrooptics*, 1996, pp. 517-518.

[19] S. Matsuo, T. Nakahara, Y. Kohama, Y. Ohiso, S. Fukushima, and T. Kurokawa, "Monolithically Integrated Photonic Switching Device Using an MSM PD, MESFET's, and a VCSEL", *IEEE Photonics Technology Letters*, vol. 7, no. 10, Oct. 1995, pp. 1165-1167.

[20] Semiconductor Industry Association, "The National Technology Roadmap for Semiconductors", SIA, San Jose, 1997.

[21] Shell, M. Vaughn, M D. Wang, A. Blumenthal, D J. Rigole, P -J. Nilsson, S., "Experimental demonstration of an all-optical routing node for multihop wavelength routed networks", *IEEE Photonics Technology Letters*, vol. 8, no. 10 Oct 1996, p 1391- 1393.

[22] S. Sherif, T.H. Szymanski, and H.S. Hinton, "Design and Implementation of a Field Programmable Smart Pixel Array", *Proc. 1996 IEEE/LEOS Summer Topical Meeting*, 1996.

[23] B. Supmonchai and T.H. Szymanski, "High Speed VLSI Concentrators for Terabit Intelligent Optical Backplanes", *Optics in Computing 1998*, June, 1998.

[24] T.H. Szymanski and H.S. Hinton, "Reconfigurable Intelligent Optical Backplane for Parallel Computing and Communications", *Applied Optics*, vol. 35, no. 8, Mar. 1996, pp. 1253-1268.

[25] T.H. Szymanski and H.S. Hinton, "Architecture of a Terabit Free-Space Intelligent Optical Backplane", *Journal of Parallel and Distributed Computing*, to appear, 1998/99.

[26] T.H. Szymanski, A. Au, M. Lafrenire-Roula, V.Tyan, B. Supmonchai, J. Wong, B. Zerrouk, and S.T. Obenaus, "Terabit Optical Local Area Networks for Multiprocessing Systems", *Applied Optics, Special Issue on Massively Parallel Optical Interconnects for Multiprocessor Systems*, vol. 37, no. 2, Jan 1998, pp.264-275.

[27] T.H. Szymanski, "Design Principles for Practical Self- Routing Nonblocking Switching Networks with $O(N \log N)$ Bit Complexity", *IEEE Transactions on Computers*, vol. 46, no. 10, Oct. 1997, pp. 1057-1069.

[28] T.H. Szymanski, "Hypermeshes - Optical Interconnection Networks for Parallel Computing", *Journal of Parallel and Distributed Computing*, vol. 26, Apr. 1995, pp. 1-23.

[29] T.H. Szymanski and B. Supmonchai, "Reconfigurable computing with Optical Interconnects - an Economic Argument for Optics", *Proceedings of the Third Int. Conf. on Massively Parallel Processing Using Optical Interconnections (MPPOI'96)*, 1996.

[30] S.K. Tweksbury, K. Devabattini, and V. Gandikota "A Parallel DSP testbed with a Heterogeneous and Reconfigurable Network Fabric", *Proceedings of the Int. Conf. Innovative Systems on Silicon, ISIS 1997*, 1997, pp. 310-322.

3 SCALABLE OPTICAL INTERCONNECTION NETWORKS FOR LARGE-SCALE PARALLEL COMPUTERS

Ahmed Louri and Brent Weech

Department of Electrical and Computer Engineering
University of Arizona
Tucson, Arizona 85721

{louri, weech}@ece.arizona.edu

Abstract: A new scalable interconnection topology called the Spanning Multichannel Linked Hypercube (SMLH) is proposed along with an optical implementation methodology that combines both the advantages of free space optics with those of wavelength division multiplexing techniques. The SMLH uses the hypercube topology as a basic building block and connects such building blocks using two-dimensional multichannel links (similar to spanning buses). In doing so, the SMLH combines positive features of both the hypercube (small diameter, high connectivity, symmetry, simple routing, and fault tolerance) and the spanning bus hypercube (SBH) (constant node degree, scalability, and ease of physical implementation), while at the same time circumventing their disadvantages. The SMLH topology supports many communication patterns found in different classes of computation, such as bus-based, mesh-based, and tree-based problems as well as hypercube-based problems. A very attractive feature of the SMLH network is its ability to support a large number of processors with the possibility of maintaining a constant degree and a constant diameter. Other positive features include symmetry, incremental scalability, and fault-tolerance. It is shown that the SMLH network provides better average message distance, average traffic density, and queueing delay than many similar networks, including the binary hypercube, the SBH, etc. Additionally, the SMLH has comparable performance to other high-performance hypercubic networks, including the Generalized Hypercube and the Hypermesh.

3.1 INTRODUCTION

Progress in VLSI technology combined with the escalating demands for more processing power and speed have recently produced a technological environment in which Massively Parallel Processors (MPPs) with hundreds or even thousands of processing elements (PEs) are becoming commonplace (examples include Intel Paragon, Cray T3D and T3E, IBM SP-1 and 2, MasPar MP-1 and 2, Stanford Dash, etc.). The interconnection network, not the PEs or their speed, is proving to be the decisive and determining factor in terms of cost and performance [11, 36, 37, 21]. Ideally, a network topology intended for massively parallel computation should have the following characteristics: (1) small or bounded degree (a small degree implies low design complexity and low cost, a bounded or fixed degree implies scalability and fixed PE complexity), (2) ability to incrementally add PEs to an existing network with minimal or no changes at all to the existing configuration, (3) a large number of PE-disjoint paths between any two pairs of PEs for increased reliability and fault-tolerance, (4) the message routing should be simple to implement and flexible to route around faulty PEs in the network, (5) the diameter should be small to minimize communication delays, (6) must have a large communication bandwidth, and (7) must possess good embedding capabilities (ability to efficiently emulate a large number of commonly required interconnection patterns). We should note that the diameter of a network remains an important criterion in characterizing the network delay (latency) even with the wide use of wormhole routing [30, 7]. Wormhole routing does not reduce the time required for intermediate PEs to process small packets (flits). Larger diameter networks must therefore keep a larger number of PEs busy for a single packet transfer which leads to a reduction in throughput for the entire system [13].

To this end, several topologies have been proposed to fit different styles of computation. Examples include crossbars, multiple buses, multistage interconnection networks, and hypercubes, to name a few. Unfortunately most of the presently known network topologies do not satisfy all the above characteristics. Among these, the hypercube has received considerable attention due mainly to its good topological characteristics (small diameter, regularity, high connectivity, simple control and routing, symmetry and fault tolerance), and its ability to efficiently permit the embedding of numerous topologies such as rings, trees, meshes, and shuffle-exchange, among others [34]. However, a drawback of the hypercube is its lack of scalability which limits its use in building large size systems out of smaller size systems. The lack of scalability of the hypercube stems from the fact that the node degree is not bounded and varies as $\log_2 N$, where N is the total number of nodes. This property makes the hypercube cost prohibitive for large N. Most hypercube-based interconnection networks proposed in the literature [9, 2, 24, 3] suffer from similar size scalability problems.

Recently some networks have been introduced that are a product of hypercube topology with fixed degree networks such as the mesh, the tree, and the deBruijn [21, 3, 10] in the quest of preserving the properties of the hypercube while improving its scalability characteristics. In a previous study the authors introduced the Optical Multi-Mesh Hypercube (OMMH) [22, 23, 18]. The OMMH is a network that combines the positive features of the hypercube with those of a mesh (constant node degree and size scalability). The OMMH can be viewed as a two-level system: a local connection level representing a set of hypercube modules and a global connection level representing the mesh network connecting the hypercube modules. The OMMH network has been physically demonstrated using a combination of free-space and fiber optics technologies, and has shown good performance characteristics [19] for a reasonable size network. However, for very large networks (greater than one thousand PEs), the OMMH experiences a logarithmic increase in diameter and requires a large amount of fiber which makes the implementation complicated and expensive.

We propose a novel network that improves the topological characteristics as well as the implementation and performance aspects of the OMMH network. The new network topology proposed is called *Spanning Multichannel Linked Hypercube* (SMLH) and possesses a constant degree and a constant diameter while preserving many properties of the hypercube. The SMLH, similar to the OMMH, employs the hypercube topology at the local connection level. The global connection level connecting the hypercube modules is called a *Spanning Multichannel Network* (SMN) and resembles a Spanning Bus Hypercube (SBH) network [43], where each bus is replaced with a multichannel link. The SBH is a D-dimensional lattice of width w in each dimension. Each node is connected to D buses, one in each of the orthogonal dimensions; w nodes share a bus in each dimension which permits only one data transfer at a time. The spanning bus hypercube offers small node degree, small diameter, and scalability. It can be scaled up by expanding the size of the spanning buses [43]. However, expanding the size of the buses leads to an $O(w)$ increase in traffic density [43] which in turn leads to bus congestion problems [16]. In the SMN, each multichannel link, spanning a dimensional axis, represents w distinct logical channels allowing up to w simultaneous data transfers. The advantages of the SMLH network are that it utilizes both the hypercube local interconnection level and the SMN multiple logical channels to decrease traffic density which alleviates the bus congestion problems encountered in pure SBH networks. This feature allows larger systems to be built since it allows the SMLH multichannel links to support a larger number of processors than the SBH network. Another attribute of the SMLH is incremental scalability with a high degree of connectivity and a low diameter.

3.2 STRUCTURE OF THE SPANNING MULTICHANNEL LINKED HYPERCUBE NETWORK

In this section, we formally define the structure of the SMLH network and discuss its properties.

3.2.1 Topology of the SMLH Interconnection Network

The topology of the SMLH can be described as an undirected graph, $G_{\text{SMLH}} = (V, E)$, where V represents a set of nodes and E represents a set of edges. The SMLH can also be viewed as a product hybrid graph because it combines a D-dimensional spanning multichannel graph and a binary hypercube graph in such a way that if $G = G1 \times G2$, where $G1$ represents the spanning multichannel graph and $G2$ the Binary Hypercube (BHC) graph, then the Cartesian product of their vertices is $V1 \times V2 = [(u2, u1)|u2 \in V2 \text{ and } u1 \in V1]$ [3]. This resultant graph relationship and other performance properties for product networks are shown in Table 3.1 [1]. The SMN resembles the SBH topology introduced by Wittie [43], with a major improvement in performance and complexity. Each multichannel link with w PEs represents w distinct logical channels which allow up to w simultaneous data transfers. In contrast, in the SBH, each bus can only allow one data transfer at a time.

The SMLH can be characterized by a three-tuple (w, n, D), where w, n and D are positive integers. The first parameter, w, defines the number of nodes attached to each multichannel link. The second parameter, n, is the degree of the point-to-point n-cube (hypercube). The third parameter, D, identifies the number of dimensions spanned by a node in the SMN. To this end, the SMLH can be viewed as a product hybrid network combining two subnetworks: a D-dimensional SMN, and a hypercube.

For an SMLH(w, n, D), the number of nodes, $|V|$, is equal to $w^D 2^n$. A node address in the SMLH is denoted by a $(w+1)$-tuple $(a_1, a_2, ..., a_w, a_{w+1})$ using a mixed radix system, where for $i = 1$ to $i = w$, $0 \le a_i < w$, and $0 \le a_{w+1} < 2^n$.

Given the set of nodes (V), the set of edges (E) is constructed as follows: For two nodes $(a_1, a_2, ..., a_w, a_{w+1})$ and $(b_1, b_2, ..., b_w, b_{w+1})$ where for $i = 1$ to $i = w$, $0 \le a_i < w$, for $j = 1$ to $j = w$, $0 \le b_j < w$, $0 \le a_n < 2^n$, and $0 \le b_{w+1} < 2^n$:

1. The two nodes span the same SMN if (1) $a_{w+1} = b_{w+1}$ and if (2) for $i = 1$ to $i = w$ there are only two components, a_j and b_j, that are different while all the other components are identical.

Table 3.1 Properties of Product Networks.

Network Property	Relationship
Resultant Graph	$G_1 \times G_2$
Size (N)	$N_1 \times N_2$
Degree (k)	$k_1 + k_2$
Diameter (d)	$d_1 + d_2$
Number of Links (L)	$N_1 L_2 + N_2 L_1$

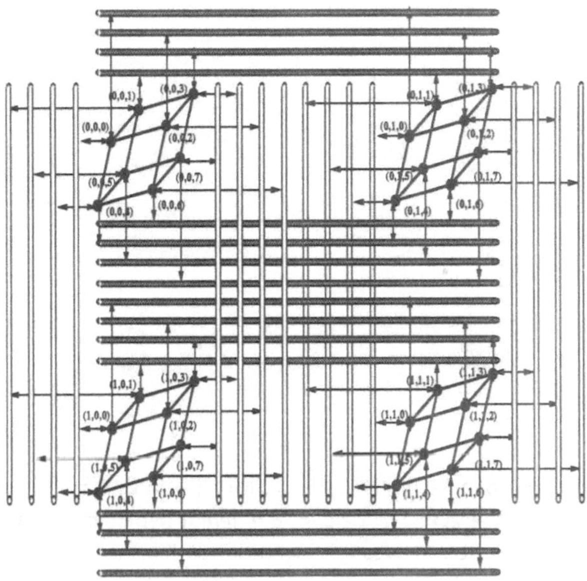

Figure 3.1 An example of the spanning multichannel linked hypercube network: a SMLH($w = 2, n = 3$) (32 nodes) interconnection is shown. Thick lines represent SMN connections while bold, thin lines represent point-to-point hypercube connections.

2. There is a link (called a hypercube link) between two nodes if and only if, for $i = 1$ to $i = w$, (1) $a_i = b_i$, and (2) a_{w+1} and b_{w+1} differ by one bit position in their binary representation (Hamming distance of one).

Fig. 3.1 shows an SMLH($w = 2, n = 3, D = 2$) interconnection where the bold, thin lines represent point-to-point hypercube links, and thick lines represent SMNs. Small dark circles represent nodes of the SMLH network which are, in this chapter, abstractions of processing elements, memory modules, or switches. Note that, because $D = 2$ each node spans two SMNs, one along each dimension. Furthermore, there are three bidirectional point-to-point links attached to a node which correspond to the hypercube links. A careful observation of Fig. 3.1, shows that the node addresses satisfy the connection rules outlined earlier.

In this chapter, we only consider SMLH networks with $D = 2$ and leave higher dimensioned SMLH network analysis for future work. Therefore in the notation, the third parameter, D, will be dropped. Consequently, an SMLH($w, n, D = 2$) network will be referred to as SMLH(w, n). As can be seen in Fig. 3.1, the SMLH($w = 2, n = 3$) consists of $2^2 \times 2^3 = 32$ nodes, and it can be viewed as 8 concurrent 2D SMNs. Note that w horizontal multichannel links and w vertical multichannel links are needed to form one $w \times w$, 2D SMN. Fig. 3.2 shows one such 2D SMN formed by nodes with the same hypercube addresses and belonging to different hypercube modules. Sim-

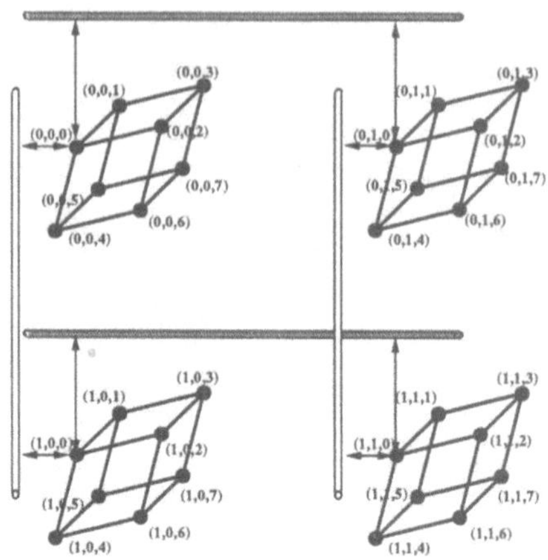

Figure 3.2 An example of a 2D SMN within a SMLH($w = 2, n = 3$) network. Note that the nodes that construct the 2D SMN belong to different hypercube modules but they possess the same binary hypercube address representation within their corresponding hypercube modules. Eight such 2D SMNs co-exist in the SMLH($w = 2, n = 3$) interconnection.

ilar considerations take place for the other seven 2D SMNs in Fig. 3.1. The SMLH($w = 2, n = 3$) network can also be viewed as 4 concurrent 3-dimensional hypercubes in which 4 nodes having identical hypercube addresses form a 2×2 SMN. The SMLH($w = 2, n = 3$) in Fig. 3.1 looks like a hypercube-clustered spanning multichannel network. In general, there are 2^n 2D SMNs and w^2 hypercube modules. Note than when w is equal to one, the SMLH becomes a pure hypercube network, while when n is equal to zero it becomes a pure spanning bus network if the multichannel link is constrained to operate on a single channel. This implies that both the hypercube and the SBH can be thought of as special cases of the SMLH network.

The choice of two parameters, w and n, completely determines the size of the network, the resources and implementation requirements, and the scaling complexity. The w parameter determines the size of the multichannel link lattice, while the n parameter defines the size of the hypercubes. From a scaling viewpoint, two scaling rules can be applied for an SMLH(w, n) network. The first rule which we call "fixed-w" rule keeps the size of the multichannel links constant and increases the size of the network by increasing n. The second rule which, we call "fixed-n" rule, keeps the size of the hypercube constant and increases the size of the network by increasing w. Clearly, the advantage of the SMLH(w, n) network is its flexibility to scale up using either, or a combination of, the two scaling rules.

For instance, the size of the SMLH can grow without altering the number of links per node by expanding the size of the multichannel links. For example, 3-dimensional hypercubes can be added on the perimeter of the 2D SMN of Fig. 3.1. Fig. 3.3 illustrates an SMLH($w = 3, n = 3$) which is constructed by

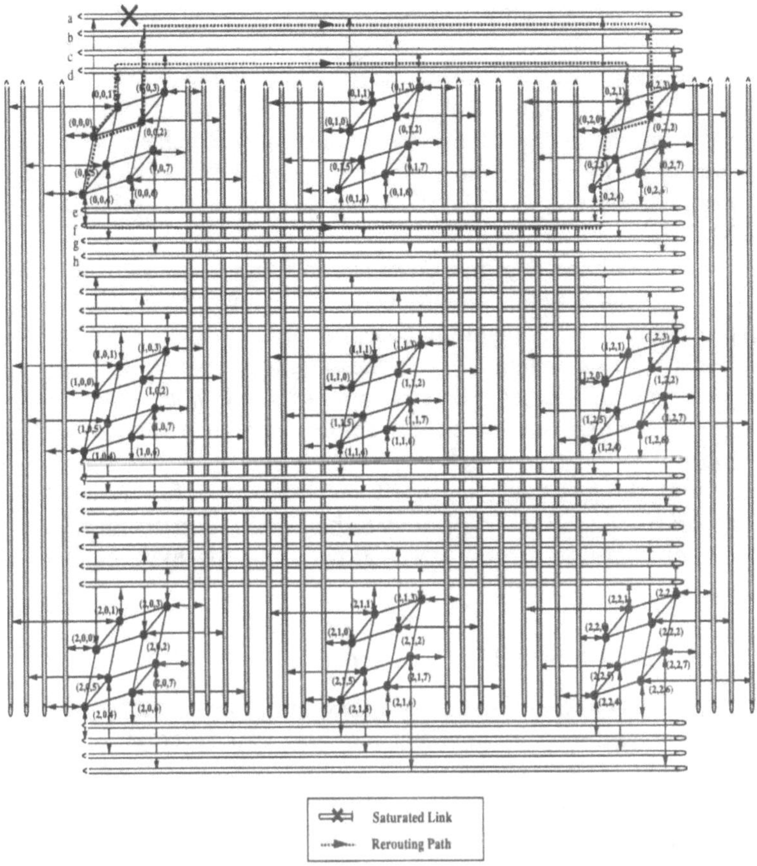

Figure 3.3 A SMLH($w = 3, n = 3$) (72 nodes) interconnection. This SMLH network can be constructed by adding hypercube modules along a row and a column.

expanding the SMLH($w = 2, n = 3$) network by adding hypercube modules along an outer row and an outer column. The existing configuration of the nodes of the SMLH($w = 2, n = 3$) network did not change because each node still spans two multichannel links and still has three bidirectional point-to-point links for the hypercube connections. This option allows the SMLH to be size scalable. Some discussion of the relationship between the two scaling rules and tradeoffs in performance are discussed in Section 3.2.3.

3.2.2 Message Routing in the SMLH Interconnection Network

Due to the regularity and symmetry of the SMLH architecture, a distributed routing scheme can be implemented without global information. At the source node, the message is formatted with the source address, the destination address, message length, and a few control bits such as semaphore bits. The interprocessor message traffic of a node gets redistributed into two categories, i.e., the hypercube communication and the spanning multichannel communication. If the source and the destination of the message are within the same hypercube subnetwork of the SMLH network, the routing procedure is exactly the same as that of the regular hypercube network. Similarly, if the source and the destination of the message are within the same SMN of the SMLH network, the routing procedure is exactly the same as that of a regular bus connected network [23].

If the source and the destination of the message share neither a hypercube nor a 2D SMN, the routing scheme uses the hypercube routing scheme until the message arrives at the same hypercube node as the destination, and then uses the SMN routing scheme for the message to arrive at the destination. Or the SMN routing scheme can first be applied to forward the message to the same hypercube where the destination resides, and then the message can reach the destination using the hypercube routing scheme. We can also mix the hypercube and the SMN routing until the message is forwarded to the same hypercube or to the same SMN where the destination resides, and then we can forward the message to the destination using the hypercube or the SMN routing scheme, respectively.

3.2.3 Properties of the SMLH Interconnection Network

3.2.3.1 Diameter and Link Complexity. The diameter of a network is defined as the maximum distance between any two processors in the network. Thus, the diameter determines the maximum number of hops that an average message may have to take. The diameter of a 2D SMN is two, as it is for a 2D SBH, and the diameter of a binary hypercube with N_{BHC} nodes is $n = \log_2 N_{BHC}$. Therefore the diameter of SMLH(w, n) is $(n+2)$. For the SMLH(w, n) network with N nodes, $N = w^2 2^n$, therefore $n = \log_2 \frac{N}{w^2}$. Consequently the diameter of the SMLH(w, n) network can be written as $\log_2 \frac{N}{w^2} + 2$. Using the fixed-w scaling rule, the diameter of the SMLH(w, n) network experiences a logarithmic increase $(O(\log_2 N))$ when the network size increases. However, using the fixed-n scaling rule would make the diameter constant, $n+2$, for any network size.

Link complexity or node degree is defined as the number of physical links per node. For a regular network where all nodes have the same number of links, the node degree of the network is that of a node. The node degree of a hypercube with N_{BHC} nodes is $n = \log_2 N_{BHC}$, and that of a 2D SMN is two, since there are only two physical SMNs links connected to each node. A node of an SMLH(w, n) network possesses links for both the hypercube connections and the SMN connections. Consequently, the node degree of the SMLH network is $(n+2)$ or $\left(\log_2 \frac{N}{w^2} + 2\right)$. Again, when using the fixed-w scaling rule the SMLH network experiences a logarithmic increase in degree $(O(\log_2 N))$; however, when the network is expanded using the fixed-n scaling rule, the degree becomes constant, $n+2$.

3.2.3.2 Bisection Width. The bisection width of a network is defined as the minimum number of links that have to be removed to partition the network into two equal halves [15]. The bisection width indicates the volume of communication allowed between any two halves of the network with an equal number of nodes. The bisection width of a n-dimensional hypercube is $2^{n-1} = N_{BHC}/2$ since that many links are connected between two $(n-1)$-dimensional hypercubes to form an n-dimensional hypercube. The equation of bisection width for the SMLH varies depending on whether w is even or odd, since w odd dictates cutting at least one hypercube in half. For w even, there are w such n-dimensional hypercubes connecting 2^n 2D SMNs, the bisection width of an SMLH(w even, n) is equal to $w \times 2^n = N/w$. For w odd, there is one such n-dimensional hypercube that must be cut in half in addition to cutting all multichannel links in one dimension and 2^{n-1} multichannel links in the second dimension. Therefore, the bisection width of an SMLH(w odd, n) is equal to $w \times 2^n + 2^n + 2^{n-1} = N/w + 3N/2w^2$.

3.2.3.3 Granularity of Size Scaling. The granularity of size scaling is the ability of the system to increase in size with minor or no change to the existing configuration, and with an expected increase in performance proportional to the extent of the increase in size. For a 2D SMN, the granularity of size scaling is only $2w+1$ since, at a minimum, one multichannel link per dimension could be added to the network in order to increase its size. Therefore the granularity of the size scaling in a $w \times w$ 2D SMN of $N_{SMN} = w^2$ nodes is $2N_{SMN}^{1/2} + 1$. However, the size of a hypercube can only be increased by doubling the number of nodes; that is, the granularity of size scaling in an n-dimensional hypercube is 2^n. When the fixed-w scaling rule is applied, the granularity of size scaling follows the hypercube size scaling. Therefore, the granularity of size scaling for the SMLH, using the fixed-w rule, is $w^2 \times 2^n = N$. When the fixed-n scaling rule is used, the granularity of size scaling follows that of the SMN. Therefore, the granularity of size scaling following the fixed-n rule is $2^n(2w+1) = 2(N/w)+2^n$. Note that the granularity of size scaling using the fixed-w rule is $O(N)$ while for the fixed-n rule is $O(N/w)$. SMLH is competitive with other networks in this smooth scalability property [39].

3.2.3.4 Average Message Distance. The average message distance in a network is defined as the average number of links that a message should travel between any two nodes. Let N_i represent the number of nodes at a distance i, then the average distance $\bar{l}$ is defined as [2]:

$$\bar{l} = \frac{1}{N-1}\sum_{i=1}^{n} iN_i, \tag{3.1}$$

where N is the total number of nodes, and n is the degree. Since the SMLH is a product network of the BHC and SMN, the average message distance can be derived from the individual average message distances for the BHC and SMN.

For the BHC, the number of nodes at a fixed distance, $N_{i,BHC}$, is given as [2]:

$$N_{i,BHC} = \binom{n}{i}(w-1)^i, \tag{3.2}$$

since for i differing digits, which are each able to vary in $(w-1)$ ways, the enumeration is $(w-1)^i$. Substituting the equation into Eq. 3.1 and computing the summation yields the common form of the BHC average message distance $\bar{l}_{BHC}$ [2]:

$$\bar{l}_{BHC} = \frac{n}{2}\left(\frac{N_{BHC}}{N_{BHC}-1}\right). \tag{3.3}$$

For a 2D SMN, the number of nodes at distance 1, $N_{1,SMN}$ is $2(w-1)$ since there are $w-1$ nodes in each direction, not including the source node. Since the diameter of the 2D SMN is 2, the number of nodes at distance 2, $N_{2,SMN}$, is equal to all nodes not at distance 1, minus the source node, i.e.,

$$N_{2,SMN} = w^2 - 2(w-1) - 1. \tag{3.4}$$

Since the SMLH network is not required to use time division multiple access (TDMA) as the SBH is to access a shared medium, the above equations for $N_{i,SMN}$ will result in an average message distance for a 2D SMN, $\bar{l}_{SMN}$, which differs from the traditional average message distance equation for a SBH of [43]:

$$\bar{l}_{SMN} = \frac{D(w-1)}{2}\left(\frac{N_{SMN}}{N_{SMN}-1}\right). \tag{3.5}$$

Assuming a collision free environment, an average message can use the hypercube routing scheme until the message arrives at the same hypercube position and then use the SMN routing scheme to arrive at the destination, as described in Sec. 3.2.2. Or the SMN routing scheme can first be applied to forward the message to the same hypercube where the destination resides, and then use the hypercube routing scheme. Therefore, and average message has the potential of encountering any of the $(N_{BHC}-1)$ nodes in a given hypercube and any of the $(N_{SMN}-1)$ SMN nodes, where N_{BHC} is the total number of nodes in a single binary hypercube and N_{SMN} is the number of nodes in the SMN network. Therefore, the average message distance in the SMLH, $\bar{l}_{SMLH}$, can be calculated as:

$$\bar{l}_{SMLH} = \frac{1}{(N_{BHC}-1)+(N_{SMN}-1)}\left(\sum_{i=1}^{n} iN_{i,BHC} + \sum_{i=1}^{2} iN_{i,SMN}\right), \tag{3.6}$$

which can be rewritten using the above equations as:

$$\bar{l}_{SMLH} = \frac{2w(w-1)+n2^{n-1}}{(w^2-1)+(2^n-1)}. \tag{3.7}$$

3.3 COMPARISONS OF SMLH WITH POPULAR NETWORKS

In this section we compare the SMLH network with existing, well known topologies. These include the Binary Hypercube (BHC) [34], the Generalized Hypercube (GHC) [2], the Torus [33], the Spanning Bus Hypercube (SBH) [43], the Hierarchical Cubic Network (HCN) [9], the Cube-Connected-Cycle (CCC) [32], the Hyper-deBruijn (HdB) [8], the Folded Peterson (FPT) [31], the Hypermesh (HM) [39], and the Optical Multi-Mesh Hypercube (OMMH) [22]. The comparison parameters include diameter, degree, number of links, average message distance, average traffic density and queueing delay. The topological characteristics of the above networks are indicated in Table 3.2. The results of the comparison are shown in Figs. 3.4, 3.5, 3.6, 3.7, 3.8, 3.9, 3.10, and 3.11.

In the figures, the SMLH($w = 32, n$) notation denotes that the network is expanded following the fixed-w ($w = 32$) rule; that is, the size of the multichannel links is kept constant (32 PEs each) and the size of the hypercube module is changed to have the same network size for comparison purposes. The SMLH($w, n = 4$) notation denotes that the network is expanded following the fixed-n rule which means that the size of the hypercube module (4) is kept fixed and the size of the multichannel links are increased. Note that when expanding the SMLH network some mathematical constraints exist. In Subsection 3.2.3 the degree and diameter of the SMLH network were derived; they are both equal to $n + 2$, or rewritten as $\log_2 \frac{N}{w^2} + 2$. The first term of the equation is a factor of both the number of nodes in the entire network (N) and the size of the multichannel links (w). The constraint is that $N \geq w^2$ because otherwise the $\log_2 \frac{N}{w^2}$ factor of the degree/diameter equation will give a negative number which would be unacceptable. The notation Torus($w = 3, n$) means the number of nodes per ring is fixed while the hypercube dimension varies. The

Table 3.2 Topological Characteristics of Several Popular Networks.

Network	Size (N)	Degree (k)	Diameter (d)	Number of Links (L)
BHC	2^n	n	n	$\frac{N}{2}n$
Torus(w, n)	w^n	$2n$	$\frac{w}{2}n$	Nn
GHC(w, n)	w^n	$(w-1)n$	n	$\frac{N}{2}(w-1)n$
SBH(w, D)	w^D	D	D	$\frac{N}{w}D$
HCN	2^{2n}	$n+1$	$2n$	$\frac{N}{2}(n+1)$
CCC(c, n)	$c2^n$	3	$\lfloor\frac{(5c-2)}{2}\rfloor$	$\frac{3}{2}N$
HdB(c, n)	$2^{(n+c)}$	$n+4$	$n+c$	$\frac{N}{2}(n+4)$
FPT	10^n	$3n$	$2n$	$\frac{3N}{2}n$
HM(w, n)	w^n	n	n	Nn
OMMH(l, l, n)	$l^2 2^n$	$n+4$	$l+n$	$\frac{N}{2}(n+4)$
SMLH(w, n)	$w^2 2^n$	$n+2$	$n+2$	$\frac{N}{2}(\frac{4}{w}+n)$

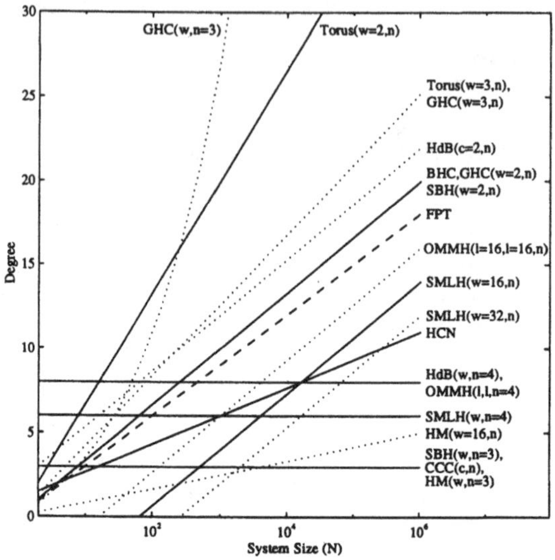

Figure 3.4 Network comparisons for degree.

notation SBH($w, D = 3$) means that the dimension of the SBH network is kept constant and the size of the buses is changed. The notation GHC($w, n = 4$) denotes that the diameter of the network is fixed while the number of nodes per dimension is varied. Similarly, the notation HM($w, n = 2$) means that the node dimension is fixed while the number of nodes per dimension is varied. The notation OMMH($l = 16, l = 16, n$) denotes that the size of the mesh network in the OMMH is fixed while the size of the hypercube is varied. Similarly the OMMH($l, l, n = 4$) notation denotes that the size of the hypercube is fixed and the mesh size is varied. The notation used for the other networks is similarly given.

3.3.1 Degree and Diameter

Figs. 3.4 and 3.5 show the graph comparisons in terms of degree and diameter as network size is increased. At the key mark of 10,000 nodes (desirable for MPPs), SBH($w, D = 3$), HM($w, n = 3$), HM($w = 16, n$), SMLH($w = 32, n$), SMLH($w, n = 4$) and SMLH($w = 16, n$) exhibit very good performances in terms of diameter and degree with values 3, 3, 3.3, 5.3, 6 and 7, respectively. The CCC(c, n) and HCN reveal good degree values (3 and 7.6) but they also exhibit fairly large diameters (17 and 13). The OMMH($l, l, n = 4$) experiences the worst diameter (29) and the GHC($w, n = 3$) the worst degree (61). Even though at 10,000 nodes the SMLH($w = 32, n$) reveals better characteristics than the SMLH($w, n = 4$) the later is more desirable because it possesses constant degree and diameter, features that allow it to be scalable. The SMLH($w = 32, n$), on the other hand, experiences a logarithmic increase in degree and

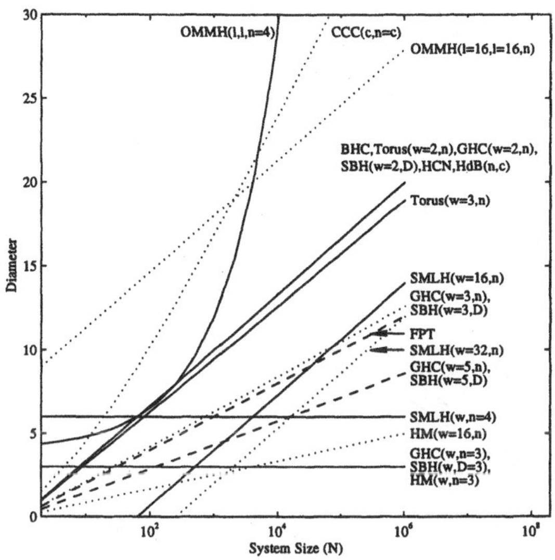

Figure 3.5 Network comparisons for diameter.

diameter, features that make it difficult to scale up to a larger number of processors. In general, from Fig. 3.4, the hybrid networks show a logarithmic increase in their degree which makes them difficult to scale them up in size.

3.3.2 Number of Links

Fig. 3.6 shows the graph comparisons in terms of the number of links as the networks scale up in size. The SBH$(w, D = 3)$, CCC(c, n), SMLH$(w, n = 3)$, SMLH$(w = 32, n)$, HM$(w, n = 2)$, SMLH$(w, n = 4)$, SMLH$(w = 16, n)$ and SBH$(w = 3, n)$ reveal the best performance characteristics in terms of number of links while the GHC$(w, n = 4)$ and Torus seem to require a larger number of links. We should note that of the SMLH links, $2N/w$ of them are actually multichannel links.

3.3.3 Average Message Distance

Fig. 3.7 illustrates graph comparisons between the BHC, SBH, SMLH, GHC, and HM networks in terms of average message distance. The average message distance of the SMLH is given in Eq. 3.7, the BHC in Eq. 3.3, and the SBH in Eq. 3.5, while equations for the other topologies of interest are [2, 39]:

$$\bar{l}_{GHC} = \frac{n(w-1)}{w}\left(\frac{N_{GHC}}{N_{GHC}-1}\right). \tag{3.8}$$

$$\bar{l}_{HM} = n\left(\frac{N_{HM}}{N_{HM}-1}\right). \tag{3.9}$$

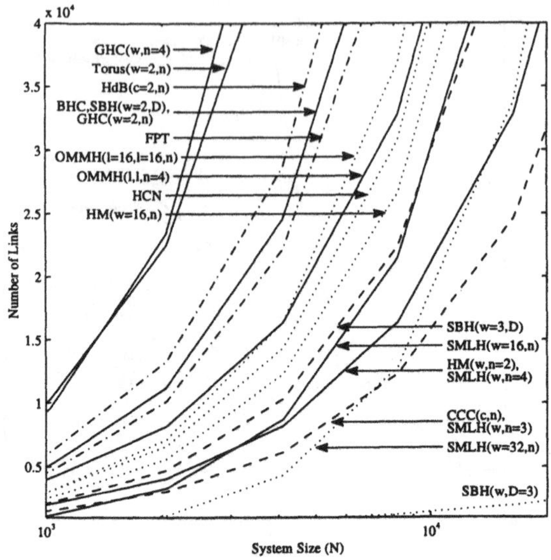

Figure 3.6 Network comparisons for number of links.

At 10,000 nodes the SMLH($w = 16, n$), SMLH($w = 32, n$), SMLH($w, n = 3$), SMLH($w, n = 4$), SMLH($w, n = 5$), HM($w, n = 2$), and GHC($w, n = 2$) exhibit the best average message distances with values close to 2. The SBH($w, D = 3$), SBH($w = 3, D$) and BHC experience the worst average message distances with values close to 31, 13 and 6.6, respectively. Fig. 3.7 indicates the average message distance of the SMLH and GHC, both having fixed size hypercubes, as well as the HM($w, n = 2$) approach a constant value for large N. However, the other networks (excluding the SBH with fixed size hypercubes) grow logarithmically with respect to the network size.

3.3.4 *Average Traffic Density*

As emphasized in Sec. 3.2, the advantages of the SMLH(w, n) network over the SBH network is its ability to use the multichannel links and point-to-point hypercube links to alleviate the spanning bus congestion. A very good measure of that is the average traffic density. The average traffic density is defined as the product of the average distance and the total number of nodes, divided by the total number of communication links [6]. Using the definition stated above, the average traffic density of the SMLH(w, n) network can also be calculated:

$$\rho = \frac{\bar{l}_{SMLH} N}{\frac{N}{2}\left(\frac{4}{w} + n\right)}. \tag{3.10}$$

Therefore the average traffic density can be rewritten as

$$\rho = \frac{2\bar{l}_{SMLH}}{\left(\frac{4}{w} + n\right)}. \tag{3.11}$$

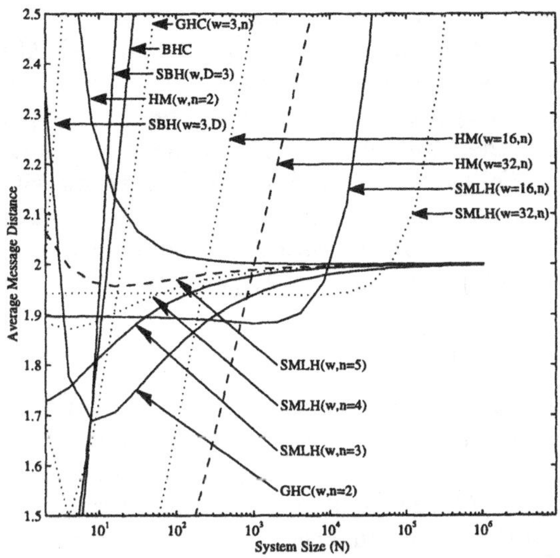

Figure 3.7 Network comparisons for average message distance.

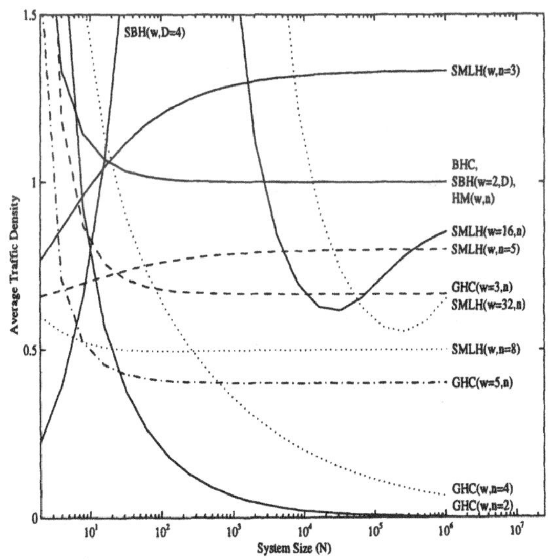

Figure 3.8 Network comparisons for average traffic density.

where $\bar{l}_{SMLH}$ can be obtained from Eq. 3.7. Eq. 3.11 reveals that when the fixed-n rule is followed to expand the network, the average traffic density of the SMLH(w, n) is essentially independent of w. This feature allows the network to utilize a much large number of nodes along the multichannel links.

Fig. 3.8 presents the average traffic density comparisons between the BHC, SBH, SMLH, GHC and HM networks. The HM, BHC, GHC with fixed w, and the SMLH with fixed sized hypercubes, have low traffic density and exhibit no sensitivity to size for larger networks. As the size of the hypercubes within the SMLH is increased, the traffic density approaches zero, as it also does for the GHC with fixed n. On the other hand, the SBH network with fixed dimension ($\text{SBH}(w, D = 4)$) shows an increase in traffic density. Therefore for larger networks, the SBH network would most likely experience severe bus congestion problems which would lead to large message delays.

3.3.5 Queueing Delay Analysis

For an exact analysis of the queueing delay of the SMLH network, we based our model on similar analysis of hypercubes [2] and optical hypercubes, meshes and hypermeshs [39]. The model is a communication net with the ith channel represented as an $M/M/1$ system with Poisson arrivals at a rate λ_i and exponential service time of mean $1/\mu c_i$. $1/\mu$ is the average packet size and c_i is the capacity of the ith channel. Additionally, the following assumptions are made [14]:

- Each node is equally likely to send a message to every other node in a fixed time period.
- The routing algorithm traverses dimensions in a static order.
- The load is evenly distributed, i.e., λ_i is the same for all i.
- The link capacities of the network have been optimally assigned.
- The cost per capacity per link is unity.

Under theses assumptions, an exact expression for the queueing delay is an M-channel N-node network is given by [14]

$$T = \frac{\bar{l}\left(\sum_{i=1}^{M}\sqrt{\frac{\lambda_i}{\lambda}}\right)^2}{\mu C\left(1 - \bar{l}\gamma\right)}, \tag{3.12}$$

where M = total number of directed links, $\lambda = \sum_{i=1}^{M} \lambda_i = M\lambda_i$, γ = the utilization factor, and $C = \sum_{i=1}^{M} c_i$ = aggregate bandwidth of the topology. Removing constants μ, C, and N, the above delay can be normalized to

$$T = \frac{\bar{l}M}{1 - \bar{l}\gamma}. \tag{3.13}$$

Figs. 3.9, 3.10, and 3.11 graph the average queueing delay in terms of increasing utilization for the BHC, SBH, SMLH, GHC, and HM with 256, 1,024, and 4,096 nodes, respectively. The delay increases exponentially with increasing

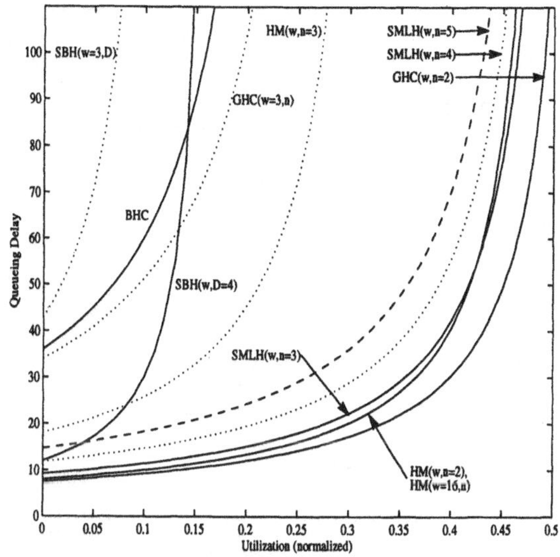

Figure 3.9 Queueing delay for networks of size $N = 256$.

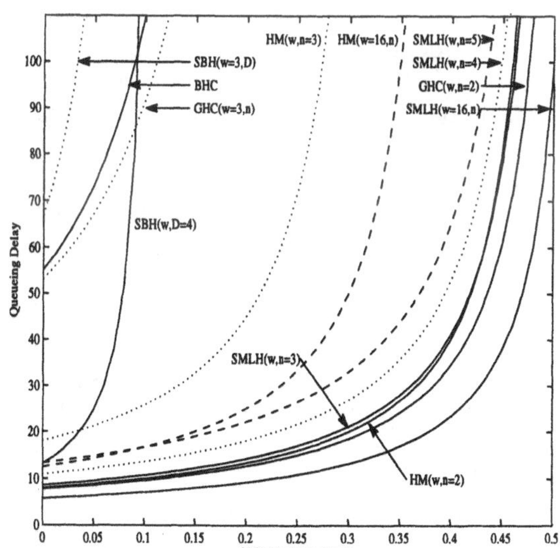

Figure 3.10 Queueing delay for networks of size $N = 1{,}024$.

utilization and saturates at a particular load, given by $\gamma_{sat} = 1/\bar{l}$. The graphs illustrate the maximum traffic capability of each topology, with the SMLH, GHC with fixed n, and HM with fixed n having higher utilization and lower queueing times than the other networks.

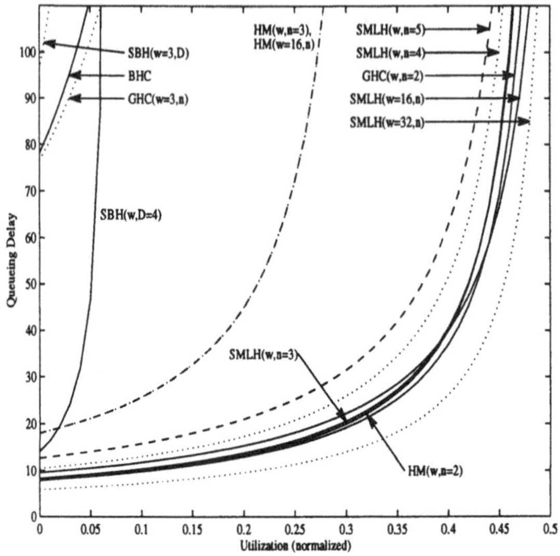

Figure 3.11 Queueing delay for networks of size N = 4,096.

The graphs further prove that the SBH network, despite its better topological characteristics (diameter, degree, etc.), has severe bus saturation problems for large number of PEs. The advantage of the SMLH network is that in addition to its good topological characteristics, it demonstrates insensitivity to traffic density when the network scales up in size. This feature allows the SMLH network to grow up in size while maintaining good traffic capability. Nevertheless, even if saturation problems appear, the SMLH network does not experience the same message delays as the SBH because it can utilize the point-to-point hypercube links to redirect the packets from another path. In the SMLH network, the saturated multichannel link can be side stepped by only two additional hops as long as that particular saturated multichannel link is not involved in the communication. Fig. 3.3 demonstrates a re-routing scheme assuming a multichannel link saturation problem. Assume that in Fig. 3.3 node $(0,0,0)$ wants to send a package to node $(0,2,0)$ via the horizontal multichannel link a. In case of a multichannel link saturation problem, the package will need to follow a different routing path. Three different routing paths are available and are all shown in Fig. 3.3 with thick dotted lines. The packet can utilize three hypercube links to access multichannel link b or multichannel link d or multichannel link f. By using one of the three multichannel links the packet will arrive at a PE which is one hop away from the destination link.

3.4 OPTICAL IMPLEMENTATION OF THE SMLH NETWORK

Obviously an electronic implementation of the proposed SMLH network is feasible. One methodology would be to use multiprocessor board technology (e.g. Multichip Module technology) for the hypercube module connections and backplanes for the multichannel link connections. To limit the number of boards required, k hypercube modules can be clustered together on a single multiprocessor board. However, for a large number of PEs, and a greater bandwidth and interconnection density, conventional backplanes have major limitations [37, 21, 38]. These include signal skew, wave reflection, impedance mismatch, skin effects, interference, and many others. A possible alternative is the use of optical interconnects. Optical interconnects offer many communication advantages over electronics, including gigahertz transfer rates in an environment free from capacitive loading effects and electromagnetic interference, high interconnection density, low power requirements, and possibly a significant reduction in design complexity through the use of multiple access techniques and the third dimension of free-space optics. The effectiveness of optical interconnects has been extensively examined [37, 21, 4, 6, 17, 25]. In the following we propose an all optical implementation of the SMLH(w, n) network where the hypercube modules are implemented using free-space, space-invariant optics [22], and the multichannel link modules are implemented using Wavelength Division Multiple Access (WDMA) techniques.

3.4.1 *Optical Implementation of the Hypercube Modules using Holographic Optical Elements (HOEs)*

The free space optical implementation of the hypercube network has been rigorously studied and analyzed [21, 22, 23]. The main objective is to exploit the third dimension and the communication advantages of free-space optics to provide efficient and adequate implementation of the hypercube network. This implementation is for illustration only and may not handle the simultaneous transfer of multiple, different packets, as a pure hypercube can. However, optics is quite capable of handling this function and a subsequent paper will address this in more detail.

The design methodology is based on an observation that PEs in a bipartite interconnection network can be partitioned into two different sets of PEs such that any two PEs in a set do not have a direct link. This is a well-known problem of bipartitioning a graph if the interconnection network is represented as a graph. For a binary n-cube, PEs whose addresses differ by more than one in Hamming distance can be in the same partition, since no link exists between two PEs if their Hamming distance is greater than one. Besides bipartitioning the graph, we arrange the PEs in each partition onto the plane such that interconnection between two planes becomes space-invariant (the connection pattern is identical for every PE in the plane). This self-imposed requirement reduces the design complexity of the optical setup [20]. The two partitions of PEs are called $plane_l$ and $plane_r$. Optical sources and detectors are assumed to be resident on processor-memory boards located on $plane_l$ and $plane_r$. Free-space holographic optical elements (HOEs) are used to implement the connection patterns required between PEs of the two planes [21, 22, 23].

Fig. 3.12 depicts the conceptual 3D free-space, space-invariant optical implementation of a 3-cube network. A 3-cube network consisting of 8 PEs is

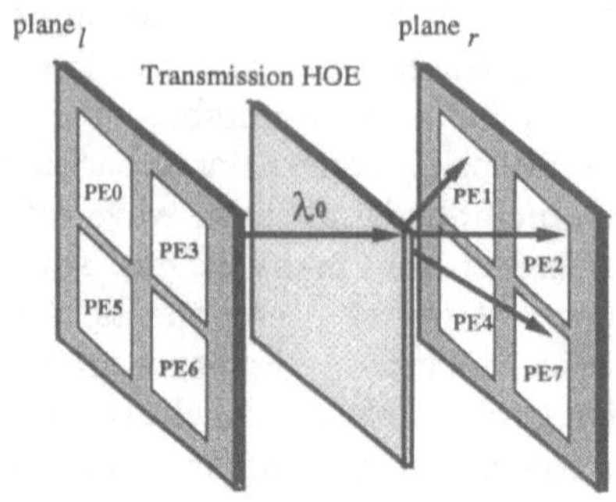

Figure 3.12 Space-invariant optical implementation of a three-cube network with a HOE.

bipartitioned into $plane_l$ and $plane_r$. Each plane consists of 4 PEs which are arranged in a 2 by 2 square configuration. The HOE provides the connection patterns required for the 3-cube. The hypercube modules have been experimentally implemented in the lab [19]. Issues relating to actual fabrication of the hologram, size, misalignment, power budget, components used, actual BER and more can be found in [19]. We should note that this is one way of implementing the Hypercube module in free space and several other means exist that are currently under study.

3.4.2 Implementation of the Spanning Multichannel Network using WDMA techniques

In this subsection the implementation of the spanning multichannel network using WDMA techniques is presented. To exploit the large communication bandwidth of optics, WDMA techniques that enable multiple multi-access channels to be realized on a single physical channel can be utilized. In a WDMA system, the optical spectrum is divided into many different logical channels, each channel corresponding to a unique wavelength. These channels are carried simultaneously on a small number of physical channels, e.g. a fiber. Additionally, each network node is equipped with a small number of transmitters and receivers, some of which are dynamically tunable to different wavelengths.

The SMLH(w, n) network consists of w^2 hypercube modules and 2^n $w \times w$ 2D SMNs. From the discussion above, every hypercube module is bipartitioned into two planes called $plane_l$ and $plane_r$. In the SMLH network, all $plane_l$s are grouped together to form a plane called $Plane_L$ while all $plane_r$s are grouped to form another plane called $Plane_R$. The multichannel links can be implemented by interconnecting the individual $plane_l$s of $Plane_L$ and $plane_r$s of $Plane_R$. The hypercube modules are implemented using free space optics to provide the connectivity between $plane_l$s and $plane_r$s. Additionally, 2^{n-1} 2D spanning multichannel networks per plane ($Plane_L$ or $Plane_R$) need to be implemented. Each 2D SMN consists of $2w$ multichannel links, therefore a total of $2w \times 2^{n-1}$ multichannel links per Plane are required.

A trivial implementation of the SMN is to assign a distinct wavelength for every PE in $Plane_L$ and $Plane_R$ and then perform WDMA techniques to implement the multichannel links. However, such a straightforward method requires a prohibitively large number of different wavelengths and fibers. For example, for an SMLH($w = 4, n = 3$) consisting of 128 PEs, a total of 64 wavelengths would be necessary. A wavelength assignment technique [22, 17] can be employed to reduce the number of wavelengths used in the system. Let's take a running example, an SMLH($w = 4, n = 3$). Fig. 3.13 shows how wavelengths are assigned for each PE of $Plane_L$. The following wavelengths are assigned to the first row: λ_1, λ_2, ..., λ_8. Then, λ_2, ..., λ_8, λ_1 are assigned as wavelengths in the second row. In general, wavelength assignment in a row is achieved by rotating the wavelength assignment of the previous row by one column. This wavelength assignment results in no two PEs in the same row or column of $Plane_L$ having an identical wavelength. Similar considerations take place for PEs of $Plane_R$. With this wavelength assignment technique, the total number of wavelengths required to implement the SMLH($w = 4, n = 3$) network is reduced from 64 to 8. In general, for an SMLH(w, n) the following wavelength assignment for the first row must be performed: λ_1, λ_2, ..., $\lambda_{w2^{\lceil (n-1)/2 \rceil}}$, and then, λ_2, ..., $\lambda_{w2^{\lceil (n-1)/2 \rceil}}$, λ_1, are assigned to the PEs of the second row and so on. Thus an implementation of an SMLH(w, n) with the above wavelength assignment requires no more than $w \times 2^{\lceil (n-1)/2 \rceil}$ wavelengths.

Referring to Fig. 3.13, the wavelengths assigned to the PEs of the first row are divided into two groups of four wavelengths each. The groups are: (λ_1, λ_3, λ_5, λ_7) and (λ_2, λ_4, λ_6, λ_8). Each of these groups correspond to the implementation of a row-wise multichannel link. Every PE in the group should be capable of tuning in to any of the wavelengths assigned to that group. For example, the node of group 1 with wavelength λ_1 must be able to tune to wavelengths λ_3, λ_5, λ_7 which correspond to wavelengths that were assigned to the other PEs of that group. Rotating the wavelength assignments of the previous rows will form the new wavelength groups that correspond to every row. Similarly, each column of Fig. 3.13 must be divided into two groups of four wavelengths each. For example, for the second column of Fig. 3.13, the following groups are formed: (λ_2, λ_4, λ_6, λ_8) and (λ_3, λ_5, λ_7, λ_1). Each of these wavelength groups correspond to the implementation of a column-wise multichannel link. Again, rotation of the column-wise wavelength assignment will result in the formation of the wavelength groups for the other columns.

We now consider the overall optical implementation of an SMLH(w, n). For simplicity and without loss of generality, we consider the implementation of an example network of size SMLH($w = 4, n = 3$). Fig. 3.13 shows an example $Plane_L$ of the SMLH($w = 4, n = 3$) network. We assume that each PE has three light sources: one fixed source, S_h, which illuminates the HOE to generate the required hypercube links, and the other two relatively tunable sources, S_r and S_c, coupled into optical fibers to implement the two multichannel links. It should be noted that full tunability is not required as explained above. In fact, each source should be tunable for a single wavelength group only. This

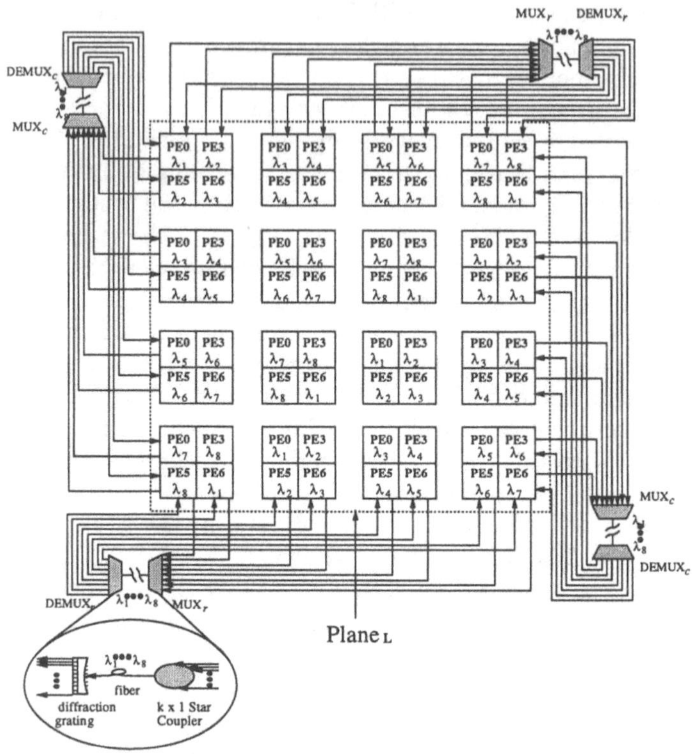

Figure 3.13 An optical implementation of $Plane_L$ of an SMLH($w = 4, n = 3$) network using WDMA. For an SMLH($w = 4, n = 3$) network 16 MUX-DEMUX pairs are need per plane. For clarity of the figure we show only four of them.

reduced range increases the efficiency, the yield, and the tuning speed of the light sources. Furthermore, each PE is equipped with three receivers; one fixed receiver, R_h, receives light from the free-space optics implementing the hypercube, and the other two receivers, R_r and R_c, receive light from fibers coming from demultiplexers. The key component that provides multichannel link connectivity here is the tunable-transmitter, fixed-receiver scheme. The wavelength assignment shown in Fig. 3.13 corresponds to the receiver wavelength assignment of every PE. Other PEs can communicate with a particular PE by simply tuning in to the wavelength assigned to that PE. Rapid progress is being made in the development of tunable devices, both in the range over which they are tunable, and their tuning times [5, 29]. Current tuning ranges are in the 4-10 nm and the tuning times vary from nanoseconds to milliseconds [5].

We also assume the availability of the following optical components: A $k \times 1$ passive optical star coupler, and a diffraction grating. The $k \times 1$ coupler acts as an optical multiplexer by funneling light from k light beams with k different wavelengths onto the same waveguide. The diffraction grating acts as an optical demultiplexer by separating and incoming combined optical signal into many signals, each corresponding to a different wavelength. Many examples that use a conventional lens, GRIN rod lenses [26], and no lens have been reported as optical demultiplexers. Recently, the idea of free-space concave grating demultiplexers has been reported [41, 42]. These demultiplexers use a wavelength range of 16 nm (40 channels) and channel spacing of 0.4 nm. The fiber-to-fiber loss was 6-9 dB and channel crosstalk was better than 25 dB. For clarity, in the rest of the chapter, we denote the $k \times 1$ star coupler as a multiplexer (MUX) and the concave grating as a demultiplexer (DEMUX), bearing in mind that these are passive devices.

We should note that the functionality of a MUX-DEMUX pair can also be realized using a single $k \times k$ passive optical star coupler. The purpose of an $k \times k$ star coupler is to couple light from one of its k input ports to all the k output ports uniformly. Star couplers with 128×128 ports and the capability of handling more than one hundred different wavelengths are feasible with currently available technology. An experimental ISDN switch architecture using eight 128×128 multiple star couplers to handle over ten thousand input port lines has been reported [5].

In Fig. 3.13, MUX_r and DEMUX_r represent the MUX and DEMUX that implement row-wise multichannel links. Similarly, MUX_c and DEMUX_c represent the MUX and DEMUX that implement column-wise multichannel links. A MUX_r multiplexes light signals from S_r sources emanating from PEs belonging to the same row of $Plane_L$, and DEMUX_r demultiplexes a light beam into $w \times 2^{\lceil (n-1)/2 \rceil} = 8$ wavelengths that are destined through other fibers to their corresponding destinations. Similarly, light beams emanating from S_c sources and belonging to PEs along the same column of $Plane_L$ or $Plane_R$ are multiplexed by MUX_c and demultiplexed by DEMUX_c. For clarity of Fig. 3.13 only the implementation of the multichannel links for two rows and two columns is shown. Similar connections exist for the other rows and columns of $Plane_L$ of the SMLH($w = 4, n = 3$) network. Each row-wise (or column-wise) multichannel link in Fig. 3.13 can be implemented using one pair of MUX-DEMUX. Fig. 3.14 shows a top view of both planes of the SMLH($w = 4, n = 3$) network. In the middle of the figure, the HOEs that implement the hypercube modules are shown. Only two MUX-DEMUX pairs are shown. The top MUX-DEMUX pair implements the two row-wise multichannel links of the first row of $Plane_R$ while the bottom MUX-DEMUX pair implements the respective multichannel links of the first row of $Plane_L$.

Limitations on fan-out capabilities of MUXs and DEMUXs dictate the number of multichannel links and the number of PEs attached to a multichannel link using this implementation method. On the other hand, if one uses optical star couplers for implementing the multichannel links, the fan-out will no longer be a limitation, but power losses associated with star couplers become the limiting factor. These issues are analyzed in the next section.

In order to alleviate multichannel link collisions (e.g. different messages destined to the same PE at the same time), the time domain along each sub-channel

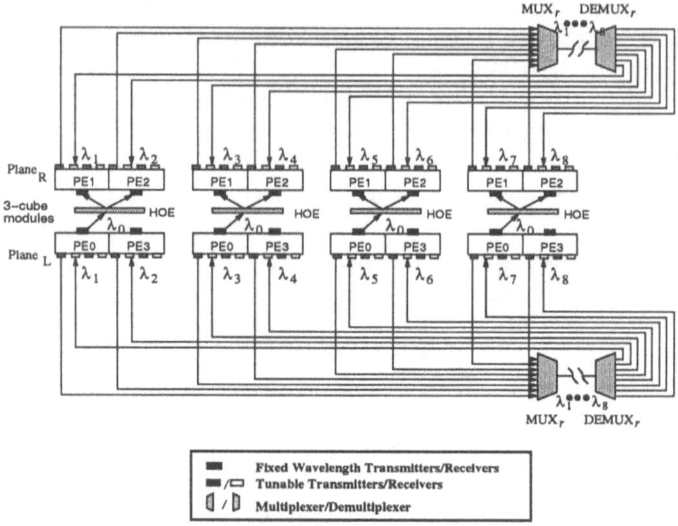

Figure 3.14 Top view of an SMLH($w = 4, n = 3$) network. The figure shows the implementation of the first row-wise multichannel links of $Plane_L$ and $Plane_R$. Similar connections exist for the other rows and columns of the SMLH($w = 4, n = 3$).

can be utilized. Time division multiple access techniques can be combined with the proposed WDMA scheme. This issue is beyond the scope of this chapter and will not be treated any further.

3.5 POWER AND DELAY ANALYSIS OF THE OPTICAL IMPLEMENTATION

3.5.1 *Bit-Error Rate Estimation*

In this section we present some system noise calculations to investigate the bit-error rate (BER) capabilities and the delay of the proposed optical implementation of the SMLH network. Calculation of BER of an optical system requires estimation of the Signal-to-Noise ratio (SNR). Estimation of total power losses, leading into the receiver sensitivity calculation is required for the SNR calculation. In what follows, the optical power loss of the implementation methodology is calculated. Then the receiver sensitivity is estimated, and consequently the BER of the proposed implementation is evaluated.

The number of nodes that an optical system can support is determined by the emitting power of the transmitter, the required receiver sensitivity, and the losses incurred between the transmitter and the receiver. Let L_{MUX} be the total insertion loss of the WDMA multiplexers, and L_{DEMUX} be the total insertion loss of the WDMA demultiplexers. Let L_{sf} be the source to fiber coupling loss, and L_{fd} be the fiber to detector coupling loss. Assume that the fiber loss is L_f and that all nodes are equidistant, meaning that if we denote

$d(i, i+1)$ as the fiber distance from node i to node $i+1$, then d is the same for all $i's$. The total transmission loss using WDMA techniques for the multichannel link implementation is then given by the following equation:

$$L_{total} = L_{MUX} + L_{DEMUX} + dL_f + L_{sf} + L_{fd} - 3. \quad (3.14)$$

To estimate the total loss of the optical system values from commercially available components were considered. We assume laser diodes sources with characteristics of +7 dBm. A dBm is defined as 10 times the logarithmic ratio of the power with a reference power quantity of 1 mW. For example, for a value of 5 mW the ratio in dBm is $10 \times log(\frac{5\ mW}{1\ mW})$ =+7 dBm. Positive values in dBm indicate that the required power is greater than the reference power (1 mW) and negative values show that the required power is less than 1 mW.

The insertion loss for a commercially available fiber coupler is taken as -1 dB while fiber to detector losses are -0.46 dB. The fiber loss is taken as 0.3 dB/Km, but since their lengths are in the order of centimeters the total fiber loss is negligible. The insertion loss of the $k \times 1$ star coupler is taken as -2 dB [6], while for the concave grating demultiplexer the insertion loss is -9 dB. In addition, a -3 dB loss is being added for engineering errors. The total power loss is -15.5 dB which is equivalent to an optical system efficiency of 2.8%. If we consider the detectors to be GaAs Metal-Semiconductor FET Transimpedance [28] with a quantum efficiency of 60% at 670 nm, and if the data rate is 2 Gbps, and if the required error probability after amplification is less than 10^{-17}, then the receiver sensitivity required is determined to be 12 μW or -19.2 dBm. Given the total power loss estimated above (-15.5 dB), and considering optical sources of 5 mW, the receiver sensitivity should be 140 μW or -8.5 dBm. This is well over the receiver sensitivity requirement for 10^{-17} BER. Therefore, the WDMA implementation proposed for the SMN subnetwork is feasible with BERs better than 10^{-17}. Depending on the optical components used, the size of the multichannel links is bounded either by the fan-out of the diffraction grating demultiplexers if the MUX-DEMUX scheme is used, or by the power loss of the star couplers if the optical star couplers are used. For the diffraction grating components, a fan-out of up to 40 channels has been reported [41, 42]. Taking this value as an upper limit, a fairly large SMLH network can be realized. For example, a SMLH($w = 20, n = 3$) network consisting of $N = 3,200$ PEs and a total of 160 MUX-DEMUX pairs is feasible. Larger networks are also feasible by implementing every multichannel link in the network with a different MUX-DEMUX pair instead of utilizing a single pair to implement $2^{\lceil (n-1)/2 \rceil}$ multichannel links. This would increase the number of PEs in the network, but it would also increase the hardware complexity. For example, a SMLH($w = 40, n = 3$) network consisting of $N = 12,800$ PEs and a total of 320 MUX-DEMUX pairs is feasible. Therefore, when designing the optical implementation of a SMLH network, trade-offs between the number of PEs desired and the hardware cost should be considered.

If a MUX-DEMUX pair were to be realized by a single optical star coupler, then Eq. 3.14 would have to include the excess (L_e) and splitting(L_{sp}) losses of the star coupler. Let P_{in} be the power entering the coupler from an input channel, and P_{out} be the power obtained from an output channel. Then, $L_{sp} = -10 \log \frac{P_{out}}{P_{in}}$, and the total transmission loss is:

$$L_{total} = L_e + L_{sp} + dL_f + L_{sf} + L_{fd} - 3. \tag{3.15}$$

P_{out} is equal to P_{in}/k where k is the fan-out of the star coupler. For the SMLH k is $w \times 2^{\lceil (n-1)/2 \rceil}$ (number of PEs on a row or column of $Plane_L$ or $Plane_R$). Consequently, Eq. 3.15 can be rewritten as

$$L_{total} = L_e - 10logk + dL_f + L_{sf} + L_{fd} - 3. \tag{3.16}$$

The excess losses of the star coupler are taken as -1 dB [12]. Rearranging Eq. 3.16 and using the above losses, the number of PEs supported by the star coupler implementation method given a desirable BER can be determined. For a BER of 10^{-17} and for 5 mW sources, the total loss in the system should be -26.2 dB yielding a star coupler fan-out of $k = 118$. This value is well within the capabilities of current star coupler technology. The optical fanout of star couplers reported to date is 128 × 128 [5]. For $k = 118$, large SMLH networks are feasible. For example, a SMLH($w = 30, n = 5$) network supporting about 28,800 PEs could be realized.

3.5.2 Communication Delay

The data transfer delay in the SMLH network is a function of the transmitter, signal propagation, receiver delays, packet collisions and channel bandwidth. The signal propagation delay includes the propagation time of the signal through the medium (i.e., fiber or free-space) and the delay when passing through optical components such as star couplers and free space optics. In this chapter we estimate the communication delay without considering contention and collisions. Estimating the delay including contention and collisions would require the inclusion of access protocols and conflict resolution schemes which is beyond the scope of this chapter.

Let τ_{tx} be the delay of the transmitter, τ_{rx} be the delay of the receiver, and τ_{prop} be the delay due to signal propagation. Therefore the total delay of a single link can be expressed as

$$\tau_d = \tau_{tx} + \tau_{rx} + \tau_{prop} + W/B, \tag{3.17}$$

where W is the packet size in bits, and B is the bandwidth of the communication medium. The value for the τ_{tx} delay includes driver turn on, laser turn on and tuning delays, while the value for the τ_{rx} delay includes the photodetector circuit delay and skew. The photodetector circuit delay consists of the preamplifier delay, decision circuit delay as well as the decision circuit skew. The propagation delay of an optical link (without any components) is

$$\tau_{prop} = \frac{L}{c/\eta}, \tag{3.18}$$

where L is the length of the link, and c/η is the speed of light in the medium. Hence the total delay of an optical link is

$$\tau_d = \tau_{tx} + \tau_{rx} + \frac{L}{c/\eta} + \tau_m + W/B, \tag{3.19}$$

where τ_m is the propagation delay through optical components.

Delays for commercially available optical fixed transmitters and receivers are in the range of 1 ns and 1.8 ns, respectively [27, 40], while those of tunable transmitters are 50 ns [35]. In SMLH the optical star couplers and the grating demultiplexers are all passive components and therefore their time delays are very small and can be ignored in this estimation.

The worst case scenario for a collisionless SMLH network is when a packet travels through both multichannel links and through n hypercube links in order to reach its destination. In this case, the maximum collision free delay is:

$$\tau_d^{max} = 2\tau_{tx}^{tune} + n\tau_{tx}^{fixed} + (n+2)\tau_{rx} + n\tau_{prop}^{BHC} + 2\tau_{prop}^{SMN} + W^{max}/B^{min}, \tag{3.20}$$

where τ_{tx}^{tune} is the delay of the tunable transmitters in the SMN subnetwork, τ_{tx}^{fixed} is the delay of the fixed transmitters in the hypercube subnetwork, τ_{rx} is the delay of the receivers, τ_{prop}^{BHC} is the propagation delay through the hypercube links, τ_{prop}^{SMN} is the propagation delay through the SMN links, W^{max} is the maximum packet size in bits, and B^{min} is the minimum bandwidth. Clearly, the communication delay using WDMA techniques is dominated by the packet size more than by any other delays.

3.6 CONCLUSIONS

We proposed a novel hybrid network which improves or is comparable to hypercube-based topologies in general, and in particular the spanning bus hypercube (SBH) and the Optical Multi-Mesh Hypercube (OMMH). The key attractive features of the proposed network include the possibility of a constant diameter and a constant degree network while it is feasible to interconnect thousands of processors at an adequate performance level. Additionally, the network is incrementally scalable and fault-tolerant. These features make SMLH very suitable for massively parallel systems. The topological characteristics of the proposed network were compared with several other well known networks and it was shown that SMLH compares extremely well with the SBH, the OMMH, the Binary Hypercube, the Torus network, the Hierarchical Cubic Network, the Cube-Connected Cycle, the Hyper-deBruijn, and the Folded Peterson, and comparable performance with the Generalized Hypercube and the Hypermesh. A WDMA technique has been proposed for the optical implementation of the SMLH network. While the structural properties of the proposed network were fully analyzed in this paper, the optical implementation was presented for illustration purposes only. Detailed analysis of the optical feasibility of the proposed network including device and optical setups is under study, and is the subject of future work.

Acknowledgment

This work was supported by the U.S. National Science Foundation under grant number MIP 93-10082.

References

[1] D.P. Agrawl, C. Chen, and J. R. Burke. A comparison of hybrid graph based topologies. In E. F. Gehringer, editor, *IEEE Computer Society Techinal Committee on Computer Architecture Newsletter*, pages 5–8, Winter 1994-95.

[2] L. N. Bhuyan and D. P. Agrawal. Generalized Hypercube and Hyperbus Structures Constructing Massively Parallel Computers. *IEEE Trans. Computers*, C-33:323–333, 1984.

[3] C. Chen, D. P. Agrawal, and J. R. Burke. dBCube: A New Class of Hierarchical Multiprocessor Interconnection Networks with Area Efficient Layout. *IEEE Trans. Parallel and Distributed Systems*, 4:1332–1344, Jan. 1993.

[4] D. M. Chiarulli, S. P. Levitan, R. G. Mehlen, M. Bidnurkar, R. Ditmore, G. Gravenstreter, Z. Guo, C. Qiao, M. F. Sakr, and J. P. Teza. Optoelectronic Buses for High-Performance Computing. *Proc. of the IEEE*, 82(11):1701–1710, Nov 1994.

[5] A. Cisneros and C. A. Brackett. A large ATM switch Based on Memory Switches and Optical Star Couplers. *IEEE J. Selected Areas in Communications*, 9:1348–1360, Oct. 1991.

[6] P. W. Dowd. Wavelength Division Multiple Access Channel Hypercube Processor Interconnection. *IEEE Trans. on Computers*, 41:1223–1241, Oct 1992.

[7] S.A. Felperin, L. Gravano, G.D. Pifarre, and J.L.C. Sanz. Fully-adaptive routing: Packet switching performance and wormhole algorithms. In *Supercomputing '91 Proceedings*, pages 654–663, 1991.

[8] E. Ganesan and D. K. Pradhan. The Hyper-deBruijn Networks: Scalable Versatile Architecture. *IEEE Trans. on Parallel and Distributed Systems*, 4:962–978, Sep 1993.

[9] K. Ghose and K. R. Desai. Hierarchical Cubic Networks. *IEEE Trans. on Computers*, 6:427–435, Apr. 1995.

[10] J. R. Goodman and C. H. Sequin. Hypertree: a Multiprocessor Interconnection Topology. *IEEE Trans. on Computers*, C-30:923–933, 1981.

[11] K. Hwang. *Advanced Computer Architecture: Parallelism, Scalability, Programmability*. McGraw-Hill, New York, NY, 1993.

[12] D. Israel, R. Baets, M. J. Goodwin, N. Shaw, M. D. Salik, and C. J. Groves-Kirkby. Comparison of Different Polymetric Multimode Star Couplers for Backplane Optical Interconnect. *IEEE J. Lightwave Technology*, 13:1057–64, June 1995.

[13] J. Kim and C. R. Das. Hypercube Communication Delay with Wormhole Routing. *IEEE Trans. on Computers*, 43:806–814, July 1994.

[14] L. Klienrock. *Queueing Systems: Vol. II, Computer Applications.* Wiley, New York, NY, 1976.

[15] V. Kumar, A. Grama, A. Gupta, and G. Karypis. *Introduction to Parallel Computing: Design and Analysis of Algorithms.* Benjamin/Cummings Publishing Co., New York, NY, 1994.

[16] G. Lerman and L. Rudolph. *Parallel Evolution of Parallel Processors.* Plenum Press, New York, 1993.

[17] Y. Li, A. W. Lohmann, and S. B. Rao. Free-space Optical Mesh-connected Bus Networks Using Wavelength-division Mutiple Access. *Applied Optics*, 32:6425–6437, 1993.

[18] A. Louri and S. Furlonge. Feasibility Study of a Scalable Optical Interconnection Network for Massively Parallel Processing Systems. *Applied Optics*, 35:1296–1308, Mar. 1996.

[19] A. Louri, S. Furlonge, and C. Neocleous. Experimental Demonstration of the Optical Multi-Mesh Hypercube: A Scalable Interconnection Network for Multiprocessors and Multicomputers. *Applied Optics*, 35:6909–6920, Dec 1996.

[20] A. Louri and H. Sung. Efficient Implementation Methodology for Three-dimensional Space-invariant Hypercube-based Free-space Optical Interconnection Networks. *Applied Optics*, 32:7200–7209, Dec. 1993.

[21] A. Louri and H. Sung. 3D Optical Interconnects for High-Speed Interchip and Interboard Communications. *Computer*, 27:27–37, Oct. 1994.

[22] A. Louri and H. Sung. A Scalable Optical Hypercube-based Interconnection Network for Massively Parallel Computing. *Applied Optics*, 33:7588–7598, Nov. 1994.

[23] A. Louri and H. Sung. An Optical Multi-Mesh Hypercube: A Scalable Optical Interconnection Network for Massively Parallel Computing. *IEEE J. Lightwave Technology*, 12:704–716, Apr. 1994.

[24] Q. M. Malluhi and M. A. Bayoumi. The Hierarchical Hypercube: A New Interconnection Topology for Massively Parallel Systems. *IEEE Trans. Parallel and Distributed Systems*, 5:17–30, Jan. 1994.

[25] A. D. McAulay. *Optical Computer Architectures: the Application of Optical Concepts to Next Generation Computers.* Wiley, New York, NY, 1991.

[26] B. D. Metcalf and J. F. Providakes. High-capacity Wavelength Demultiplexer with a Large-diameter GRIN Rod Lens. *Applied Optics*, 21:794–796, 1982.

[27] R. A. Morgan. Advances in vertical cavity surface emitting lasers. *Proc. of the SPIE*, 2147:97–119, 1994.

[28] T. V. Moui. Receiver design for high-speed optical fiber systems. *IEEE J. Lightwave Technology*, pages 243–267, 1984.

[29] P. F. Moulton. Tunable Solid State Lasers. *Proc. of the IEEE*, 80:348–364, Mar. 1992.

[30] L. M. Ni and P. K. McKinley. A Survey of Wormhole Routing Techniques in Direct Networks. *Computer*, 26:62–76, Feb. 1993.

[31] S. Ohring and S.K. Das. Folded Petersen Cube Networks: New Competitors for the Hypercubes. *IEEE Trans. Parallel and Distributed Systems*, 7:151–168, Feb. 1996.

[32] F. P Preparata and J. Vuillemin. The cube-connected cycles: a versatile network parallel computation. *Communications of the ACM*, 24:300–309, May 1981.

[33] D. A. Reed and R. M. Fujimoto. *Multicomputer Networks: Message-Based Parallel Processing.* MIT Press, Cambridge, MA, 1987.

[34] Youcef Saad and Martin H. Schultz. Topological Properties of Hypercubes. *IEEE Trans. on Computers*, 37:867–872, July 1988.

[35] N. K. Shankaranarayanan, U. Koren, B. Glance, and G. Wright. Two-section DBR laser transmitters with accurate channel spacing and fast arbitrary-sequence tuning for FDMA networks. In *Technical Digest of Fiber Optical Communication.*, pages 36–7. Washington, DC: Optical Society of America, 1994.

[36] H. J. Siegel. *Interconnection Networks for Large-scale Parallel Processing.* McGraw-Hill, New York, NY, 1990.

[37] H. S. Stone and J. Cocke. Computer Architecture in the 1990s. *Computer*, pages 30–38, Sep. 1991.

[38] P. Sweazey. Limits of performance of backplane buses. In *Digital Bus Handbook.* New York: McGraw-Hill, 1990.

[39] T. Szymanski. Hypermeshes: Optical Interconnection Networks for Parallel Computing. *J. Parallel and Distribted Computing*, 26:1–23, 1995.

[40] A. Takai, T. Kato, S. Yamashita, S. Hanatani, Y. Motegi, K. Ito, H. Abe, and H. Kodera. $200 - Mb/s/ch$ 100m Optical Subsystem Interconnections Using 8-channel 1.3μ Laser Diode Arrays and Single-Mode Fiber Arrays. *IEEE J. Lightwave Technology*, 12:260–269, Feb 1994.

[41] F. N Tinofeev, P. Bayvel, J. E. Midwinter, and M. N. Sokolskii. High-Performance, free-space ruled concave grating demultiplexer. *Electronics Letters*, 31:2200–1, Dec 1995.

[42] F. N Tinofeev, P. Bayvel, J. E. Midwinter, M. N. Sokolskii, E. G Churin, and A. Stavdas. Free-Space aberration-corrected grating demultiplexer for application in densely-spaced, subnanometer wavelength-routed optical networks. *Electronics Letters*, 31:1368–70, Aug 1995.

[43] L. D. Wittie. Communication Structures for Large Networks of Microcomputers. *IEEE Trans. on Computers*, C-30:264–273, Apr. 1981.

4 THE COMMUNICATION CAPABILITIES OF PARTITIONED OPTICAL PASSIVE STARS NETWORKS

Rami Melhem, Greg Gravenstreter,
Donald Chiarulli, and Steven Levitan

Department of Computer Science
University of Pittsburgh
Pittsburgh, Pennsylvania 15261
{melhem,greg,don,levitan}@cs.pitt.edu

Abstract: This chapter analyzes the communication capabilities of Partitioned Optical Passive Stars (POPS) networks. POPS is a topological approach to providing reconfigurable multiple data channels using multiple non-hierarchical couplers. POPS topologies provide an architecture which supports flexible design trade-offs between system complexity, network throughput, power budget, and control overhead. It is shown that high performance and high coupler utilization are achieved for random, as well as regular communication patterns. These qualities are maintained as the POPS topologies are scaled to large systems.

4.1 INTRODUCTION

The incorporation of optical fiber link technology into existing multi-hop network designs solves many problem that are associated with electronic interconnections, including driving power, capacitive/inductive loading, and relatively low transmitter rates implied by sensitivity to noise [22, 1, 14]. This simple technological replacement provides at least an order of magnitude increase in link throughput. Further, the lack of reactive factors and a high noise immunity is well suited to the demands of faster, larger, and more distributed environments. Such networks, however, still suffer from throughput bottlenecks and high latencies resulting from electronic/optical conversions and processing at intermediate hops. These disadvantages can be avoided in "all-optical" networks, where electronic technology is only present at the beginning and the end of the communications pathway. Certain topologies implemented exclusively with passive optical technology provide enormous potential throughput with very low latencies. Communication protocols incorporating time (TDM) [21, 23], wavelength (WDM) [12], or code (CDM) [16] division multiplexing have been developed to increase utilization and provide more efficient access to this tremendous capacity.

However, existing "all-optical" networks make very poor use of the fiber optic physical bandwidth. It is estimated that fiber optic links provide as much as four orders of magnitude more of usable bandwidth than what the best commercial systems make use of today [3]. Any network which provides a single communication channel for each physical data channel is inherently limited by the electronic capabilities at its input and output ports. New networking designs are needed that present multiple data channels which are physically concurrent [11]. Wavelength division multiplexing (WDM) is one method for achieving concurrent multiple channels. A large amount of research has been accomplished on WDM-based systems [8, 7]. The majority of this has been to develop multiple-access multiple-channel protocols on star architectures using tunable transmitters and receivers [13]. For example, IBM's Rainbow-1 demonstrated the feasibility of these types of WDM networks supporting up to 32 nodes. With individual throughput of 300 megabits per second, the network capacity is on the order of 10 gigabits per second [17].

Theoretically, WDM has remarkable potential, yet serious problems remain. First, one nanometer channel spacing would provide for thousands, perhaps tens of thousands, of high capacity channels. But, much more progress is needed to achieve acceptable tuning speeds (nanoseconds) over necessary ranges (hundreds of nanometers). If the link wavelengths are not tunable, then either unique channels must be assigned to both transmitters and receivers, or arbitration must be performed to allocate channels. Second, optical power budgets become a major issue for large WDM systems, since each transmitter must broadcast to all receivers. Erbium-doped optical amplifiers may relieve some of these problems, but again, the range of amplification is limited. Finally, the protocols which control WDM-systems require substantial quantities of both network bandwidth and processing power.

Several studies have been done on innovative topological approaches to address some of these specific problems. Power optimization in multiple star networks, both hierarchical and non-hierarchical, was investigated by Birk [2]. Networks of multiple passive stars, both totally and partially connected, as well as the reconfigurability of WDM systems operating on them, was studied by Ganz [6]. A multi-hop WDM method applied to multiple star topologies was presented by Hluchyj [15]. Multihop routing in multiple star topologies that are not fully connected are studied in [5].

This chapter investigates a purely topological approach to providing multiple physical data channels. The Partitioned Optical Passive Stars (POPS) topology is an interconnection architecture that uses multiple non-hierarchical stars to achieve single-hop networks [4, 10, 9]. It is an "all-optical" topology constructed exclusively with passive optical components, and benefits from all the corresponding characteristics discussed. That is, no intermediate electronic/optical conversions, no reactive factors and high noise immunity.

A POPS topology is configured at design time to provide a fixed number of physically concurrent data channels, each of which is capable of high capacity in a circuit-switched system. The number of such channels is not absolutely limited, and is a key engineering tradeoff. This design flexibility provides for a customized optimization between lower total system complexity versus the combination of higher system throughput, lower power budgets and lower network control overheads. The analysis presented confirms that, independent of the number of channels configured, POPS network data channels can be efficiently utilized for both random permutation-based communication patterns and several important application environments.

All data channels in a POPS network may use the same fixed wavelength. Fixed transmitters and fixed receivers using current technologies may be chosen based on cost, mechanical, or other factors. The design flexibility of a POPS topology may be used to avoid power budget problems, eliminating the need for optical amplification. Similarly, simple network control mechanisms can be achieved. Additionally, established multiplexing methods, such as TDM or WDM are particularly appropriate and highly efficient on a POPS network.

The rest of this chapter is organized as follows. Section 4.2 discusses the components, parameters, and notation for a POPS topology. Section 4.3 presents the routing and scheduling of messages in a POPS topology. Section 4.4 illustrates the performance of POPS topologies in detail. Absolute bounds for delivery times of message sets are determined. An analytical examination is performed, and simulation results are obtained, for several performance metrics. Section 4.5 examines the capabilities of POPS topologies in several common application environments. We conclude with a summary of the capabilities of POPS topologies.

4.2 DESCRIPTION OF THE POPS TOPOLOGY

A small POPS network is shown in Figure 4.1. Source nodes relay messages through optical links to passive optical couplers. These couplers send the message through other optical links to the destination nodes. A minimal set of two independent parameters completely determines a POPS network implementation. The first parameter, a measure of the system size, is the number of nodes and is denoted by n. The second parameter, a measure of the coupler complexity, is the degree of each coupler and is denoted d. Each coupler is a $d \times d$ passive optical star which equally distributes the optical power on any of its d inputs to all of its d outputs. The small example POPS network in Figure 4.1 is for $n = 16$ and $d = 8$.

The set of n source nodes are partitioned into $g = n/d$ equal-sized source groups. Similarly, the set of n destination nodes are partitioned into g destination groups. Each source group and each destination group consists of $d = n/g$ nodes. Typically, a source node and the corresponding destination node are the same processing element. Consequently, the source groups and the corre-

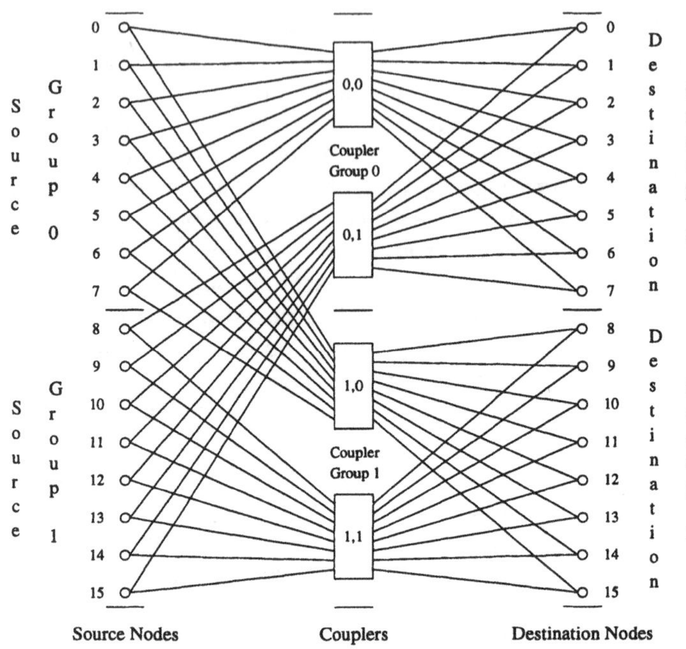

Figure 4.1 A $n = 16$ High-Degree ($d = 8$) POPS network ($g = 2$).

sponding destination groups contain the same set of nodes. These g groups of nodes are denoted $GN_0, ..., GN_{g-1}$.

A set of $c = g^2 = n^2/d^2$ couplers is partitioned into g groups of g couplers each. These g groups of couplers are denoted by $GC_0, ..., GC_{g-1}$. The g couplers in coupler group GC_i are denoted by $C_{i,0}, ..., C_{i,g-1}$, where $0 \leq i < g$. The d inputs for a given coupler, $C_{i,j}$, are connected to the d nodes in source group GN_j, and the d outputs of $C_{i,j}$, are connected to the d nodes in destination group GN_i.

Each source node, i, where $0 \leq i < n$, has g transmitters, denoted $T_{i,0}, T_{i,1}$, ..., $T_{i,g-1}$. Transmitter $T_{i,j}$ is connected to a coupler in GC_j, and thus is used to communicate with nodes in destination group GN_j. Similarly, each destination node i has g receivers, denoted $R_{i,0}, ..., R_{i,g-1}$, and receiver $R_{i,j}$ is connected to a coupler which provides communication with nodes in source group GN_j.

The optical links in a POPS network fall into one of two classes. In the first class, called source links, each link connects a transmitter at a source node to an input of a coupler. For each coupler, at most one source link should be active at a given time to prevent collisions. In the second class, called destination links, each link connects an output of a coupler to a receiver at a destination node. For each coupler, all destination links are either simultaneously active or inactive. A summary of all components in a POPS network implementation is presented in the following table.

Parameter	Per Node/Coupler	Per Group	In Total
Nodes	—	d	n
Groups	—	—	$g = n/d$
Couplers	—	$g = n/d$	$c = g^2 = n^2/d^2$
Transmitters	$g = n/d$ (/node)	n	n^2/d
Receivers	$g = n/d$ (/node)	n	n^2/d
Links	—	—	$l = 2n^2/d$

Given a fixed system size n, the choice of coupler degree d allows for a wide range of system characteristics. As the coupler degree approaches n, the system assumes the nature of a single passive star which includes low system cost and complexity, restricted throughput, and increased power dissipation. As the coupler degree approaches one, the system becomes a completely-connected topology, with very high system cost and optimal performance. The example POPS topology shown in Figure 4.1 has a relatively large coupler degree ($d = 8$). In Figure 4.2, another example POPS topology, with the same number of nodes as in Figure 4.1, but a smaller coupler degree ($d = 4$), is given for comparison.

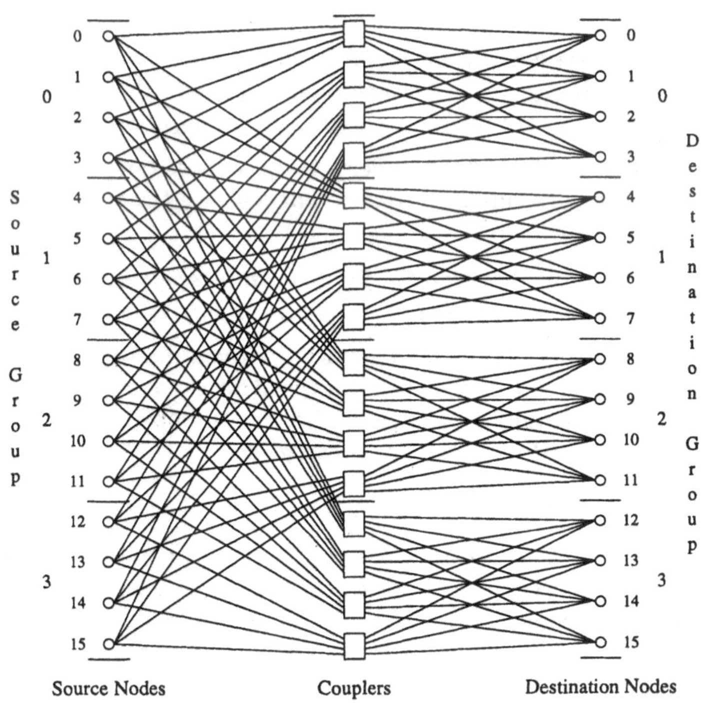

Figure 4.2 A $n = 16$ Low-Degree ($d = 4$) POPS network ($g = 4$).

4.3 COMMUNICATION USING POPS NETWORKS

In a POPS network, the determination of the route for a specified message is a simple process. The source node determines the appropriate transmitter for communication with the required destination group, and the destination node determines the appropriate receiver for communication with the required source group. Specifically, consider a single message, denoted $M_{x,y}$, which originates at source node x in source group $GN_{x'}$, where $0 \leq x < n$ and $x' = \lfloor x/d \rfloor$. This message terminates at destination node y in destination group $GN_{y'}$, where $0 \leq y < n$ and $y' = \lfloor y/d \rfloor$. Source node x uses transmitter $T_{x,y'}$ to send the message. Coupler $C_{y',x'}$ transports the message, and destination node y uses receiver $R_{y,x'}$ to get the message.

Lemma 1. *In a POPS network, any specific message $M_{x,y}$ has a unique path composed of a transmitter, a coupler, and a receiver, $(T_{x,y'}, C_{y',x'}, R_{y,x'})$.* ∎

Lemma 2. *For any two given messages, M_{x_i,y_i} and M_{x_j,y_j}, where $x_i \neq x_j$ and $y_i \neq y_j$, the only possible point of conflict in the path of the two messages is the coupler. No conflict occurs if either $x'_i \neq x'_j$ or $y'_i \neq y'_j$.* ∎

Thus, the conflicts during the delivery of a set of messages in a POPS network can be determined. Subsequently, the minimal schedule of delivery for a set of messages can also be readily calculated as described next.

Definition 1. *A* **Message Set** *of size m is a set of m messages, denoted by $\mathcal{M} = \{M_{x_0,y_0}, ..., M_{x_{m-1},y_{m-1}}\}$, where $x_0, ..., x_{m-1}$ are the m message source nodes, and $y_0, ..., y_{m-1}$ are the m message destination nodes.*

Given a message set $\mathcal{M}$, let a coupler's usage, denoted by u_i , where $0 \leq i < c$, be the number of individual messages that will use the ith coupler, that is, $C_{\lfloor i/g \rfloor, i \bmod g}$.

Definition 2. *A* **Coupler Profile** *for a given message set $\mathcal{M}$ is an ordered c-tuple of coupler usages, denoted $C(\mathcal{M}) = (u_0, u_1, ..., u_{c-1})$.*

Lemma 1 states that the coupler used by a given message is a function of the message's source and destination. Applied to the elements of a message set, Lemma 1 yields a unique coupler profile. For example, consider the POPS of Figure 4.1 ($n = 16$, $d = 8$) and the message set

$$\mathcal{M}_1 = \{M_{0,0}, M_{3,6}, M_{6,12}, M_{9,2}, M_{12,8}, M_{15,14}\}.$$

The paths for both $M_{0,0}$ and $M_{3,6}$ use coupler $C_{0,0}$, the paths for both $M_{12,8}$ and $M_{15,14}$ use coupler $C_{1,1}$, the path for $M_{6,12}$ uses $C_{1,0}$ and the path for $M_{9,2}$ uses $C_{0,1}$. Thus the coupler profile $C(\mathcal{M}_1) = (2, 1, 1, 2)$.

Only a single message per coupler can be delivered in a fixed time period, which we call a time slot. Thus, a maximum of c messages can be delivered by a POPS network implementation in a single time slot. When multiple messages require the same coupler, a series of consecutive time slots, called a sequence, is required. During each time slot in a sequence, some non-zero number of couplers will deliver one message each.

A scheduling algorithm chooses, for each coupler, the order and timing of delivery for the pending messages requiring that coupler. A greedy scheduling algorithm is defined to be one that, by some selection criteria in each time slot, assigns one message to each coupler where messages are pending. For example, a random greedy scheduling algorithm might randomly select from the messages pending for a specific coupler. The following theorem and its corollary show that scheduling messages in POPS networks is straightforward.

Theorem 1. *In a given POPS network, for any given message set, the scheduling of any specific coupler is independent of all other couplers.*

Proof. From Lemma 1, any given message set yields a unique coupler profile. From Lemma 2, it follows that the subset of messages requiring a given coupler all have conflicting paths. However, no element of this subset of the messages conflicts with any other message. Thus, the subset can be independently scheduled. ∎

Corollary 1. *Any greedy scheduling algorithm has the minimum possible sequence length,* $u_{max} = max\{u_0, u_1, ..., u_{c-1}\}$. ∎

For example, the message set $\mathcal{M}_1$, whose coupler profile is $(2, 1, 1, 2)$, may be delivered in two time slots. Many message schedules may achieve this minimum sequence length, including a greedy schedule which schedules $\{M_{0,0}, M_{6,12}, M_{9,2}, M_{12,8}\}$ in the first time slot and $\{M_{3,6}, M_{15,14}\}$ in the second time slot.

Network traffic can be either static or dynamic. A static communications pattern is known a priori, while a dynamic pattern is not. In a POPS network, as seen in Lemma 1, route determination is trivial. Further, Corollary 1 demonstrates that, as a result of the independent nature of the couplers, scheduling is both easy and efficient. For dynamic message traffic, it is still necessary to specify an arbitration protocol on a per coupler basis. Established wavelength-division multiplexing protocols, such as surveyed in [20], might be adaptable. In [4], a specially optimized time-division multiplexing protocol is presented for the POPS topology.

The primary focus of this chapter is the study of performance in a POPS network for static communications patterns. That is, a predetermined message set , $\mathcal{M}$, of size m is generated and delivered. Two types of static patterns are considered. In the next section, random message sets that are permutations or subsets of permutations are studied, while the following section studies some regular communication patterns.

4.4 PERMUTATION CAPABILITIES OF THE POPS TOPOLOGY

Certain key aspects concerning the performance of a POPS network are examined. Given a message set size, absolute bounds for the sequence length are determined. Analytical methods, as well as results from general simulations, are used to survey several performance metrics. Sequence length distribution, average sequence length, and coupler utilization are analyzed. First, we define permutation-based random message sets as follows:

Definition 3. *A* **Permutation-Based Message Set** *is a message set of m messages, $1 \leq m \leq n$, where the m source nodes $x_0, x_1, ..., x_{m-1}$ are distinct, and the m destination nodes $y_0, y_1, ..., y_{m-1}$ are also distinct.*

4.4.1 *Absolute Bounds for Message Sets*

Assume a fixed POPS topology and a specific message set $\mathcal{M}$ of size m, with coupler profile $C(\mathcal{M})$, and a maximum coupler usage of u_{max}. If s is the sequence length that a greedy algorithm yields, then, from Corollary 1, $s = u_{max}$. Clearly, for a given $\mathcal{M}$, u_{max} depends on m and on the message distribution in $\mathcal{M}$. For a given m, however, the value of u_{max} is bounded as defined below.

Definition 4. *The* **Greatest Lower Bound** *for a given message set size m, denoted by $glb(m)$, is the minimum value of s over the sample space of all permutation-based message sets of size m. Similarly, the* **Least Upper Bound** *for a given message set size m, denoted by $lub(m)$, is the maximum value of s over the sample space of all permutation-based message sets of size m.*

The greatest lower and least upper bounds represent absolute limits to the sequence length for a greedy algorithm. These absolute bounds to the sequence length are functions of the message set size and the POPS topology parameters, as specified by the following theorem.

Theorem 2. *For a given POPS network and message set $\mathcal{M}$ containing m messages, $glb(m) = \lceil m/c \rceil$, and $lub(m) = min\{m, d\}$.*

Proof. (Greatest Lower Bound) From Euclid's Division Algorithm [19], there exist unique non-negative integers x and y, such that $m = xc + y$, where $0 \leq y < c$. Consider a message set where the first x time slots utilize the maximum network throughput c, and a final time slot delivers y messages. This requires a total of $s = x + \lceil y/c \rceil = \lceil m/c \rceil$ time slots. This is a lower bound, since the couplers in a POPS network are 100% utilized for a throughput of c, and thus, xc messages can not be delivered in less than x time slots. Further, these x time slots can not deliver any of the y additional messages. Thus, any $s < \lceil m/c \rceil$ can not deliver the m messages.

In order to show that $\lceil m/c \rceil$ is the greatest lower bound for s, we construct a message set $\mathcal{M}$ of size m which requires a sequence of length $s = \lceil m/c \rceil$. Let the source nodes of the m messages in $\mathcal{M}$ be evenly distributed among the g node groups such that either $\lfloor m/g \rfloor$ or $\lceil m/g \rceil$ messages originate in each node group, GN_i, $i = 0, \ldots, g-1$. For simplicity, take these nodes to be the first r_i nodes in GN_i, where r_i is either $\lfloor m/g \rfloor$ or $\lceil m/g \rceil$, and $r_0 + \cdots + r_{g-1} = m$. The set $\mathcal{M}$ is then constructed such that the message originating at the kth node in GN_i is destined to the $(i +_d k)^{th}$ node in $GN_{k \bmod g}$, where $+_d$ denotes the addition modulo d. Specifically,

$$\mathcal{M} = \{M_{id+k,(k \bmod g)d+(i+_d k)} : i = 0, \ldots, g-1 \ \& \ k = 0, \ldots, r_i - 1\}.$$

The utilization of any coupler $C_{j,i}$ is equal to the number of messages in $\mathcal{M}$ that originate at a node in GN_i and terminates at a node in GN_j. This is

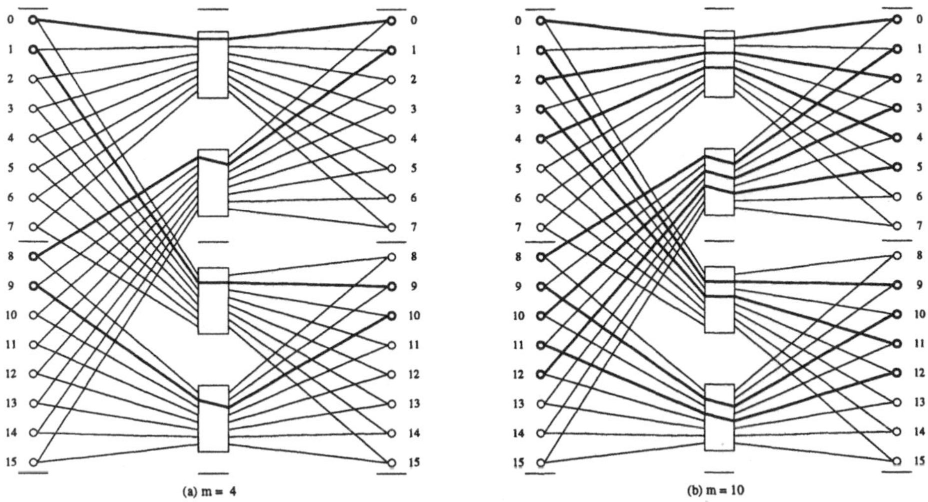

Figure 4.3 The Greatest Lower Bound for a $n = 16, d = 8$ POPS network.

equal to $\lceil (r_i - j)/g \rceil$. The maximum coupler utilization for $\mathcal{M}$ is thus $\lceil m/g^2 \rceil = \lceil m/c \rceil$. To illustrate the construction, we consider the POPS of Figure 4.1 ($n = 16, d = 8, g = 2$) and show in Figure 4.3 the paths (in bold face) for the message sets constructed for the two cases $m = 4$ and $m = 10$. In the former case $r_0 = r_1 = 2$, while in the latter case $r_0 = r_1 = 5$.

(Least Upper Bound) First, $min\{m, d\}$ is an upper bound for s. If $m \leq d$, then all of the m messages is the most that any coupler can deliver, and $s \leq m$. If $d \leq m$, the most messages that a coupler can deliver is d, since all message source nodes and all message destination nodes are distinct, and $s \leq d$. Thus, $s \leq min\{m, d\}$. Next, we show that $min\{m, d\}$ is the least upper bound for s. If $m > d$, there are message sets such that $s = m$. For example, where m nodes in a single node group send messages to themselves. If $d \leq m$, there are messages sets such that $s = d$. For example, where all d nodes in a single node group send messages to themselves. Thus, any number less than $min\{m, d\}$ is not an upper bound. ∎

Figure 4.4 shows the behavior of both $glb(m)$ and $lub(m)$ for a fixed POPS network as the message set size ranges from one to n. Combining the plots for greatest lower bound and least upper bound, plus the line segment connecting the values of these bounds for $m = n$, yields a polygon which we call the sequence envelope.

For any given message set $\mathcal{M}$ of size ranging from $m = 1$ (a single message) to $m = n$ (a permutation message set), possible sequence lengths are represented by integral values of s that lie on or within the sequence envelope in the appropriate vertical line. This range of sequence lengths, $glb(m) \leq s \leq lub(m)$, is a function of m, for a given n and d. The probability distribution of s in

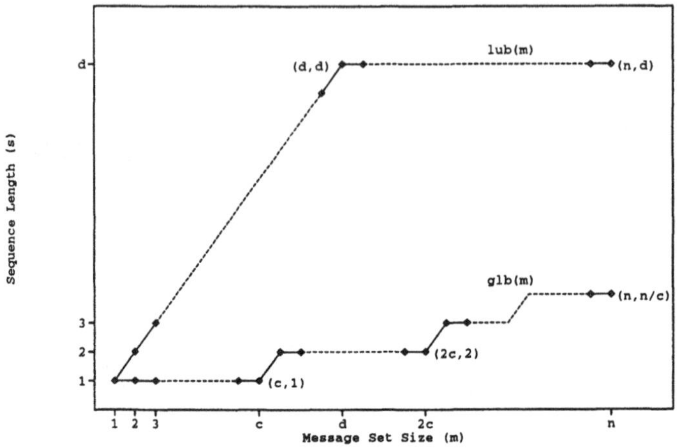

Figure 4.4 Greatest Lower and Least Upper Bounds.

that range is an important measure of the performance of a POPS network. It determines the average and variance of the communications delay for sets of messages of size m.

4.4.2 *Sequence Length Probability*

Assume a POPS topology with fixed n and d, and consider a set $\mathcal{M}$ of m messages with distinct source nodes and distinct destination nodes. Denote by $\aleph$, the set of all possible message sets containing m messages. Further, denote by $\aleph_s$, the set of all message sets containing m messages that require a sequence of length s for delivery. Thus, $\aleph = \bigcup_{glb(m) \le s \le lub(m)} \aleph_s$.

Lemma 3. *The probability that a set $\mathcal{M}$ of m randomly selected messages will require a sequence of length s for delivery is $\rho(m,s) = |\aleph_s| \,/\, |\aleph|$.* ∎

The number of ways to choose the source nodes for a message set $\mathcal{M}$ is $C^n_m = n!/(n-m)!m!$, the number of unordered ways to select m nodes from n nodes. For each of these choices, the number of ways to select the destination nodes for the message set is $P^n_m = n!/(n-m)!$, the number of ordered ways to select m nodes from n nodes. Thus, $|\aleph| = P^n_m C^n_m = (n!)^2/((n-m)!)^2 m!$.

To determine $\rho(m,s)$, the calculation of $|\aleph_s|$ is still required. It is possible to enumerate all the message sets in $\aleph_s$, but typically, the number of these message sets makes this impractical. A more efficient approach, based on the enumeration of coupler profiles is given.

Denote by $\mathcal{C}(\aleph) = \{C(\mathcal{M}) | \mathcal{M} \in \aleph\}$, the set of all possible coupler profiles for message sets containing m messages. Also, denote by $\mathcal{C}(\aleph_s) = \{C(\mathcal{M}) | \mathcal{M} \in \aleph_s\}$, the set of all possible coupler profiles for message sets containing m messages that require a sequence of length s for delivery. It follows that the POPS topology imposes several constraints on the elements of $\mathcal{C}(\aleph_s)$.

Lemma 4. *Let $C(\mathcal{M}) = (u_0, u_1, ..., u_{c-1})$ be an element of the set of coupler profiles, $\mathcal{C}(\aleph_s)$, for sets of m messages requiring a sequence of length s for delivery. The following conditions should be satisfied in $C(\mathcal{M})$:*

1. $u_i \leq s \leq d$, *for* $0 \leq i \leq c-1$;
2. $\exists i, 0 \leq i \leq c-1$, *such that* $u_i = s$;
3. $\sum_{j=0}^{g-1} u_{ig+j} \leq d$, *for* $0 \leq i \leq g-1$;
4. $\sum_{i=0}^{g-1} u_{ig+j} \leq d$, *for* $0 \leq j \leq g-1$;
5. $\sum_{i=0}^{g-1} \sum_{j=0}^{g-1} u_{ig+j} = m$. ∎

The first condition says that the number of messages through any coupler is at most s, else a sequence length of s could not deliver that coupler's messages. The second condition says that at least one coupler must handle s messages, else a sequence length of s would not be required for the message set. The third condition says that the total number of messages through couplers connected to any specific source group is at most d, since each node can initiate at most one message. Similarly, the fourth condition says that the total number of messages through couplers connected to any specific destination group is at most d. Finally, the fifth condition says that the total number of messages through all couplers must be m.

In general, each coupler profile $C \in \mathcal{C}(\aleph_s)$ supports more than one message set $\mathcal{M} \in \aleph_s$. For example, consider a message set where two of the messages use a single given coupler to reach two different destinations. Another message set supported by the same coupler profile, would have the same two destinations reversed. The set of message sets supported by a specific element, C, of the set of coupler profiles, $\mathcal{C}(\aleph_s)$, is denoted $\mathcal{M}_C$. The elements of $\mathcal{C}(\aleph_s)$ yield a partitioning of $\aleph_s$ by the message sets $\mathcal{M}_C$. That is $\aleph_s = \bigcup_{C \in \mathcal{C}(\aleph_s)} \mathcal{M}_C$. Further, $|\aleph_s| = \left|\bigcup_{C \in \mathcal{C}(\aleph_s)} \mathcal{M}_C\right| = \sum_{C \in \mathcal{C}(\aleph_s)} |\mathcal{M}_C|$. The determination of $|\aleph_s|$, and thus of $\rho(m, s)$, can be achieved by summing the values of $|\mathcal{M}_C|$ during an enumeration of the C in $\mathcal{C}(\aleph_s)$. The value of $|\mathcal{M}_C|$ is now specified.

Lemma 5. *Given n and d for a POPS topology, and a coupler profile $C = (u_0, u_1, ..., u_{c-1}) \in \mathcal{C}(\aleph_s)$. The number of message sets supported by C is,*

$$|\mathcal{M}_C| = \prod_{i=0}^{g-1} \prod_{j=0}^{g-1} \left[C_{u_{ig+j}}^{d-\alpha} P_{u_{ig+j}}^{d-\beta} \right],$$

where $\alpha = \sum_{l=0}^{i-1} u_{lg+j}$, *and* $\beta = \sum_{l=0}^{j-1} u_{ig+l}$.

Proof. The enumeration of the message sets is derived as follows:

- $\prod_{i=0}^{g-1}$ considers each coupler group in the POPS network.

- $\prod_{j=0}^{g-1}$ considers each coupler in the specific coupler group.
- The large brackets enumerate the contribution of a single coupler.
  - $C^{d-\alpha}_{u_{ig+j}}$ considers the ways to choose the u_{ig+j} inputs from the d source group nodes, minus the total of previous usage of that group.
  - $P^{d-\beta}_{u_{ig+j}}$ considers the ways to map the u_{ig+j} inputs selected to the d destination group nodes, minus the total of previous usage of that destination group. ∎

For example, in the network of Figure 4.1, the coupler profile $C = (2, 1, 1, 2)$ can result from $C^8_2 P^8_2 * C^8_1 P^6_1 * C^6_1 P^8_1 * C^7_2 P^7_2 \approx 3 \times 10^9$ possible message sets of size 6. First, the two messages using coupler $C_{0,0}$ can be from any two of the eight sources in GN_0 to any two of the eight destinations in GN_0, thus the factor $C^8_2 P^8_2$. For each choice for the two messages using $C_{0,0}$, the one message using $C_{0,1}$ can be from any source in GN_1 to any of the six remaining destinations in GN_0, thus the factor $C^8_1 P^6_1$. For each choice of the first three messages, the message using $C_{1,0}$ can be from any of the six remaining sources in GN_0 to any of the destinations in GN_1, thus the factor $C^6_1 P^8_1$. Finally, the last two messages using $C_{1,1}$ can be from any two of the seven remaining sources in GN_1 to any two of the seven remaining destinations in GN_1, thus the factor $C^7_2 P^7_2$.

Combining the above Lemmas, the probability that a random message set of size m will require a sequence of length s for delivery, $\rho(m, s)$, can be specified.

Theorem 3. (The Probability Calculation Theorem)

$$\rho(m,s) = \left[\sum_{C \in \mathcal{C}(\aleph_s)} \prod_{i=0}^{g-1} \prod_{j=0}^{g-1} \left[C^{d-\alpha}_{u_{ig+j}} P^{d-\beta}_{u_{ig+j}}\right]\right] \Big/ \left[P^n_m C^n_m\right],$$

where $\alpha = \sum_{l=0}^{i-1} u_{lg+j}$, *and* $\beta = \sum_{l=0}^{j-1} u_{ig+l}$. ∎

4.4.3 Sequence Envelope Special Points

The Probability Calculation Theorem specifies the probability that a random message set $\mathcal{M}$ of size m will require a sequence of length s for delivery. The coupler profile approach, which enumerates the profiles in $\mathcal{C}(\aleph_s)$, is more efficient than a complete enumeration of all message sets in $\aleph_s$. Moreover, probabilities for certain special positions on the sequence envelope can be calculated by the coupler profile approach. This is due to various additional restrictions being appropriate at these positions. A couple of these special degenerate cases will be considered.

As shown in Figure 4.4, the greatest lower bound increases slowly, holding steady except for unit increases as the message set size traverses multiples of the number of couplers ($glb(m) = \lceil m/c \rceil$). When the message set size is a multiple of the number of couplers, the greatest lower bound occurs when the messages are evenly distributed among the couplers. In each of these cases, where all coupler usages are identical, only one coupler profile is possible. This restriction provides for a major simplification of Lemma 5, and consequently, The Probability Calculation Theorem.

Corollary 2. *Given n and d for a POPS topology, $\forall m$, such that $m = xc$, where $1 \leq x \leq n/c$,*

$$\begin{aligned}\rho(m,x) &= \left[\prod_{i=0}^{g-1}\prod_{j=0}^{g-1}\left[C_x^{d-ix}P_x^{d-jx}\right]\right] \Big/ \left[P_m^n C_m^n\right] \\ &= \left[\prod_{i=0}^{g-1}\prod_{j=0}^{g-1}\left[\frac{(d-ix)!(d-jx)!}{(d-ix-x)!(d-jx-x)!x!}\right]\right] \Big/ \left[P_m^n C_m^n\right].\end{aligned}$$

Proof. First, when considering the contribution of a single coupler (within the inner brackets), the total of the previous usage (α in Lemma 5) for the required source group is ix. This represents the x messages from that source group for each of the i previous coupler groups. Second, and similarly, the total previous usage (β in Lemma 5) for the required destination group is jx. This represents the x messages to that destination group for each of the j previous coupler groups. Third, all coupler usages $u_{ig+j} = x, \forall 0 \leq i < g, \forall 0 \leq j < g$. The corollary then follows from Lemma 3. ∎

For small message sets, $m \leq d$, the least upper bound increases with the message set size, as shown in Figure 4.4. Under these conditions, all the messages in a message set pass through a single coupler. In each of these cases, where only one coupler usage is not zero, c coupler profiles are possible. This restriction provides for another major simplification of Lemma 5, and consequently, The Probability Calculation Theorem.

Corollary 3. *Given n and d for a POPS topology, $\forall m, 1 \leq m \leq d$, then* $\rho(m,m) = \left[(n/d)^2 C_m^d P_m^d\right] / \left[P_m^n C_m^n\right] = \left[n^2(d!)^2/d^2((d-m)!)^2 m!\right] / \left[P_m^n C_m^n\right]$.

Proof. For each of the $c = (n/d)^2$ coupler profiles, there are C_m^d ways to choose the m inputs from the d source group nodes. Also, there are P_m^d ways to map the m inputs selected to the d destination group nodes. The corollary then follows from Lemma 3. ∎

For example, consider, once more, the POPS of Figure 4.1 and message sets of size 8. The minimum sequence length to deliver 8 messages is 2. This specifically occurs when the utilization of each coupler is 2, which from Corollary 2 has a probability $\rho(8,2) \approx 0.07$. The maximum sequence length to deliver 8 messages is 8, which occurs when the 8 messages use one coupler. From Corollary 3, this occurs with probability $\rho(8,8) \approx 2.4 \times 10^{-8}$.

4.4.4 *Sequence Length Probability Experiments*

For POPS topologies of low complexity (relatively small n or large d), a program based on The Probability Calculation Theorem is practical. For a given POPS topology, this calculator program enumerates all coupler profiles of a specific message set size, and is more efficient than an enumeration of all possible message sets. After the profile enumeration, $\rho(m,s), \forall s, glb(m) \leq s \leq lub(m)$ are tabulated.

The exhaustive nature of the calculation-based program is not practical for more complex systems (for example, when $c > 32$). A program which estimates the sequence length probability was also developed. For a given POPS topology, random message sets of a specific size (permutations, or subsets thereof) are generated, then resulting coupler profiles are used to estimate $\rho(m,s), \forall s$ such that $glb(m) \leq s \leq lub(m)$.

4.4.4.1 Probability Distribution. The nature of the distribution of sequence length probabilities is extremely consistent in the range of POPS topologies considered. Three characteristics are observed across a wide range of system sizes, coupler degrees, and message set sizes. First, the sequence length extremes, *glb* and *lub*, have extremely low probabilities. Second, all non-trivial probabilities occur in a small range of sequence lengths. Third, sequence lengths with non-trivial probabilities constitute a small fraction of the possible sequence lengths, these being among the very shortest.

Corollary 2 calculates the probability that, given a random permutation-based message set whose size is some multiple of the number of couplers, the minimum possible sequence length for delivery is required. This represents the likelihood that a random message set of that size will make the most efficient use of the POPS network. Similarly, Corollary 3 calculates the probability that, given a random permutation-based message set of some fixed size ($m \leq d$), the maximum possible sequence length for delivery is required. This represents the likelihood that a random message set of that size will make the least efficient use of the POPS network. A short survey of the probability that the sequence length will be the greatest lower bound or the least upper bound for several POPS topologies, is presented in the following table.

		$n = 32$	$n = 64$	$n = 128$
$d = 8$	lub when $m = n$	$1.4 \cdot 10^{-13}$	$3.3 \cdot 10^{-18}$	$1.3 \cdot 10^{-22}$
$d = 8$	glb when $m = d$	$4.1 \cdot 10^{-04}$	$3.8 \cdot 10^{-16}$	$2.6 \cdot 10^{-216}$
$d = 16$	lub when $m = n$	$1.1 \cdot 10^{-17}$	$6.7 \cdot 10^{-29}$	$7.3 \cdot 10^{-39}$
$d = 16$	glb when $m = d$	$2.8 \cdot 10^{-04}$	$2.4 \cdot 10^{-05}$	$1.9 \cdot 10^{-22}$

This table indicates that, though not impossible, it is very unlikely that the messages will all require the same coupler. Likewise, it is also highly unlikely, though not impossible, that the messages will be evenly distributed among the couplers. In fact, as the number of couplers increases (there are 256 couplers for $n = 128$ and $d = 8$) the greatest lower bound probability quickly approaches zero.

In general, estimator program data shows that all non-trivial probabilities occur in a very narrow range of sequence lengths. Further, the distribution involves relatively few of the total number of possible sequence lengths, and those involved are the shortest ones. For example, the small POPS network where $n = 32, d = 16$, and $m = 32$, delivers more than 98% of the possible message sets in four sequence lengths (8 through 11). These represent the lowest possible sequence lengths ($glb = 8 \leq s \leq 16 = lub$).

These distribution characteristics become even more pronounced in larger POPS networks. For example, when $n = 256, d = 64$ and $m = 128$, possible sequence lengths range from the greatest lower bound of 8 to the least upper bound of 64. The most probable sequence length of 13 delivers over 25% of the possible message sets. In fact, over 88% of the message sets can be delivered by one of only five sequence lengths (11 through 15), more than 98% in ten sequence lengths (8 through 17). This represents the shortest 18% of the valid sequence lengths ($glb = 8 \leq s \leq 64 = lub$).

4.4.4.2 Average Sequence Length. For a given POPS topology and a message set $\mathcal{M}$ of size m, let the average sequence length be the probability-weighted arithmetic mean of all possible sequence lengths, $s_{avg} = \sum_{s=glb(m)}^{lub(m)} s \cdot \rho(m, s)$. The consistent nature of the non-trivial part of the probability distribution, as described above, makes the average sequence length a suitable and adequate measure of the performance of a POPS topology.

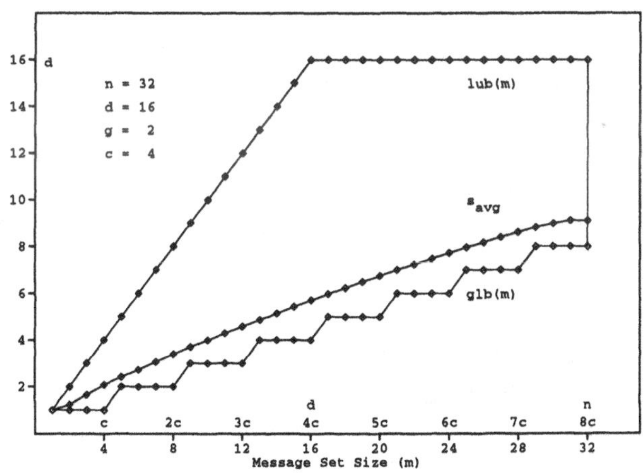

Figure 4.5 Average Sequence Length on a Sequence Envelope.

Figure 4.5 shows the sequence envelope for a small POPS network ($n = 32, d = 16$), as well as the average sequence length $s_{avg}, \forall m$ such that $1 \leq m \leq n$. In this example, for any size message set, the average sequence length is always less than two, and often only one, sequence length unit larger than the greatest lower bound. Typically, this is only about 10% of the sequence length range (least upper bound minus greatest lower bound).

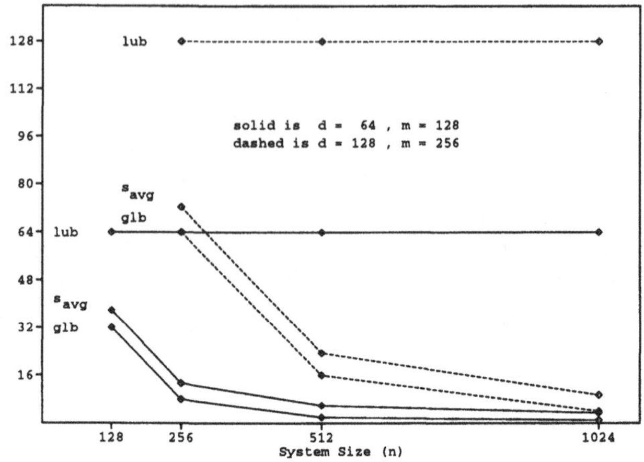

Figure 4.6 Average Sequence Length for Various POPS Topologies.

The efficient delivery of a random permutation-based message set in a POPS topology is not dependent on small system size. In fact, similar and even more pronounced behavior is observed in larger systems. Figure 4.6 shows the average sequence length for seven large POPS networks involving $n = 128, 256, 512, 1024$ and $d = 64, 128$.

As an example of high performance in a very large POPS topology, consider a system with 1024 nodes and a coupler degree of 64 (16 groups, 256 couplers), and message sets containing 512 messages. In this system, message sets require sequence lengths of at least 2 and at most 64. The most probable sequence length is found to be 7 and delivers 45.1% of the possible message sets. The eight shortest sequence lengths (2 through 9) will deliver 98.8% of random message sets. The average sequence length is only $s_{avg} = 6.992$. Figure 4.7 shows the average number of messages delivered (couplers utilized) per time slot in the sequence. In the figure, individual averages are shown by the solid curve, and cumulative averages appear as the dashed line. The horizontal line for 256 messages delivered represents 100% coupler utilization.

The low average sequence lengths observed in POPS topologies is even more notable given the static (that is, a single message set generated at initialization) nature of the analysis. This type of analysis tends towards a message delivery *tail* with very low utilization, as seen during time slots 4 through 7 in Figure 4.7. The effect of this low tail utilization will be sharply reduced in a dynamic environment where the transmission of different message sets may be overlapped.

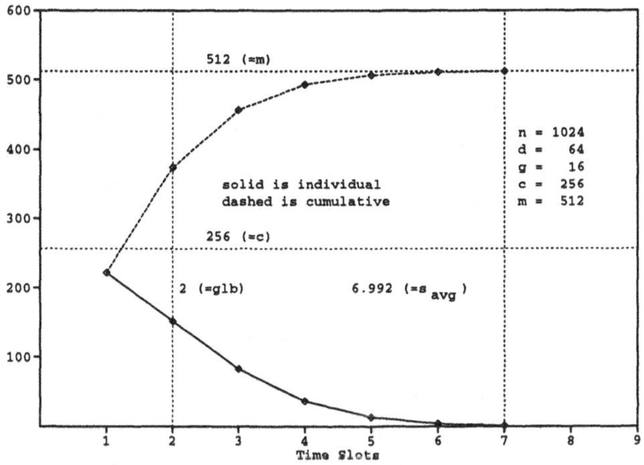

Figure 4.7 Average Messages Delivered per Time Slot.

4.5 REALIZATION OF SPECIFIC COMMUNICATION PATTERNS

The performance of the POPS topology is examined with respect to several common communication patterns. The effect of the efficiency and throughput observed in the previous section is determined. Group and multicasting patterns are briefly examined. The all-to-all personalized communications pattern is then shown to be optimally supported. Terminology is consistent with [18].

4.5.1 Group and Casting Patterns

The nature of the communications between node groups in a POPS topology has a powerful effect on the optimization of some of the most common communication patterns. Strictly on the node group level, a POPS network provides an all-to-all personalized communications capability in a single time slot. In other words, during each time slot and for every pair of source node group $GN_{x'}, 0 \leq x' < g$ and destination node group $GN_{y'}, 0 \leq y' < g$, one unique message $M_{x,y}$ from a node x in $GN_{x'}$ to a node y in $GN_{y'}$ is possible. Multiple messages M_{x_0,y_0} and M_{x_1,y_1} between a single source and a single destination node group pair ($x'_0 = x'_1$ and $y'_0 = y'_1$), must be handled one message per time slot.

Short sequence lengths are achieved by those communication patterns where each step emphasizes both inter-group communications, and no more than one message between any source/destination node group pair. High coupler utilizations are achieved by those communication patterns where each step approaches an all-to-all personalized communications on the group level. Some hierarchical or intra-group patterns may not make efficient use of the available POPS network throughput. Nonetheless, many of the most important of these patterns may be ideally implemented.

Several of the most common communication patterns in tightly-coupled multi-processor environments are multicasting variations. Generically, these occur when identical messages are sent from one node to a set of nodes. That is, a multicasting communication pattern has a message set of the form $\mathcal{M} = \{M_{x,y_0}, ..., M_{x,y_{m-1}}\}$, where $M_{x,y_i} = M_{x,y_j}$, for any $0 \leq i,j < m$. These patterns are frequently important to both computational and control aspects of a system's operation. The POPS topology has rich support for these demands.

Immediate (single time slot) system-wide one-to-all broadcast capability ($\mathcal{M} = \{M_{x,0}, ..., M_{x,n-1}\}$, where $M_{x,i} = M_{x,j}$, for $0 \leq i,j < n$) is inherent in a POPS topology. During any single time slot, any specific node may send a common message to all nodes. All transmitters for the source node send the message, and each of the g couplers connected to the source node group fans the message out to all nodes in a specific destination node group. Further, an all-to-all broadcast, where each node sends a single message to all nodes, is trivial to schedule in n time slots. If multiple receivers for a given node can be active in a single time slot, then g nodes, one from each node group, can broadcast g messages during the same time slot. Hence, an all-to-all broadcast can be done in $n/g = d$ time slots.

Practical network designs often partition the nodes into groups. Both the computation algorithm and the control protocol are frequently optimized for this group-oriented design. In any POPS network, one-to-all broadcasting within a group ($\mathcal{M} = \{M_{x,x'd}, M_{x,x'd+1}, ..., M_{x,x'd+(d-1)}\}$, where $M_{x,i} = M_{x,j}$, for $x'd \leq i,j < x'd + (d-1)$) requires a single time slot. All-to-all broadcasting within the d nodes in a group is easily scheduled in d time slots. A key aspect to group multicasting in a POPS topology is independence. In a single time slot, all or any number of the node groups may engage in separate multicasting operations. Importantly, node groups not involved in these multicasting operations may participate in normal point-to-point communications.

4.5.2 The All-to-All Personalized Communication Pattern

A major communications pattern is all-to-all personalized (complete) communications. Here, each node sends a unique message to every node in the system. This complete communications requires a total of $m = n^2$ messages. If all couplers are utilized during every time slot required to deliver these messages, the sequence length is the minimum possible, so the all-to-all personalized communications is considered optimal.

For a POPS topology, an optimal complete communications requires all c couplers to be used during each time slot needed to deliver the n^2 messages. If $d < \sqrt{n}$, then $n < c$. Thus, since each processor can generate one message every time slot, for a total of n messages, not all couplers can be utilized simultaneously, and all-to-all personalized communications can not be optimal.

If $d \geq \sqrt{n}$, then $n \geq c$, and an optimal complete communications is always possible as stated in the following theorem.

Theorem 4. *In any POPS topology where n and d are powers of two, if $d \geq \sqrt{n}$, then all-to-all personalized communications can be performed in $s = d^2$ time slots, which is optimal since $n^2/c = d^2$.*

Proof. Partition the n nodes into $p = n/c$ subsets, denoted $\mathcal{N}_i, \forall 0 \leq i < p$, where each subset contains $n/p = c$ nodes evenly distributed among the node groups. Specifically, each $\mathcal{N}_i$ contains $c/g = g$ nodes from each of $GN_0, GN_1, ..., GN_{g-1}$. For example, we may define $\mathcal{N}_i = \bigcup_{j=0}^{g-1} \mathcal{N}_i^j$, where $\mathcal{N}_i^j = \{ig + jd + k; 0 \leq k < g\}$ contains g nodes from node group GN_j. Since $d^2/n \geq 1$ from $d \geq \sqrt{n}$, and both d^2 and n are powers of two, then p is a positive integer.

The all-to-all personalized communication pattern will be accomplished in p^2 phases denoted $\mathcal{P}_r$, $r = 0, \ldots, p^2 - 1$. One source subset $\mathcal{N}_{i_s}, 0 \leq i_s < p$, completely communicates with one destination subset $\mathcal{N}_{i_d}, 0 \leq i_d < p$, in a single phase. That is, consecutive time slots in a phase are exclusively used to deliver all messages originating from any node in the source subset, $\mathcal{N}_{i_s}$, and terminating at any node in the destination subset, $\mathcal{N}_{i_d}$. One possible choice for i_s and i_d for phase $\mathcal{P}_r$ is $i_s = \lfloor r/p \rfloor$ and $i_d = r \bmod p$.

Further, divide the time slots in each phase $\mathcal{P}_r$ into g equal partitions, which we shall call sub-phases and denote $\mathcal{P}_r^u, 0 \leq u < g$. In a single sub-phase $\mathcal{P}_r^u$, any given source node in any given $\mathcal{N}_{i_s}^{j_s}, 0 \leq j_s < g$, completely communicates with an entire single destination $\mathcal{N}_{i_d}^{j_d}$. Specifically, the kth node, $0 \leq k < g$, in $\mathcal{N}_{i_s}^{j_s}$ communicates with all the nodes in $\mathcal{N}_{i_d}^{j_d}$ where $j_d = (k + u) \bmod g$. This communication can be arranged such that in the tth time slot of $\mathcal{P}_r^u$, $0 \leq t < g$, the kth node, in $\mathcal{N}_{i_s}^{j_s}$ communicate with the yth node in $\mathcal{N}_{i_d}^{j_d}$, where $y = (j_s + t) \bmod g$.

Hence, in any time slot, regardless of the sub-phase and the phase, each of the g nodes in any $\mathcal{N}_{i_s}^{j_s}, 0 \leq j_s < g$ communicates with a unique $\mathcal{N}_{i_d}^{j_d}, 0 \leq j_d < g$. Thus, in a single time slot, all c nodes in $\mathcal{N}_{i_s}$ communicate utilizing all c couplers. Further, in the g time slots within any sub-phase, regardless of the phase, any given node in $\mathcal{N}_{i_s}$ communicates with every node in a particular $\mathcal{N}_{i_d}^{j_d}$. Finally, in the g sub-phases within any phase, any given node in $\mathcal{N}_{i_s}$ communicates with every node in $\mathcal{N}_{i_d}$.

In summary, within the g time slots per sub-phase, the g sub-phases per phase, and the p phases, an all-to-all personalized communication pattern can be achieved. Thus, in any POPS topology where $d \geq \sqrt{n}$, an optimal complete communications can be constructed, and requires a sequence length of $s = g * g * p^2 = d^2$. ∎

For example, in the POPS of Figure 4.1, in which $n = 16$ and $d = 8$ ($c = 4$ and $g = 2$), the four subsets defined in Theorem 4 can be formed as follows:

$$\begin{aligned}
\mathcal{N}_0 &= \{0,1\} \cup \{8,9\},\\
\mathcal{N}_1 &= \{2,3\} \cup \{10,11\},\\
\mathcal{N}_2 &= \{4,5\} \cup \{12,13\},\\
\mathcal{N}_3 &= \{6,7\} \cup \{14,15\}.
\end{aligned}$$

In order to illustrate the communication within each of the sixteen phases needed to accomplish all-to-all communications, consider the third phase in which the nodes in $\mathcal{N}_0$ communicate with the nodes in $\mathcal{N}_2$. This phase is divided into two sub-phases. In the first sub-phase nodes 0 and 8 communicate with nodes in $\{4,5\}$ and nodes 1 and 9 communicate with nodes in $\{12,13\}$. In the second sub-phase, nodes 0 and 8 communicate with nodes in $\{12,13\}$ and nodes 1 and 9 communicate with nodes in $\{4,5\}$. In each sub-phase, eight messages are delivered in two time slots. For example, in the first sub-phase, messages $M_{0,4}$, $M_{8,5}$, $M_{1,12}$ and $M_{9,13}$ may be delivered in the first time slot, and messages $M_{0,5}$, $M_{8,4}$, $M_{1,13}$ and $M_{9,12}$ may be delivered in the second time slot. The complete all-to-all personalized communication can thus be completed in $16 * 2 * 2 = 64$ time slots, which is optimal since it is the minimum required for the delivery of 256 messages using four couplers.

4.6 CONCLUSION

The POPS topology is an all-optical non-hierarchical star-based topology. Route determination in POPS is trivial and message scheduling is simple. This latter fact is due to the independence of individual coupler scheduling. These characteristics promote a very low control overhead. Absolute bounds of the sequence length for delivery of a random permutation-based message set can be easily calculated.

For random permutation patterns, results consistently show a high and narrow sequence length distribution centered relatively close to the theoretical minimum. Surveys of system size, coupler degree, and message set size show that this results in an average sequence length close to the greatest lower bound under almost any conditions. Low system complexity, high utilization and very high relative throughput are all possible in a POPS topology. A wide range of POPS topologies can provide optimal all-to-all personalized communications.

The results presented in this chapter estimate the minimum number of time slots, s, needed to deliver the messages in a message set $\mathcal{M}$. The same results directly apply if $\mathcal{M}$ is a set of connections that are to be established in a time-multiplexed manner in the POPS. In this case, s represents the minimum degree of multiplexing which allows all the connections in $\mathcal{M}$ to be established in the POPS. Finally, if WDM is to be used, then s would be the minimum number of wavelengths needed to simultaneously establish the connections in $\mathcal{M}$.

References

[1] E. Arnould, F. Bitz, E. Cooper, R. Sansom, and P. Steenkiste. The design of nectar: A network backplane for heterogeneous multicomputers. In *Proceedings of the Third International Conference on Architectural Support for Programming Languages and Operating Systems*, pages 205–216. IEEE, 1989.

[2] Y. Birk. Power-optimal layout of passive, single-hop, fiber-optic interconnections whose capacity increases with the number of stations. In *Infocom '93: 12th Joint Conference of the Computer and Communications Societies*, pages 565–572. IEEE, 1993.

[3] C. Brackett. Dense wavelength division multiplexing networks: Principles and applications. *IEEE Journal on Selected Areas in Communications*, 8(6):948–964, August 1990.

[4] D. Chiarulli, S. Levitan, R. Melhem, J. Teza, and G. Gravenstreter. Multiprocessor interconnection networks using partitioned optical passive star (POPS) topologies and distributed control. In *Proceedings of the First International Workshop on Massively Parallel Processing Using Interconnections*, pages 70–80. IEEE, April 1994.

[5] D. Coudert, A. Ferreira, and X. Munoz. Multiprocessor architectures using multi-hop multi-ops lightwave networks and distributed control. In *Proceedings of the Ninth International Parallel Processing Symposium*. IEEE, April 1998.

[6] A. Ganz, B. Li, and L. Zenou. Reconfigurability of multi-star based lightwave LANs. In *Globecom '91*, volume 3, pages 1906–1910. IEEE, 1992.

[7] R. Gidron. Teranet: A multi-gigabits per second ATM network. *Network Architecture*, 1992.

[8] M. Goodman, H. Kobrinski, M. Vecchi, R. Bulley, and J. Gimlett. The LAMBDANET multiwavelength network: Architecture, applications, and demonstration. *IEEE Journal of Selected Areas in Communications*, 8(6):995–1004, August 1990.

[9] G. Gravenstreter and R. Melhem. Realizing common communication patterns in patitioned optical passive stars (pops) networks. *IEEE Trans. on Computers*, To appear - 1998.

[10] G. Gravenstreter, R. Melhem, D. Chiarulli, S. Levitan, and J. Teza. The partitioned optical passive stars (POPS) topology. In *Proceedings of the Ninth International Parallel Processing Symposium*. IEEE, April 1995.

[11] P. Green. An all-optical computer network: Lessons learned. *IEEE Network*, pages 56–60, March 1992.

[12] P. Green. *Fiber Optic Networks*. Prentice Hall, Englewood Cliffs, NJ, 1993.

[13] I. Habbab, M. Kavehrad, and C. Sundberg. Protocols for very high-speed optical fiber local area networks using a passive star topology. *IEEE Journal of Lightwave Technology*, LT-5(12):1782–1793, December 1987.

[14] R. Handel and M. Huber. *Integrated Broadband Networks: An introduction to ATM-based Networks.* Addison-Wesley, Reading, MA, 1990.

[15] M. Hluchyj and M. Karol. ShuffleNet: An application of generalized perfect shuffles to multihop lightwave networks. In *Infocom '88*, pages 379–390. IEEE, 1988.

[16] A. Holmes and R. Syms. All-optical CDMA using quasi-prime codes. *Journal of Lightwave Technology*, 10(2):279–286, February 1992.

[17] F. Janniello, R. Ramaswami, and D. Steinberg. A prototype circuit-switched multi-wavelength optical metropolitan-area network. In *Proceedings of the IEEE International Communications Conference.* IEEE, 1992.

[18] V. Kumar, A. Grama, A. Gupta, and G. Karypis. *Introduction to Parallel Computing.* Benjamin/Cummings, Redwood City, CA, 1994.

[19] S. Lay. *Analysis - An Introduction to Proof.* Prestice-Hall, Englewood Cliffs, NJ, 1986.

[20] B. Mukherjee. WDM-Based local lightwave networks, Part I: Single-hop systems. *IEEE Networks*, pages 12–27, May 1992.

[21] C. Qiao and R. Melhem. Reconfiguration with time division multiplexing MINs for multiprocessor communications. *IEEE Transactions on Parallel and Distributed Systems*, 5(4):337–352, 1994.

[22] M. Sexton and A. Reid. *Transmission Networking: SONET and the Synchronous Digital Hierarchy.* Artech House, Norwood, MA, 1992.

[23] R. Thompson, R. Anderson, J. Camlet, and P. Giordano. Experimental modular switching system with a time-multiplexed photonic center stage. In *Proceedings of OSA Topical Meeting on Photonic Switching*, 1992.

5 OTIS OPTOELECTRONIC COMPUTERS

Chih-fang Wang and Sartaj Sahni

Department of Computer and Information Science and Engineering
University of Florida
Gainesville, Florida 32611
{wang,sahni}@cise.ufl.edu

Abstract: OTIS computers employ the optical transpose interconnect system for inter package communications and a regular electronic interconnect system (*e.g.*, mesh, hypercube, mesh of trees) for intra package communications. In this chapter, we explore the characteristics of OTIS computers and summarize some of the algorithmic developments that have been made.

5.1 INTRODUCTION

It is well known that when communication distances exceed a few millimeters, optical interconnects provide speed (bandwidth) and power advantages over electronic interconnects [1, 5]. Therefore, in the construction of very large multiprocessor computers it is prudent to interconnect physically close processors using electronic interconnect and to use optical interconnect for pairs of processors that are distant. We shall assume that physically close processors are in the same physical package (chip, wafer, board) and processors that are not physically close are in different packages. As a result, electronic interconnects are used for intra package communications while optical interconnect is used for inter package communication.

Various combinations of interconnection networks for intra package (*i.e.*, electronic) communications and inter package (*i.e.*, optical communications) have been proposed. In OTIS computers [3, 6, 13], optical interconnects are realized via a free space optical interconnect system known as the optical transpose interconnection system (OTIS).

In this chapter, we begin by describing the OTIS. Next, we describe the OTIS-Mesh and OTIS-Hypercube parallel computers that result, respectively, when the OTIS optical interconnect system is used for inter package communication and a mesh or hypercube is used for intra package communication. Properties of both the OTIS-Mesh and OTIS-Hypercube computer are also developed. We conclude by summarizing the complexities of OTIS-Mesh and OTIS-Hypercube algorithms for data routing as well as of basic operations and matrix multiplication on OTIS-Mesh computers. In this chapter we are concerned with deterministic algorithms alone. Randomized algorithms for the OTIS-Mesh can be found in [8].

5.2 OPTICAL TRANSPOSE INTERCONNECTION SYSTEM (OTIS)

The optical transpose interconnection system (OTIS) was proposed by Marsden *et al.* [6]. The OTIS connects $L = MN$ inputs to L outputs using free space optics and two arrays of lenslets. The first lenslet array is a $\sqrt{M} \times \sqrt{M}$ array and the second one is of dimension $\sqrt{N} \times \sqrt{N}$. Thus, a total of $M + N$ lenslets are used. The L inputs and outputs are arranged to form a $\sqrt{L} \times \sqrt{L}$ array. The L inputs are arranged into $\sqrt{M} \times \sqrt{M}$ groups with each group containing N inputs arranged into a $\sqrt{N} \times \sqrt{N}$ array. Figure 5.1 shows the arrangement of the $L = 64$ inputs when $M = 4$ and $N = 16$. The $M \times N$ inputs are indexed (i, j) with $0 \leq i < M$ and $0 \leq j < N$. Inputs with the same i value are in the same $\sqrt{N} \times \sqrt{N}$ block. The notation $(1, *)$, for example, refers to all inputs of the form $(1, j)$.

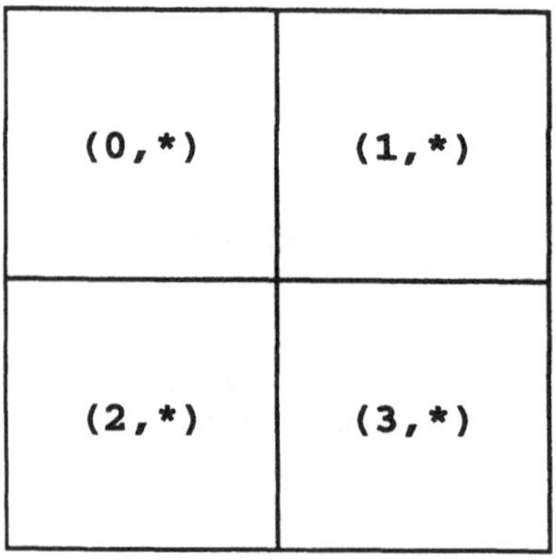

Figure 5.1(a) $\sqrt{M} \times \sqrt{M} = 2 \times 2$ grouping of inputs.

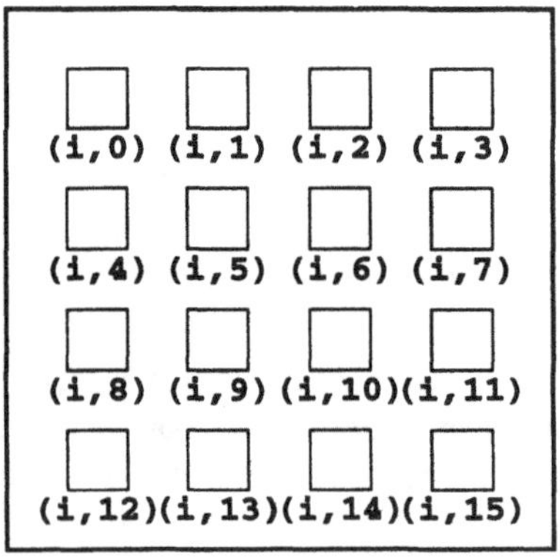

Figure 5.1(b) The $(i, *)$ group, $0 \leq i < M = 4$.

Figure 5.1 2-dimensional arrangement of $L = 64$ inputs when $M = 4$ and $N = 16$.

In addition to using the two-dimensional notation (i, j) to refer to an input, we also use a four-dimensional notation (i_r, i_c, j_r, j_c) where (i_r, i_c) gives the coordinates (row,column) of the $\sqrt{N} \times \sqrt{N}$ block that contains the input (see Figure 5.1(a)) and (j_r, j_c) gives coordinates of the element within a block (see Figure 5.1(b)). So all elements $(i, *)$ with $i = 0$ have $(i_r, i_c) = (0, 0)$; those with $i = 1$ have $(i_r, i_c) = (0, 1)$; those with $i = 2$ have $(i_r, i_c) = (1, 0)$; and those with $i = 3$ have $(i_r, i_c) = (1, 1)$. Similarly, all inputs with $j = 3$ have $(j_r, j_c) = (0, 3)$, and those with $j = 12$ have $(j_r, j_c) = (3, 0)$.

The L outputs are also arranged into a $\sqrt{L} \times \sqrt{L}$ array. This time, however, the $\sqrt{L} \times \sqrt{L}$ array is composed of $\sqrt{N} \times \sqrt{N}$ blocks with each block containing M outputs that are arranged as a $\sqrt{M} \times \sqrt{M}$ array. The $L = MN$ outputs are indexed (i, j) with $0 \leq i < N$, $0 \leq j < M$. All outputs of the form $(i, *)$ are in the same block, block i. Block i is in position (i_r, i_c) with $i = i_r\sqrt{N} + i_c$ of the $\sqrt{N} \times \sqrt{N}$ block arrangement. Outputs of the form $(*, j)$ are in position (j_r, j_c) of their block, $j = j_r\sqrt{M} + j_c$.

In the physical realization of OTIS, the $\sqrt{L} \times \sqrt{L}$ output arrangement is rotated 180°. We have 4 two-dimensional planes; the first is the $\sqrt{L} \times \sqrt{L}$ input plane; the second is a $\sqrt{M} \times \sqrt{M}$ lenslet plane, the third is a $\sqrt{N} \times \sqrt{N}$ lenslet plane, and the fourth is the $\sqrt{L} \times \sqrt{L}$ plane of outputs rotated 180°. When the OTIS is viewed from the side, only the first column of each of these planes is visible. Such a side view for the case $L = M \times N = 4 \times 16$ is shown in Figure 5.2. Notice that the first column of the input plane consists of the inputs (0,0), (0,4), (0,8), (0,12), (2,0), (2,4), (2,8), (2,12) which in 4D notation are (0,0,0,0), (0,0,1,0), (0,0,2,0), (0,0,3,0), (1,0,0,0), (1,0,1,0), (1,0,2,0), (1,0,3,0). The inputs in the same row as (0,0,0,0) are $(0, *, 0, *)$, those in the same row as (i_r, i_c, j_r, j_c) are $(i_r, *, j_r, *)$. The (i_r, j_r) values top to bottom are (0,0), (0,1), (0,2), (0,3), (1,0), (1,1), (1,2), (1,3). The first column in the output plane (after the 180° rotation) has the outputs (15,3), (15,1), (11,3), (11,1), (7,3), (7,1), (3,3), (3,1) which in 4D notation are (3,3,1,1), (3,3,0,1), (2,3,1,1), (2,3,0,1), (1,3,1,1), (1,3,0,1), (0,3,1,1), (0,3,0,1). The outputs in the same row as (3,3,1,1) are $(3, *, 1, *)$, those in the same row as (i_r, i_c, j_r, j_c) are $(i_r, *, j_r, *)$. The (i_r, j_r) values top to bottom are (3,1), (3,0), (2,1), (2,0), (1,1), (1,0), (0,1), (0,0).

Each lens of Figure 5.2 denotes a row of lenslets and each ○ a row of inputs or outputs. The interconnection pattern defined by the given arrangement of inputs, outputs, and lenslets connects input $(i, j) = (i_r, i_c, j_r, j_c)$ to output $(j, i) = (j_r, j_c, i_r, i_c)$. The connection is established via an optical ray that originates at input position (i_r, i_c, j_r, j_c), goes through lenslet (i_r, i_c) of the first lenslet array, then through lenslet (j_r, j_c) of the second lenslet array, and finally arrives at output position (j_r, j_c, i_r, i_c).

The basic connectivity provided by the OTIS is an optical connection between input (i, j) and output (j, i), $0 \leq i < M$, $0 \leq j < N$.

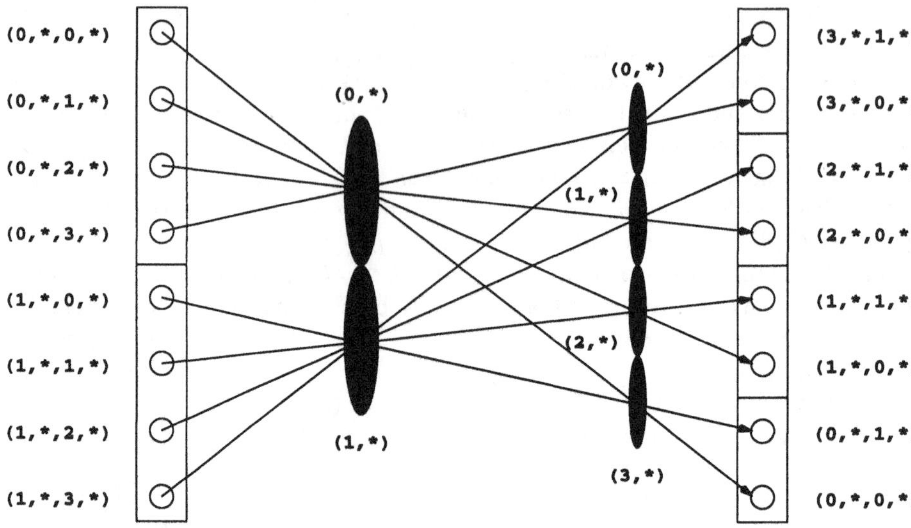

Figure 5.2 Side view of the OTIS with $M = 4$ and $N = 16$.

5.3 OTIS PARALLEL COMPUTERS

Marsden *et al.* [6] have proposed several parallel computer architectures in which OTIS is used to connect processors in different groups (packages) and an electronic interconnection network is used to connect processors in the same group. Since Krishnamoorthy *et al.* [5] have shown that bandwidth is maximized and power consumption minimized when an $L = N^2$ processor OTIS computer is partitioned into N groups of N processors each, Zane *et al.* [13] limit the study of OTIS parallel computers so that each processor group (package) has N processors and the parallel computer has a total of N groups (packages). Let (i, j) denote processor j of package i, $0 \leq i < N$, $0 \leq j < N$. Processor (i, j), $i \neq j$, is connected to processor (j, i) using free space optics (*i.e.*, OTIS). The only other connections available in an OTIS computer are the electronic intra group connections.

A generic 16 processor OTIS computer is shown in Figure 5.3. The solid boxes denote processors. Each processor is labeled (g, p) where g is the group index and p is the processor index. OTIS connections are shown by arrows. Intra group connections are not shown.

In an OTIS-Mesh, processors in the same group are connected as a 2-D mesh [6, 13, 9]; and in an OTIS-Hypercube, processors in the same group are connected using the hypercube topology [6, 13, 10]. OTIS-Mesh of trees [6], OTIS-Perfect shuffle, OTIS-Cube connected cycles, etc., may be defined in an analogous manner.

When analyzing algorithms for OTIS architectures, we count data moves along electronic interconnects (*i.e.*, electronic moves) and those along optical

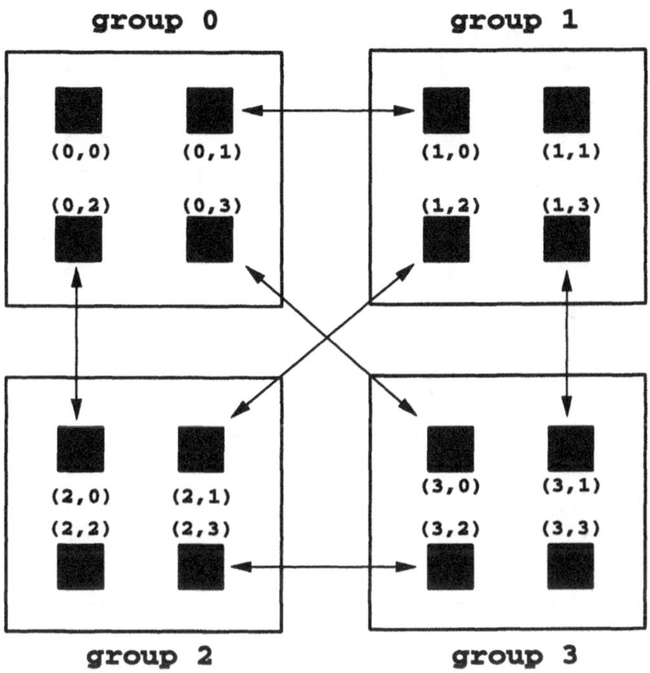

Figure 5.3 Example of OTIS connections with 16 processors.

interconnects (*i.e.*, OTIS moves) separately. This allows us to later account for any differences in the speed and bandwidth of these two types of interconnect.

5.4 OTIS-MESH

In an N^2 processor OTIS-Mesh, each group is a $\sqrt{N} \times \sqrt{N}$ mesh and there are a total of N groups. Figure 5.4 shows a 16 processor OTIS-Mesh. The processors of groups 0 and 2 are labeled using two dimensional local mesh coordinates while the processors in groups 1 and 3 are labeled in row-major fashion. We use the notation (g, p) to refer to processor p of group g.

5.4.1 Diameter of the OTIS-Mesh

Let (g_1, p_1) and (g_2, p_2) be two OTIS-Mesh processors. The shortest path between these two processors is of one of the form:

(a) The path involves only electronic moves. This is possible only when $g_1 = g_2$.

(b) The path involves an even number of optical moves. In this case the path is of the form $(g_1, p_1) \xrightarrow{E^*} (g_1, p_1') \xrightarrow{O} (p_1', g_1) \xrightarrow{E^*} (p_1', g_1') \xrightarrow{O} (g_1', p_1') \xrightarrow{E^*} (g_1', p_1'') \xrightarrow{O} (p_1'', g_1') \xrightarrow{E^*} (p_1'', g_1'') \xrightarrow{O} \cdots \xrightarrow{E^*} (g_2, p_2)$.

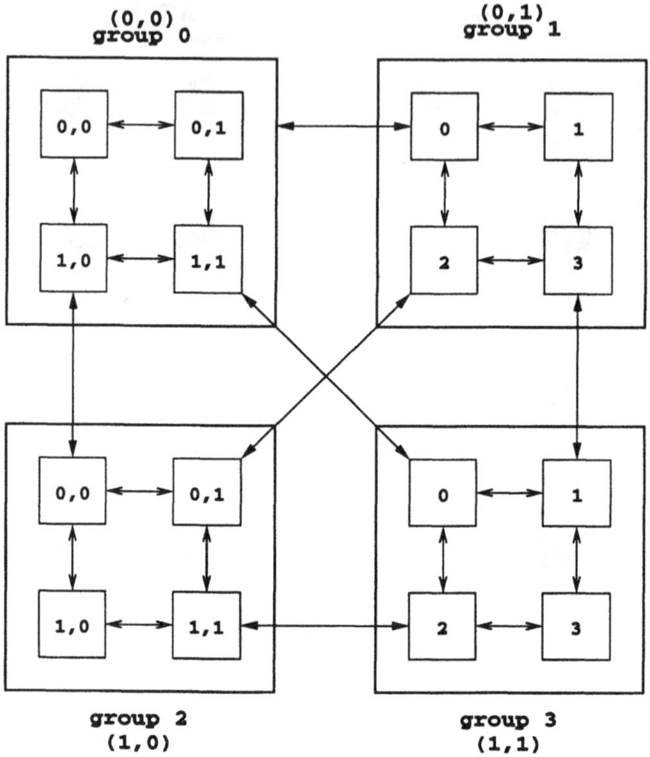

Figure 5.4 16 Processor OTIS-Mesh.

Here E* denotes a sequence (possibly empty) of electronic moves and O denotes a single OTIS move. If the number of OTIS moves is more than two, we may compress paths of this form into the shorter path $(g_1, p_1) \xrightarrow{E^*} (g_1, p_2) \xrightarrow{O} (p_2, g_1) \xrightarrow{E^*} (p_2, g_2) \xrightarrow{O} (g_2, p_2)$. So we may assume that the path is of the above form with exactly two OTIS moves.

(c) The path involves an odd number of OTIS moves. In this case, it must involve exactly one OTIS move (as otherwise it may be compressed into a shorter path with just one OTIS move as in (b)) and may be assumed to be of the form $(g_1, p_1) \xrightarrow{E^*} (g_1, g_2) \xrightarrow{O} (g_2, g_1) \xrightarrow{E^*} (g_2, p_2)$.

Let $d(i, j)$ be the shortest distance between processors i and j of a group using a path comprised solely of electronic moves. So, $d(i, j)$ is the Manhattan distance between the two processors of the local mesh group. Shortest paths of type (a) have length $d(p_1, p_2)$ while those of types (b) and (c) have length $d(p_1, p_2) + d(g_1, g_2) + 2$ and $d(p_1, g_2) + d(p_2, g_1) + 1$, respectively.

From the preceding discussion we have the following theorems:

Theorem 1 *[9] The length of the shortest path between processors* (g_1, p_1) *and* (g_2, p_2) *is* $d(p_1, p_2)$ *when* $g_1 = g_2$ *and* $\min\{d(p_1, p_2) + d(g_1, g_2) + 2, d(p_1, g_2) + d(p_2, g_1) + 1\}$ *when* $g_1 \neq g_2$.

Proof. When $g_1 = g_2$, there are three possibilities for the shortest path. It may be of types (a), (b), or (c). If it is of type (a), its length is $d(p_1, p_2)$. If it is of type (b), its length is $d(p_1, p_2)+d(g_1, g_2)+2 = d(p_1, p_2)+2$. If it is of type (c), its length is $d(p_1, g_2)+d(p_2, g_1)+1 = d(p_1, g_1)+d(p_2, g_1)+1 = d(p_1, g_1)+d(g_1, p_2)+1 \geq d(p_1, p_2)+1$. So, the shortest path has length $d(p_1, p_2)$. When $g_1 \neq g_2$, the shortest path is either of type (b) or (c). From our earlier development it follows that its length is $\min\{d(P_1, P_2) + d(G_1, G_2) + 2, d(P_1, G_2) + d(P_2, G_1) + 1\}$. ∎

Theorem 2 *[9] The diameter of the OTIS-Mesh is* $4\sqrt{N} - 3$.

Proof. Since each group is a $\sqrt{N} \times \sqrt{N}$ mesh, $d(p_1, p_2)$, $d(p_2, g_1)$, $d(p_1, g_2)$, and $d(g_1, g_2)$ are all less than or equal to $2(\sqrt{N}-1)$. From Theorem 1, it follows that no two processors are more than $4(\sqrt{N} - 1) + 1 = 4\sqrt{N} - 3$ apart. Hence, the diameter is $\leq 4\sqrt{N} - 3$. Now consider the processors (g_1, p_1), (g_2, p_2) such that p_1 is in position $(0, 0)$ of its group and p_2 is in position $(\sqrt{N} - 1, \sqrt{N} - 1)$ (*i.e.*, p_1 the top left processor and p_2 the bottom right one of its group). Let g_1 be 0 and g_2 be $N-1$. So, $d(p_1, p_2) = d(g_1, g_2) = d(p_1, g_2) = d(p_2, g_1) = \sqrt{N}-1$. Hence, the distance between (g_1, p_1) and (g_2, p_2) is $4\sqrt{N} - 3$. As a result, the diameter of the OTIS-Mesh is exactly $4\sqrt{N} - 3$. ∎

5.4.2 Simulation of a 4D Mesh

Zane *et al.* [13] have shown that the OTIS-Mesh can simulate each move of a $\sqrt{N} \times \sqrt{N} \times \sqrt{N} \times \sqrt{N}$ four-dimensional mesh by using either a single electronic move local to a group or using one local electronic move and two inter group OTIS moves. For the simulation, we must first embed the 4D mesh into the OTIS-Mesh. The embedding is rather straightforward with processor (i, j, k, l) of the 4D mesh being identified with processor (g, p) of the OTIS-Mesh. Here, $g = i\sqrt{N} + j$ and $p = k\sqrt{N} + l$.

The mesh moves $(i, j, k \pm 1, l)$ and $(i, j, k, l \pm 1)$ can be performed with one electronic move of the OTIS-Mesh while the moves $(i, j \pm 1, k, l)$ and $(i \pm 1, j, k, l)$ require one electronic and two optical moves. For example, the move $(i, j + 1, k, l)$ may be done by the sequence $(i, j, k, l) \xrightarrow{O} (k, l, i, j) \xrightarrow{E} (k, l, i, j + 1) \xrightarrow{O} (i, j + 1, k, l)$.

The above efficient embedding of a 4D mesh implies that 4D mesh algorithms can be run on the OTIS-Mesh with a constant factor (at most 3) slowdown [13]. Unfortunately, the body of known 4D mesh algorithms is very small compared to that of 2D mesh algorithms. So, it is desirable to consider a 2D mesh embedding. Such an embedding will enable one to run 2D mesh algorithms on the OTIS-Mesh. Naturally, one would do this only for problems for which no 4D algorithm is known or for which the known 4D mesh algorithms are not faster than the 2D algorithms.

5.4.3 Simulation of a 2D Mesh

There are at least two intuitively appealing ways to embed an $N \times N$ mesh into the OTIS-Mesh. One is the *group row mapping* (GRM) in which each group of the OTIS-Mesh represents a row of the 2D mesh. The mapping of the mesh row onto a group of OTIS processors is done in a snake-like fashion as in Figure 5.5(a). The pair of numbers in each processor of Figure 5.5(a) gives the (row,column) index of the mapped 2D mesh processor. The thick edges show the electronic connections used to obtain the 2D mesh row. Notice that the assignment of rows to groups is also done in a snake-like manner. Let (i, j) denote a processor of a 2D mesh. The move to $(i, j+1)$ (or $(i, j-1)$) can be done with one electronic move as (i, j) and $(i, j+1)$ are neighbors in a processor group. If all elements of row i are to be moved over one column, then the OTIS-Mesh would need one electronic move in case of a MIMD mesh and 3 in case of a SIMD mesh as the row move would involve a shift by one left, right, and down within a group. A column shift can be done with 2 additional OTIS moves as in the case of a 4D mesh embedding. GRM is particularly nice for the matrix transpose operation. Data from processor (i, j) can be moved to processor (j, i) with one OTIS and zero electronic moves.

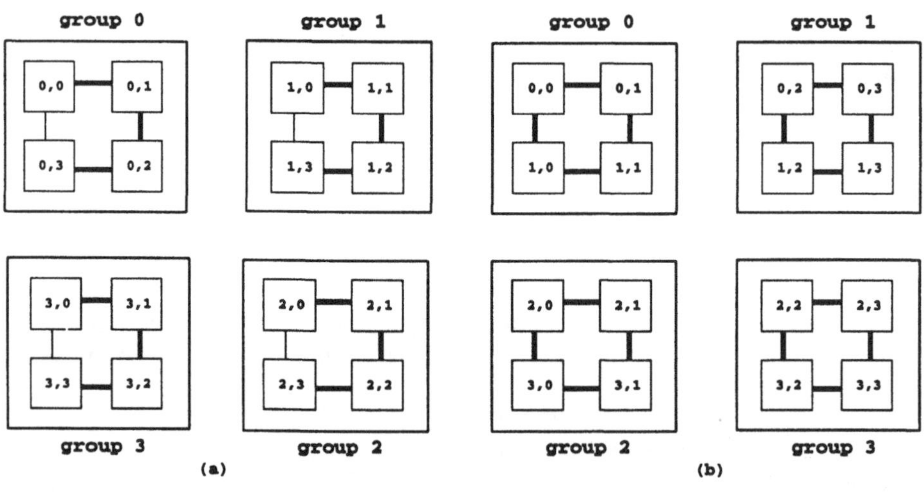

Figure 5.5 Mapping a 4×4 mesh onto a 16 processor OTIS-Mesh: (a) GRM; (b) GSM.

The second way to embed an $N \times N$ mesh is to use the *group submesh mapping* (GSM). Here, the $N \times N$ mesh is partitioned into N $\sqrt{N} \times \sqrt{N}$ submeshes. Each of these is mapped in the natural way onto a group of OTIS-Mesh processors. Figure 5.5(b) shows GSM of a 4×4 mesh. Moving all elements of row or column i over by one is now considerably more expensive. For example, a row shift by +1 would be accomplished by the following data movements (a boundary processor is one on the right boundary of a group):

Step 1: Shift data in non-boundary processors right by one using an electronic move.

Step 2: Perform an OTIS move on boundary processor data. So, data from (g, p) moves to (p, g).

Step 3: Shift the data moved in Step 2 right by one using an electronic move. Now, data from (g, p) is in $(p, g + 1)$.

Step 4: Perform an OTIS move on this data. Now data originally in (g, p) is in $(g + 1, p)$.

Step 5: Shift this data left by $\sqrt{N} - 1$ using $\sqrt{N} - 1$ electronic moves. Now, the boundary data originally in (g, p) is in the processor to its right but in the next group.

The above five step process takes $\sqrt{N}$ electronic and two OTIS moves. Note, however, that if each group is a wraparound mesh in which the last processor of each row connects to the first and the bottom processor of each column connects to the top one, then row and column shift operations become much simpler as Step 1 may be eliminated and Step 5 replaced by a right wraparound shift of 1. The complexity is now two electronic and two OTIS moves.

GSM is also inferior on the transpose operation which now requires $8(\sqrt{N}-1)$ electronic and 2 OTIS moves.

Theorem 3 *[9] The transpose operation of an $N \times N$ mesh requires $8(\sqrt{N}-1)$ electronic and 2 OTIS moves when the GSM is used.*

Proof. Let $g_x g_y$ and $p_x p_y$ denote processor (g, p) of the OTIS-Mesh. This processor is in position (p_x, p_y) of group (g_x, g_y) and corresponds to processor $(g_x p_x, g_y p_y)$ of the $N \times N$ embedded mesh. To accomplish the transpose, data is to be moved from the $N \times N$ mesh processor $(g_x p_x, g_y p_y)$ (*i.e.*, the OTIS-Mesh processor $(g, p) = (g_x g_y, p_x p_y)$) to the mesh processor $(g_y p_y, g_x p_x)$ (*i.e.*, the OTIS-Mesh processor $(g_y g_x, p_y p_x)$). The following movements do this: $(g_x p_x, g_y p_y) \xrightarrow{E^*} (g_x p_y, g_y p_x) \xrightarrow{O} (p_y g_x, p_x g_y) \xrightarrow{E^*} (p_y g_y, p_x g_x) \xrightarrow{O} (g_y p_y, g_x p_x)$. Once again E* denotes a sequence of electronic moves local to a group and O denotes a single OTIS move. The E moves in this case perform a transpose in a $\sqrt{N} \times \sqrt{N}$ mesh. Each of these transposes can be done in $4(\sqrt{N}-1)$ moves [7]. So, the above transpose method uses $8(\sqrt{N}-1)$ electronic and 2 OTIS moves.

To see that this is optimal, first note that every transpose algorithm requires at least 2 OTIS moves. For this, pick a group $g_x g_y$ such that $g_x \neq g_y$. Data from all N processors in this group are to move to the processors in group $g_y g_x$. This requires at least one OTIS move. However, if only one OTIS move is performed, data from $g_x g_y$ is scattered to the N groups. So, at least two OTIS moves are needed if the data ends up in the same group.

Next, we shall show that independent of the OTIS moves, at least $8(\sqrt{N} - 1)$ electronic moves must be performed. The electronic moves cumulatively perform one of the following two transforms (depending on whether the number of OTIS moves is even or odd, see previous section about the diameter):

(a) local moves from (p_x, p_y) to (p_y, p_x); local moves from (g_x, g_y) to (g_y, g_x);

(b) local moves from (p_x, p_y) to (g_y, g_x); local moves from (g_x, g_y) to (p_y, p_x).

For $(p_x, p_y) = (g_x, g_y) = (0, \sqrt{N} - 1)$, (a) and (b) require $2(\sqrt{N} - 1)$ left and $2(\sqrt{N} - 1)$ down moves. For $(p_x, p_y) = (g_x, g_y) = (\sqrt{N} - 1, 0)$, (a) and (b) require $2(\sqrt{N} - 1)$ right and $2(\sqrt{N} - 1)$ up moves. The total number of moves is thus $8(\sqrt{N} - 1)$. So, $8(\sqrt{N} - 1)$ is a lower bound on the number of electronic moves needed. ∎

5.5 OTIS-HYPERCUBE

In an N^2 processor OTIS-Hypercube, each group is a hypercube of dimension $\log_2 N$. Figure 5.6 shows a 16 processor OTIS-Hypercube. The number inside a processor is the processor index within its group.

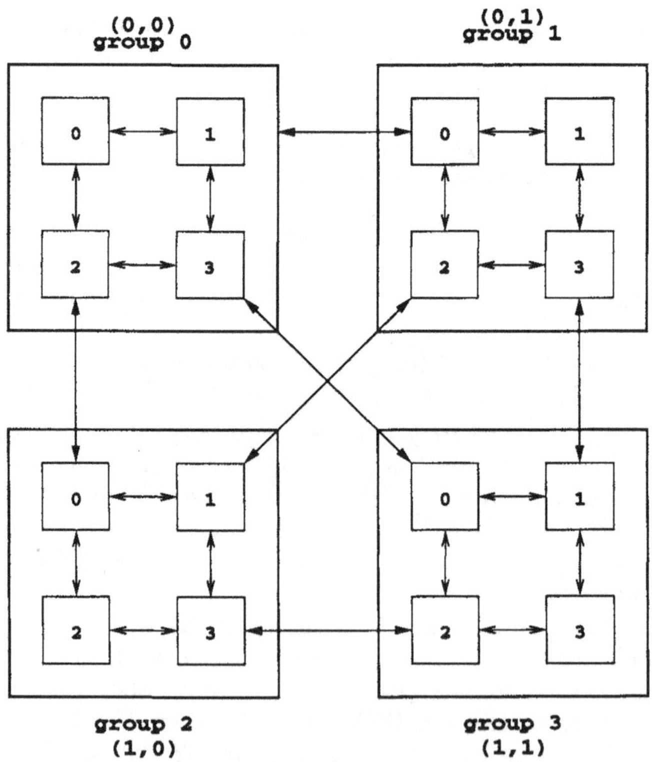

Figure 5.6 16 processor OTIS-Hypercube.

5.5.1 *Diameter of the OTIS-Hypercube*

Let $N = 2^d$ and let $D(i,j)$ be the length of the shortest path from processor i to processor j in a hypercube. Let (g_1, p_1) and (g_2, p_2) be two OTIS-Hypercube processors. Similar to the discussion of the diameter of OTIS-Mesh in the previous section, The shortest path between these two processors fits into one of the following categories:

(a) The path employs electronic moves only. This is possible only when $g_1 = g_2$.

(b) The path employs an even number of OTIS moves. If the number of OTIS moves is more than two, we may compress the path into a shorter path that uses 2 OTIS moves only: $(g_1, p_1) \xrightarrow{E^*} (g_1, p_2) \xrightarrow{O} (p_2, g_1) \xrightarrow{E^*} (p_2, g_2) \xrightarrow{O} (g_2, p_2)$.

(c) The path employs an odd number of OTIS moves. Again, if the number of moves is more than one, we can compress the path into a shorter one that employs exactly one OTIS move. The compressed path looks like: $(g_1, p_1) \xrightarrow{E^*} (g_1, g_2) \xrightarrow{O} (g_2, g_1) \xrightarrow{E^*} (g_2, p_2)$.

Shortest paths of type (a) have length exactly $D(p_1, p_2)$ (which equals the number of ones in the binary representation of $p_1 \oplus p_2$). Paths of type (b) and type (c) have length $D(p_1, p_2) + D(g_1, g_2) + 2$ and $D(p_1, g_2) + D(p_2, g_1) + 1$, respectively.

The following theorem follows from the preceding discussion:

Theorem 4 *[10] The length of the shortest path between processors (g_1, p_1) and (g_2, p_2) is $d(p_1, p_2)$ when $g_1 = g_2$ and $\min\{D(p_1, p_2) + D(g_1, g_2) + 2, D(p_1, g_2) + D(p_2, g_1) + 1\}$ when $g_1 \neq g_2$.* ∎

Theorem 5 *[10] The diameter of the OTIS-Hypercube is $2d + 1$.*

Proof. Since each group is a d-dimensional hypercube, $D(p_1, p_2)$, $D(g_1, g_2)$, $D(p_1, g_2)$, and $D(p_2, g_1)$ are all less than or equal to d. From theorem 4, we conclude that no two processors are more than $2d + 1$ apart. Now consider the processors (g_1, p_1), (g_2, p_2) such that $p_1 = 0$ and $p_2 = N - 1$. Let $g_1 = 0$ and $g_2 = N - 1$. So $D(p_1, p_2) = D(g_1, g_2) = D(p_1, g_2) = D(P_2, G_1) = d$. Hence, the distance between (g_1, p_1) and (g_2, p_2) is $2d + 1$. As a result, the diameter of the OTIS-Mesh is exactly $2d + 1$. ∎

5.5.2 *Simulation of an N^2 hypercube*

Zane *et al.* [13] have shown that each move of an N^2 processor hypercube can be simulated by either a single electronic move or by one electronic and two OTIS moves in an N^2 processor OTIS-Hypercube. For the simulation, processor q of the hypercube is mapped to processor (g, p) of the OTIS-Hypercube. Here $gp = q$ (*i.e.*, g is obtained from the most significant $\log_2 N$ bits of q and p

comes from the least significant $\log_2 N$ bits). Let $g_{d-1} \cdots g_0$ and $p_{d-1} \cdots p_0$, $d = \log_2 N$, be the binary representations of g and p respectively. The binary representation of q is $q_{2d-1} \cdots q_0 = g_{d-1} \cdots g_0 p_{d-1} \cdots p_0$. A hypercube move moves data from processor q to processor $q^{(k)}$ where $q^{(k)}$ is obtained from q by complementing bit k in the binary representation of q. When k is in the range $[0, d)$, the move is done in the OTIS-Hypercube by a local intra group hypercube move. When $k \geq d$, the move is done using the steps

$$\begin{aligned} & (g_{d-1} \cdots g_j \cdots g_0 p_{d-1} \cdots p_0) \\ \xrightarrow{\text{O}} & (p_{d-1} \cdots p_0 g_{d-1} \cdots g_j \cdots g_0) \\ \xrightarrow{\text{E}} & (p_{d-1} \cdots p_0 g_{d-1} \cdots \overline{g_j} \cdots g_0) \\ \xrightarrow{\text{O}} & (g_{d-1} \cdots \overline{g_j} \cdots g_0 p_{d-1} \cdots p_0) \end{aligned}$$

where $j = k - d$.

5.6 PERMUTATION ROUTING ON OTIS COMPUTERS

Suppose we wish to rearrange the data in an N^2 processor OTIS computer according to the permutation $\Pi = \Pi[0] \cdots \Pi[N^2 - 1]$. That is, data from processor $i = gN + p$ is to be sent to processor $\Pi[i]$, $0 \leq i < N^2$. We assume that the interconnection network in each group is able to sort the data in its N processors (equivalently, it is able to perform any permutation of the data in its N processors). This assumption is certainly valid for the mesh, hypercube, perfect shuffle, cube-connected cycles, and mesh of trees interconnections mentioned earlier.

Theorem 6 *Every OTIS computer in which each group can sort can perform any permutation* Π *using at most 2 OTIS moves.*

Proof. When 2 OTIS moves are permitted, the data movement can be modeled by a 3 stage MIN (multistage interconnection network) as in Figure 5.7. Each switch represents a processor group which is capable of performing any N input to N output permutation. The OTIS moves are represented by the connections from one stage to the next.

The OTIS inter stage connections are equivalent to the inter stage connections in a standard MIN that uses $N \times N$ switches. From MIN theory [4], we know that when $k \times k$ switches are used, $2 \log_k N^2 - 1$ stages of switches are sufficient to make an N^2 input N^2 output network that can realize every input to output permutation. In our case (Figure 5.7), $k = N$. Therefore, $2 \log_N N^2 - 1 = 3$ stages are sufficient. Hence 2 OTIS moves suffice to realize any permutation.

An alternative proof comes from an equivalence with the preemptive open shop scheduling problem (POSP) [2]. In the POSP we are given n jobs that are to be scheduled on m machines. Each job i has m tasks. The task length of the

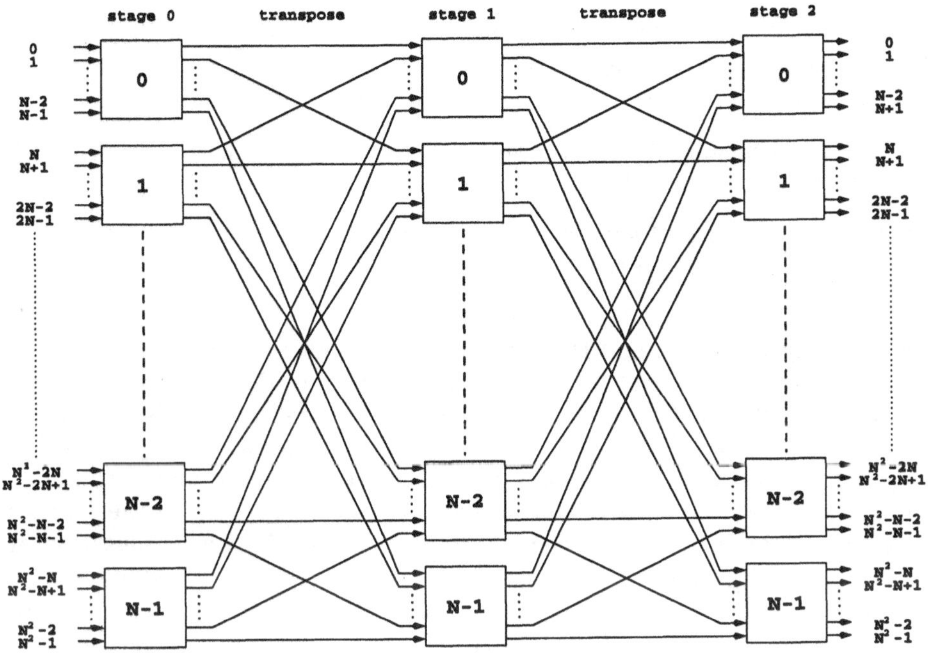

Figure 5.7 Multistage interconnection network (MIN) defined by OTIS.

jth task of job i is the integer $t_{ij} \geq 0$. In a preemptive schedule of length T, the time interval from 0 to T is divided into slices of length 1 unit each. A time slice is divided into m slots with each slot representing a unit time interval on one machine. Time slots on each machine are labeled with a job index. The labeling is done in such a way that (a) Each job (index) i is assigned to exactly t_{ij} slots on machine j, $0 \leq j < m$, and (b) No job is assigned to two or more machines in any time slice. T is the schedule length. The objective is to find the smallest T for which a schedule exists. Gonzalez and Sahni [2] have shown that the length T_{min} of an optimal schedule is

$$T_{min} = \max\{J_{max}, M_{max}\},$$

where $J_{max} = \max_i\{\sum_{j=0}^{m-1} t_{ij}\}$ (*i.e.*, J_{max} is the maximum job length) and $M_{max} = \max_j\{\sum_{i=0}^{n-1} t_{ij}\}$ (*i.e.*, M_{max} is the maximum processing to be done by any machine).

We can transform the OTIS computer permutation routing problem into a POSP. First, note that to realize a permutation Π with 2 OTIS moves, we must be able to write Π as a sequence of permutations $\Pi_0 T \Pi_1 T \Pi_2$ where Π_i is the permutation realized by the switches (*i.e.*, processor groups) in stage i and T denotes the OTIS (transpose) interstage permutation. Let (g_q, p_q) denote processor p_q of group g_q where $q \in \{i_0, o_0, i_1, o_1, i_2, o_2\}$ (i_0 = input of stage 0,

o_0 = output of stage 0, etc.). Then, the data path is $(g_{i_0}, p_{i_0}) \xrightarrow{\Pi_0} (g_{o_0}, p_{o_0}) \xrightarrow{T} (p_{o_0}, g_{o_0}) = (g_{i_1}, p_{i_1}) \xrightarrow{\Pi_1} (g_{o_1}, p_{o_1}) \xrightarrow{T} (p_{o_1}, g_{o_1}) = (g_{i_2}, p_{i_2}) \xrightarrow{\Pi_2} (g_{o_2}, p_{o_2})$.

We observe that to realize the permutation Π, the following must hold:

(i) Switch i of stage 1 should receive exactly one data item from each switch of stage 0, $0 \le i < N$.

(ii) Switch i of stage 1 should receive exactly one data item destined for each switch of stage 2, $0 \le i < N$.

Once we know which data items are to get to switch i, $0 \le i < N$, we can easily compute Π_0, Π_1, and Π_2. Therefore, it is sufficient to demonstrate the existence of an assignment of the N^2 stage 0 inputs to the switches in stage 1 satisfying conditions (i) and (ii). For this, we construct an N job N machine POSP instance. Job i represents switch i of stage 0 and machine j represents switch j of stage 2. The task time t_{ij} equals the number of inputs to switch i of stage 0 that are destined for switch j of stage 2 (*i.e.*, t_{ij} is the number of group i data that are destined for group j). Since Π is a permutation, it follows that $\sum_{j=0}^{N-1} t_{ij}$ = total number of inputs to switch i of stage 0 = N and $\sum_{i=0}^{N-1} t_{ij}$ = total number of inputs destined for switch j of stage 2 = N. Therefore, $J_{max} = M_{max} = N$ and the optimal schedule length is N. Since $\sum_{i=0}^{N-1} \sum_{j=0}^{N-1} t_{ij} = N^2$ and the optimal schedule length is N, every slot of every machine is assigned a task in an optimal schedule. From the property of a schedule, it follows that in each time slice all N job labels occur exactly once. The N labels in slice i of the schedule define the inputs that are to be assigned to switch i of stage 1, $0 \le i < N$. From properties (a) and (b) of a schedule, it follows that this assignment satisfies the requirements (i) and (ii) for an assignment to the stage 1 switches. ∎

Even though every permutation Π can be realized with just 2 OTIS moves, it takes many more OTIS moves to compute the decomposition $\Pi = \Pi_0 T \Pi_1 T \Pi_2$. Therefore, simulating the 3 stage MIN of Figure 5.7 does not result in an efficient algorithm to perform permutation routing. Consequently, Sahni and Wang [9, 10] have developed customized algorithms for specific as well as generalized BPC permutations. General permutations may be realized using the sorting algorithm of [11], which uses $o(\sqrt{N})$ OTIS moves.

5.7 SUMMARY OF OTHER RESULTS

Algorithms for various operations on OTIS computers have been developed in [9, 10, 11, 12, 8]. The complexity of these algorithms is governed by the number of data moves performed. So we summarize these algorithms by providing the number of electronic and OTIS moves they require.

Tables 5.1 and 5.2 give the number of data moves for the commonly used permutations — transpose, perfect shuffle, etc. — and BPC permutations [7] on the OTIS-Mesh [9] and OTIS-Hypercube [10], respectively. Note that in Table 5.2, $d = \log_2 N$.

Table 5.1 Performance of Permutations on OTIS-Mesh [9].

Operation	Electronic	OTIS
Transpose	0	1
Perfect Shuffle	$4\sqrt{N}+6$	2
Unshuffle	$4\sqrt{N}+6$	2
Bit Reversal	$8(\sqrt{N}-1)$	1
Vector Reversal	$8(\sqrt{N}-1)$	2
Bit Shuffle	$\frac{28}{3}\sqrt{N}-4$	$\log_2 N+2$
Shuffled Row-Major	$\frac{28}{3}\sqrt{N}-4$	$\log_2 N+2$
BPC	$12(\sqrt{N}-1)$	$\log_2 N+2$

Table 5.2 Performance of Permutations on Processor OTIS-Hypercube [10].

Operation	Electronic	OTIS
Transpose	0	1
Perfect Shuffle	$2d$	2
Unshuffle	$2d$	2
Bit Reversal	$2d$	1
Vector Reversal	$2d$	2
Bit Shuffle	$3d$	$d/2+2$
Shuffled Row-major	$3d$	$d/2+2$
BPC	$3d$	$d/2+2$

Table 5.3 gives the number of data moves for the basic operations — broadcast, prefix sum, rank, sorting, etc. — on the OTIS-Mesh with both SIMD and MIMD models [11]. Note that w in window broadcast indicates the window size, where $0 < w \leq \sqrt{N}$ and w divides $\sqrt{N}$; s in regular and circular shifts is the shift distance, $0 \leq s < \sqrt{N}/2$; and M in data accumulation, consecutive sum, and adjacent sum is the block size, $0 \leq M < \sqrt{N}/2$.

Table 5.4 gives the number of data moves for various matrix/vector multiplication operations [12]. Since the mapping of the matrix and/or vector is a determining factor on the outcome, results for both GRM and GSM mappings are listed. The big O besides the matrix × matrix operation denotes the memory capacity constraints for each processor and K denotes the maximum

Table 5.3 Performance of Basic Operations on OTIS-Mesh [11].

	SIMD		MIMD	
Operation	Electronic	OTIS	Electronic	OTIS
Broadcast	$4(\sqrt{N}-1)$	1	$4(\sqrt{N}-1)$	1
Window Broadcast	$4\sqrt{N}-2w-2$	2	$4\sqrt{N}-2w-2$	2
Prefix Sum	$7(\sqrt{N}-1)$	2	$7(\sqrt{N}-1)$	2
Data Sum	$8(\sqrt{N}-1)$	1	$4\sqrt{N}$	1
Rank	$7(\sqrt{N}-1)$	2	$7(\sqrt{N}-1)$	2
Regular Shift	s	2	s	2
Circular Shift	$\sqrt{N}$	2	s	2
Data Accumulation	$\sqrt{N}$	2	M	2
Consecutive Sum	$2(M-1)$	2	$M-1$	2
Adjacent Sum	$\sqrt{N}$	2	M	2
Concentrate	$7(\sqrt{N}-1)$	2	$4(\sqrt{N}-1)$	2
Distribute	$7(\sqrt{N}-1)$	2	$4(\sqrt{N}-1)$	2
Generalize	$7(\sqrt{N}-1)$	2	$4(\sqrt{N}-1)$	2
Sorting	$22\sqrt{N}+o(\sqrt{N})$	$o(\sqrt{N})$	$11\sqrt{N}+o(\sqrt{N})$	$o(\sqrt{N})$

amount of data that can be transferred in a single OTIS move (in an electronic move, 1 unit of data may be transferred. Since the optical bandwidth is larger than the electronic bandwidth, $K > 1$ units of data may be transferred in each OTIS move).

The results for randomized routing, sorting, and selection on the OTIS-Mesh [8] are listed in Table 5.5. These results are obtained under the assumption that the cost of OTIS and electronic moves is the same, and that the OTIS-Mesh is a MIMD machine.

5.8 SUMMARY

In this chapter, we have described the OTIS family of computer architectures. This family accounts for the fact that computers with a very large number of processors cannot be built using very short connections (*i.e.*, less than a few millimeters) alone. Therefore, at some level of the packaging hierarchy, the inter processor distance will be such as to favor an optical interconnect over an electronic one. The OTIS system is a simple and easy to implement free space optical interconnect system. Using OTIS for the long connections and

Table 5.4 Performance of Matrix Multiplication on OTIS-Mesh [12].

Scheme	GRM		GSM	
Operation*	Electronic	OTIS	Electronic	OTIS
C × R	$2\sqrt{N}$	2	$4(\sqrt{N}-1)$	2
R × C	$2(\sqrt{N}-1)$	1	$5(\sqrt{N}-1)$	2
R × M	$4(\sqrt{N}-1)$	3	$8(\sqrt{N}-1)$	2
M × C	$4(\sqrt{N}-1)$	1	$8(\sqrt{N}-1)$	2
M × M $O(N)$	$4(N-1)$	$N/K+1$	$4(N-1)$	$2\sqrt{N}/K+1$
M × M $O(1)$	$8N+O(\sqrt{N})$	$N+1$	$4N+O(\sqrt{N})$	$\sqrt{N}$

*C, R, and M denote column vector, row vector, and matrix respectively.

Table 5.5 Performance of Randomized Algorithms on OTIS-Mesh [8].

Operation	Complexity
Randomized Routing	$4\sqrt{N}+\tilde{o}(\sqrt{N})$
Randomized Sorting	$8\sqrt{N}+\tilde{o}(\sqrt{N})$
Randomized Selection	$6\sqrt{N}+\tilde{o}(\sqrt{N})$

electronic mesh (say) for the short interconnects results in a computer with a hybrid optoelectronic interconnection network. The development of algorithms for computers using a hybrid interconnection network poses a challenge. Using the simulation methods developed by Zane *et al.* [13], it is possible to obtain efficient OTIS algorithms from efficient algorithms for the 4D-mesh and hypercube architectures. As one might expect, we can do better than this simulation by developing algorithms specifically for the OTIS architecture [9, 10, 11, 12].

Acknowledgment

This work was supported, in part, by the Army Research Office under grant DAA H04-95-1-0111.

References

[1] M. Feldman, S. Esener, C. Guest, and S. Lee. Comparison between electrical and free-space optical interconnects based on power and speed considerations. *Applied Optics*, 27(9):1742–1751, May 1988.

[2] Teofilo Gonzalez and Sartaj Sahni. Open shop scheduling to minimize finish time. *Journal of the Association for Computing Machinery*, 23(4):665–679, October 1976.

[3] W. Hendrick, O. Kibar, P. Marchand, C. Fan, D. V. Blerkom, F. McCormick, I. Cokgor, M. Hansen, and S. Esener. Modeling and optimization of the optical transpose interconnection system. in Optoelectronic Technology Center, Program Review, Cornell University, September 1995.

[4] David M. Koppelman and A. Yavuz Oruç. The complexity of routing in clos permutation networks. *IEEE Transactions on Information Theory*, 40(1):278–284, January 1994.

[5] A. Krishnamoorthy, P. Marchand, F. Kiamilev, and S. Esener. Grain-size considerations for optoelectronic multistage interconnection networks. *Applied Optics*, 31(26):5480–5507, September 1992.

[6] Gary C. Marsden, Philippe J. Marchand, Phil Harvey, and Sadik C. Esener. Optical transpose interconnection system architectures. *Optics Letters*, 18(13):1083–1085, July 1 1993.

[7] David Nassimi and Sartaj Sahni. An optimal routing algorithm for mesh-connected parallel computers. *Journal of the Association for Computing Machinery*, 27(1):6–29, January 1980.

[8] Sanguthevar Rajasekaran and Sartaj Sahni. Randomized routing, selection, and sorting on the OTIS-Mesh optoelectronic computer. *IEEE Transactions on Parallel and Distributed Systems*, 1998. To appear.

[9] Sartaj Sahni and Chih-Fang Wang. BPC permutations on the OTIS-Mesh optoelectronic computer. In *Proceedings of the fourth International Conference on Massively Parallel Processing Using Optical Interconnections (MPPOI'97)*, pages 130–135, 1997.

[10] Sartaj Sahni and Chih-Fang Wang. BPC permutations on the OTIS-Hypercube optoelectronic computer. *Informatica*, 1998. To appear.

[11] Chih-Fang Wang and Sartaj Sahni. Basic operations on the OTIS-Mesh optoelectronic computer. In *Proceedings of the fifth International Conference on Massively Parallel Processing Using Optical Interconnections (MPPOI'98)*, pages 150-157, 1998.

[12] Chih-Fang Wang and Sartaj Sahni. Matrix multiplication on the otis-mesh optoelectronic computer. Technical report, CISE Department, University of Florida, 1998.

[13] Francis Zane, Philippe Marchand, Ramamohan Paturi, and Sadik Esener. Scalable network architectures using the optical transpose interconnection system (OTIS). In *Proceedings of the second International Conference on Massively Parallel Processing Using Optical Interconnections (MPPOI'96)*, pages 114–121, 1996.

6 ON WAVELENGTH ASSIGNMENT IN WDM OPTICAL NETWORKS

Hyeong-Ah Choi and Eric J. Harder

Department of Electrical Engineering and Computer Science
George Washington University
Washington, DC 20052
{choi,eharder}@seas.gwu.edu

Abstract: Routing and wavelength assignments are fundamental problems in efficient use of optical networks that employ wavelength division multiplexing of the fiber bandwidth. Three perspectives are taken in this chapter. First, the computational complexity of wavelength assignment is considered. Next, wavelength assignment with traffic restricted to all-to-all gossiping is examined, where lower bounds and an approximation algorithm are developed. Finally, results on the related problem of file transfer scheduling in optical networks are given for several network topologies.

6.1 INTRODUCTION

Many technologies exist to improve the network communications among processors in a computing network. Gigabit bitrates, low propagation latency, and low error rates are among the bandwidth enhancements provided by optical networks. By spatially partitioning the fiber bandwidth into distinct parallel optical channels, *wavelength division multiplexing* (WDM) enables the efficient use of optical resources. Combining WDM with passive or active wavelength selective switches results in a variety of wavelength-routing network topologies.

Since the number of wavelengths in a given network is limited, the careful choice of routes and wavelengths for network connection requests is fundamental to the efficient use of wavelength-routing networks. This is more precisely stated as the

Routing and Wavelength Assignment Problem (RWAP)

Given: A directed graph G modeling the physical topology of a wavelength-routing network N, where vertices and edges of G correspond to nodes and links of N, respectively, and a set C of connection requests between nodes of N. Each connection request has the form of a (source, destination) pair.

Objective: For each connection request, find an assignment of a route, or a path in G, and a wavelength in such a way that

- no two paths are assigned the same wavelength if they share a link,
- the total number of wavelengths required to perform the assignment is minimized.

A modified version *Wavelength Assignment Problem (WAP)* of the RWAP is essentially the same as the RWAP except that it presupposes the routing of the connection requests. In this case, it is assumed that a path in G is assigned to each connection request in the set C which is a route from the source to the destination. The set of paths produced by the routing algorithm is denoted by P and is called the ***path set*** for C. The maximum number of paths using any directed link is denoted by L_{max}. This figure is also known as the ***congestion*** [1] of the routing algorithm and is a useful metric for later analysis. The combination of a path and a wavelength forms a ***lightpath*** [26]. The successful formation of a set of lightpaths, in which all of the current connection requests are satisfied, provides a feasible solution to the RWAP, an optimal solution being one that minimizes the number of wavelengths required.

First, the computational complexity of wavelength assignment is studied. On close examination, it turns out that the complexity of finding the optimal routes and wavelengths even for specific topologies is not a simple matter. In particular, the problem of determining the optimal number of wavelengths needed for any set of connection requests on directed rectangular meshes that use arbitrary routing is shown to be NP-complete, leading to the conclusion that approximation algorithms achieving constant bounded performance for this problem do not exist.

Next, a specific connection request set, all-to-all gossiping, is examined. An algorithm for 12/7 optimal wavelength assignment for gossiping, a data distribution technique that is of interest to parallel algorithm designers, on square meshes using row-column routing is developed and analyzed.

Finally, the applied problem of transferring files in optical networks is examined. Lower bounds for the lengths of transfer schedules are found, and it is shown that calculating the optimal schedule is NP-complete under very restricted conditions. Approximation algorithms for scheduling 2-hop file transfers are developed under various network topologies.

6.2 COMPLEXITY

6.2.1 Arbitrary Topologies

For the problem of a set of k connection requests in the RWAP, k disjoint paths, one for each source and destination pair, are desirable. Then only one wavelength is needed over which to communicate since no paths share an edge. However, it is possible that no set of disjoint paths exists that also satisfies all of the connection requests. In the following theorem, we describe NP-completeness results for the RWAP and the WAP. The proofs can be found in [13].

Theorem 1 *Given an arbitrary network topology,*

(i) the RWAP is NP-complete even if the number of available wavelengths is onc

(ii) the WAP is NP-complete even if $L_{max} = 3$ and the length of any path is bounded by two. ∎

While wavelength assignment is thus hard for network topologies in general, it remains to be determined if the same holds true for specific topologies.

6.2.2 Directed Mesh Networks

It is first shown here that deciding if a set of paths over a directed mesh can be properly colored with k colors is at least as hard as deciding if the vertices of an undirected graph can be colored with k colors. It is then shown that no polynomial-time algorithm exists to properly color a given set of paths with k colors, where k is bounded by any constant times L_{max}, unless $P = NP$. To bridge the terminologies of wavelength assignment and graph vertex coloring, the following definitions are made.

A *path coloring* of a graph G is an assignment of colors to a set of paths in G. A *proper* path coloring of a set of paths in G is a path coloring in which no two paths of the same color share an edge. If k colors are used, the path set is said to be *k-colorable.*

Graph k-Colorability Problem (KCOL):
Given: A simple graph $G(V, E)$ and an integer $k \leq |V(G)|$.

Question: Is there a proper coloring of the vertices of G with k colors?

Directed Mesh Path Coloring Problem (DMPC):
Given: A directed mesh G', a set P of directed paths, and a positive integer $k' \leq |P|$.

Question: Is there a proper coloring of the paths in P with k' colors?

The following theorem shows a polynomial-time reduction from KCOL to DMPC.

Theorem 2 *DMPC is NP-complete.*

Proof. A non-deterministic algorithm for DMPC need only guess a coloring of size k' for the paths in P and check in polynomial time that no pair of paths share both a color and an edge of G'. Hence DMPC $\in$ NP.

In [12], Garey and Johnson list KCOL as an NP-complete problem. A polynomial reduction is given from an arbitrary instance of KCOL to an instance of DMPC. Let G be the simple graph and k be the positive integer in an arbitrary instance of KCOL. A directed graph G' and a set P of paths in G' are constructed so that G is k-colorable if and only if P is k-colorable ($k' = k$).

To each vertex of G, a path and a directed mesh are associated in the construction of an instance of DMPC. It is first shown how to construct the meshes and paths, then how to connect the meshes to form G'. The key idea is to transform the vertex adjacency information in G into path adjacency information in G'.

Let $n = |V(G)|$ and label the vertices $V(G) = \{1, 2, \ldots, n\}$ in breadth-first order, with children of a given vertex being labeled in least to greatest order. Define M^i, $1 \le i \le n$, as an $n \times n$ directed mesh. Denote a vertex of M^i by $m^i_{j,k}$. The edges of M^i are, for $1 \le j, k \le n - 1$,

$$\{(m^i_{j,k}, m^i_{j,k+1})\} \cup \{(m^i_{j,k+1}, m^i_{j,k})\} \cup \{(m^i_{j,k}, m^i_{j+1,k})\} \cup \{(m^i_{j+1,k}, m^i_{j,k})\},$$

where $(m^i_{j,k}, m^i_{j+1,k})$ is a directed edge. In essence, there are pairs of vertical edges between each vertex in a column (except the boundary vertices), and pairs of horizontal left-and-right edges between each vertex in a row(except boundary vertices). In the construction of the paths, however, no use is made of any of the right-to-left edges $\{(m^i_{j,k+1}, m^i_{j,k})\}$.

The construction of the paths in G' proceeds as the vertices of G are traversed in the labeled order determined previously, forming partial paths in each mesh, and later connecting the partial paths and meshes to form the final paths and G'. Let π_i, $1 \le i \le n$, denote the paths in the set P, and $p_{i,j}$, $1 \le i, j \le n$, denote the partial path of path π_i in mesh M^j. For instance, the path π_1 is composed of partial paths $p_{1,1} \cup p_{1,2} \cup \cdots \cup p_{1,n}$. Each partial path $p_{i,j}$ begins in mesh M^j at vertex $m^j_{i,1}$

As G is traversed, the adjacency information of the currently visited vertex, say v_k, is transformed into partial paths in the mesh M^k in the following way:

- each partial path $p_{i,k}$, $i \ne k$, consists of vertices (and the intervening directed edges) $m^k_{i,1}, m^k_{i,2}, \ldots, m^k_{i,n}$.
- for each vertex v_s that is adjacent to v_k and has not yet been visited, the partial path $p_{k,k}$ is allowed traverse vertical down edges to row s, and traverse a horizontal edge, thus sharing an edge with the partial path

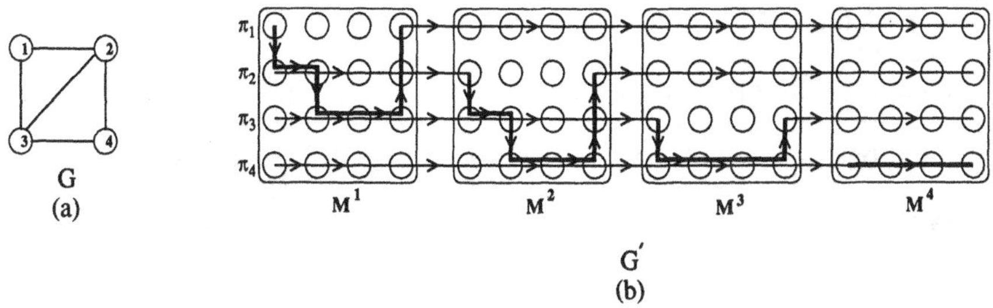

Figure 6.1 Graph G and graph G' with 4 paths.

$p_{s,k}$ (The restriction to unvisited vertices prevents us from re-encoding adjacency information). Since $V(G)$ has been labeled as described above, v_s an unvisited vertex adjacent to v_k implies $k < s$, so only down vertical edges are needed to extend the path. If v_s is the qth such unvisited vertex in G, then the partial path has been extended to a vertex in the qth column of mesh M^k. Suppose the partial path is in row $r \leq n$ at this point. Then $p_{k,k}$ is at vertex $m^k_{r,q}$. The path is now extended to vertex $m^k_{s,q}$, traversing the $r - s$ vertical edges, and then to vertex $m^k_{r,q+1}$ to cause $p_{k,k}$ to share an edge with $p_{k,q}$. Since each mesh is $n \times n$, this procedure can be repeated for every vertex adjacent to each vertex in G. When the last adjacent vertex is transformed, the path is extended to the end column and traverse vertical edges to return to row k (the originating row for the partial path $p_{k,k}$.

Finally, when each vertex in G has been visited and adjacent vertices have been transformed, mesh M^i is connected to M^{i+1}, $1 \leq i \leq n-1$, via directed edges

$$(m^i_{j,n}, m^{i+1}_{j,1}).$$

Each partial path $p_{i,j}$ is similarly joined to partial path $p_{i+1,j}$ by including the above edge.

Figure 6.1 (a) shows a simple graph G and 6.1 (b) a completed directed mesh G' and a path set. Note that the directed edges in G' have been ignored to better display the paths. The construction given in the proof can be followed by concentrating on how the path π_1 is constructed. Within mesh M^1, π_1 consists of the partial path

$$m^1_{1,1}, m^1_{2,1}, m^1_{2,2}, m^1_{3,2}, m^1_{3,3}, m^1_{3,4}, m^1_{2,4}, m^1_{1,4}.$$

By the construction, π_1 shares the edge $(m^1_{2,1}, m^1_{2,2})$ with path π_2, and edge $(m^1_{3,2}, m^1_{3,3})$ with path π_3. In the other meshes, the partial paths for π_1 remain along the first row since there are no other adjacencies involving v_1 to transform.

Observe that a path π_i shares an edge with path π_j, for $i \neq j$, if and only if vertex v_i is adjacent to vertex v_j in G. So the set P of paths can be properly colored with k colors if and only if the vertices of G can be properly colored with k colors.

To complete the proof, note that this is a polynomial time reduction since

- the labeling of $V(G)$ can be performed in polynomial time, and
- each path is wholly contained in $n \times n$ meshes, of which there are n.

This completes the proof of the theorem. ∎

In the reduction from the graph k-colorability problem (for an arbitrary value k), the parameter L_{max} of the input to DWAP always remains to be two even if G' may require more than k wavelengths. Suppose there exists an algorithm $\mathcal{A}$ for the wavelength assignment problem such that the number $\mathcal{A}(G', P, L_{max})$ of wavelengths used by $\mathcal{A}$ for an arbitrary set $\{G', P, L_{max}\}$ of input parameters is bounded by cL_{max} for a constant c. (Without loss of any generality, assume that c is an integer.) Let G be a complete graph on $2c + 1$ vertices (i.e., G has $2c + 1$ vertices and every vertex is connected to $2c$ other vertices; hence, G is not $2c$-colorable). Let G' and P be the input to the wavelength assignment problem constructed using the method discussed in the proof of Theorem 1. The value of L_{max} is then two. Applying algorithm $\mathcal{A}$ to this input results in $\mathcal{A}(G', P, 2) \leq 2c$, which implies that G is $2c$-colorable, a contradiction. The results that follow are established from the above discussion.

Theorem 3 *The sufficient number of wavelengths for the WAP has no relation to the parameter L_{max} of the input.* ∎

Remark. Note that all of the existing results in the literature attempt to approximate the number of required wavelengths for the WAP in terms of L_{max}, in contrast to this result.

The following theorem is also obtained from the proof of Theorem 2; this is the *first* complexity result addressing the approximation hardness of the WAP.

Theorem 4 *There exists no algorithm for the wavelength assignment problem that guarantees the number of wavelengths bounded by a constant times L_{max} even if L_{max} is two and the network topology is a mesh.* ∎

6.2.3 Star and Tree Networks

The following results are described without proofs. The proofs can be found in [13].

Theorem 5 *The WAP can be solved in polynomial time for star networks.* ∎

Theorem 6 *The WAP problem is NP-complete even if the length of each path in trees is bounded by three.* ∎

6.3 ALL-TO-ALL GOSSIPING ON SQUARE MESHES

An approach for combined routing and wavelength assignment is presented here. The recursive nature of certain square meshes is employed to define optimal routing and wavelength assignment. The set of communication requests that is considered is one in which each node communicates with every other node uniquely. This set is referred to as the *all-to-all gossiping set* (or simply the gossiping set). Gossiping is restricted to the use of same length messages.

It has been reported by Beauquier in a theorem in [6] that for 2-dimensional tori an optimal coloring exists for all-to-all gossiping. The proof depends, however, on a result in [16] in which row-column routing is not employed. This contrasts with our results developed in this chapter which focus on simple row-column routing.

The bisection width of communication networks has received much attention, and is considered in detail in [23], where the bisection width of an r-dimensional N-sided array is shown to be at least N^{r-1}. This characterization is appropriate for network flow problems and algorithms that require only a single channel over any edge. The communication networks considered in this chapter, however, are meant to support gossiping among nodes, and hence must consider many channels over any edge. The important measure in this case is the optimal maximum number of required channels, or L^{opt}_{max}, which is found to be 2^{3k-2} for $2^k \times 2^k$ meshes with row-column routing.

The gossiping set of connection requests on $2^k \times 2^k$ directed meshes is considered. For any routing algorithm, it is shown that the optimal maximum load across any edge, denoted as L^{opt}_{max}, is at least $\lceil \frac{n}{2} \rceil \lfloor \frac{n}{2} \rfloor (n)$. Using this lower bound, it is later shown that a $2^k \times 2^k$ directed WDM mesh has an optimal routing, i.e. the value of L_{max} using the row-column routing algorithm achieves the lower bound. The construction of a $2^k \times 2^k$ mesh is first defined, as well as a notation for its edges. Later, optimal bounds and the routing of the gossiping set are developed.

Definition 1 *M^k is defined as a $2^k \times 2^k$ directed mesh, that is, between each pair of adjacent mesh vertices there are two oppositely directed edges.*

Definition 2 *Let M^k be a directed mesh. Any edge in M^k has a location, denoted by its origination point i, j, and a direction in which it points, denoted with the letters r, l, u, d, for right, left, up, and down, respectively. Thus the directed edge from node (i,j) to node $(i, j+1)$ is denoted as $e^r_{i,j}$, and the directed edge from node (i,j) to node $(i+1, j)$ is denoted as $e^d_{i,j}$. Similarly, the directed edge from node (i,j) to node $(i, j-1)$ is denoted as $e^l_{i,j}$, and the directed edge from node (i,j) to node $(i-1, j)$ is denoted as $e^u_{i,j}$. When it is not necessary to specify the direction, the notation $e^\alpha_{i,j}$ with $\alpha \in \{r, l, u, d\}$ is used.*

For any routing of the all-to-all gossiping connection set, the optimal maximum load across any edge, denoted L^{opt}_{max}, has the following lower bound.

Theorem 7 *If M is an $n \times n$ mesh, then $L^{opt}_{max} \geq \lceil \frac{n}{2} \rceil \lfloor \frac{n}{2} \rfloor (n)$ for any routing algorithm.*

Proof. Suppose M is an $n \times n$ mesh. M is partitioned along the $\lceil \frac{n}{2} \rceil$-column into a left submesh M_L of size $n \times \lceil \frac{n}{2} \rceil$ and a right submesh M_R of size $n \times \lfloor \frac{n}{2} \rfloor$. For the $\lceil \frac{n}{2} \rceil$ vertices in row i of M_L, there are $\lfloor \frac{n}{2} \rfloor (n)$ directed paths starting at each of these vertices, and ending in M_R, for a total of $\lceil \frac{n}{2} \rceil \lfloor \frac{n}{2} \rfloor (n)$ directed paths starting at row I and ending in M_R. Since there are n rows in M_L, there are $\lceil \frac{n}{2} \rceil \lfloor \frac{n}{2} \rfloor (n^2)$ paths starting in M_L and ending in M_R. There are only n edges connecting M_L to M_R, so at least one edge connecting M_L to M_R has at least $\lceil \frac{n}{2} \rceil \lfloor \frac{n}{2} \rfloor (n)$ paths traversing it. Therefore $L^{opt}_{max} \geq \lceil \frac{n}{2} \rceil \lfloor \frac{n}{2} \rfloor (n)$. ∎

Before our row-column routing algorithm is recursively defined, the number of gossiping paths that traverse an edge is characterized, and it is found the optimal maximum load is achieved.

Let $p^k_{RC}(e^{\alpha}_{i,j})$ denote the number of all-to-all gossiping paths in a directed mesh M^k that traverse the edge $e^{\alpha}_{i,j}$ when the row-column (RC) routing algorithm is used.

Lemma 1 *Suppose $\alpha \in \{r, l\}$ and $e^{\alpha}_{i,j} \in M^k$. Then, for RC routing, the number of paths in the all-to-all gossiping set that traverse $e^{\alpha}_{i,j}$ is given by the following formula:*

$$p^k_{RC}(e^{\alpha}_{i,j}) = j(2^k - j)(2^k).$$

Proof. Let $\alpha = r$. Using RC routing for the gossiping set, each vertex to the left of $(i, j+1)$ is routed to every vertex in the block $2^k \times (2^k - j)$ with vertex $(i, j+1)$ located in the leftmost column of the block. A symmetric argument exists for edges $e^{l}_{i,j}$. ∎

Lemma 2 *Suppose $\alpha \in \{u, d\}$ and $e^{\alpha}_{i,j} \in M^k$. Then, for RC routing, the number of paths in the gossiping set that traverse $e^{\alpha}_{i,j}$ is given by the following formula:*

$$p^k_{RC}(e^{\alpha}_{i,j}) = i(2^k - i)(2^k).$$

Proof. Let $\alpha = d$. Then, under RC routing of the gossiping set, every vertex in the block of size $i \times 2^k$ block, starting at vertex $(1, 1)$, routes to each of the $(2^k - i)$ vertices below $e^{d}_{i,j}$. A symmetric argument can be made for the $e^{u}_{i,j}$ edges. ∎

With these characterizations for RC routing of the gossiping set, the following theorem is deduced:

Theorem 8 *For any $2^k \times 2^k$ directed mesh*

$$\max_{1 \le i,j \le 2^k} \{p^k_{RC}(e^\alpha_{i,j})\} = 2^{3k-2}.$$

Proof. Let $n = 2^k$ and consider $e^r_{i,j}$. Since $p^k_{RC}(e^r_{i,j}) = j(n-j)(n)$, the value

$$\max_j \{p^k_{RC}(e^r_{i,j})\} = \max_j \{j(n-j)(n)\}$$

is sought. Upon differentiating $p^k_{RC}(e^\alpha_{i,j})$ with respect to j, it is found that

$$p'^k_{RC}(e^r_{i,j}) = n^2 - 2jn.$$

So the maximum occurs at $j = n/2$. Similar reasoning applies to the other directions, with the result that $L_{max} = n^3/4 = 2^{3k-2}$. This result, together with Theorem 1, implies that $L_{max} = L^{opt}_{max}$. ∎

The row-column routing algorithm is now defined recursively.

Definition 3 *The row-column **RC** routing of the all-to-all gossiping path set on a mesh M^k is defines as:*

- *For $k = 1$ see Figure 6.2 for the paths defined on M^1 and Figure 6.3 for routes over the first row of M^2.*
- *For $k > 2$, M^k is subdivided into four submeshes: M^{k-1}_1, the upper left submesh; M^{k-1}_2, the upper right submesh; M^{k-1}_3, the lower left submesh; and M^{k-1}_4, the lower right submesh. The routes to connect vertices between the submeshes are routed by row-column.*
- *Each of the four submeshes is a mesh M^{k-1} that is recursively subdivided into four submeshes, with row-column routing between submeshes.*

The paths routed by RC can also be counted recursively. The following two lemmas prove this, which ensures that the recursive definition of RC routing does not change the bounds described in Theorem 7.

Lemma 3 *For $\alpha \in \{r, l\}$ and RC routing, the number of paths in the gossiping set that traverse $e^\alpha_{i,j}$ is given by the following formulae:*

- *For $1 \le j \le 2^{k-1}$,*

$$p^k_{RC}(e^\alpha_{i,j}) = p^{k-1}_{RC}(e^\alpha_{i,j}) + j(2^{k-1})(2^k) + j(2^{k-1})(2^{k-1} - j).$$

- *For $2^{k-1} < j \le 2^k$,*

$$p^k_{RC}(e^\alpha_{i,j}) = p^{k-1}_{RC}(e^\alpha_{i,j}) + (2^{k-1})(2^k - j)(2^k) + (j - 2^{k-1})(2^{k-1})(2^k - j).$$

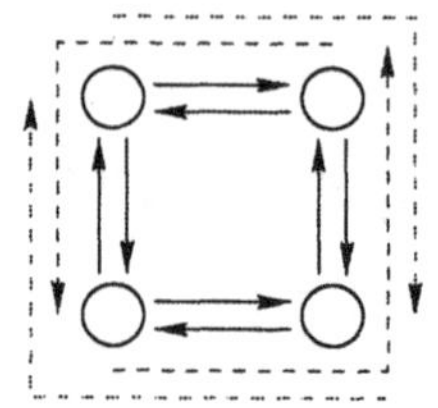

Figure 6.2 RC routing of 2 × 2 mesh.

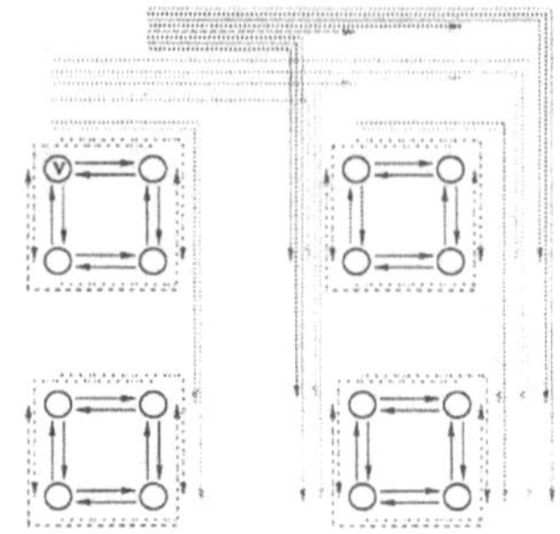

Figure 6.3 RC routing across row 1 of 4 × 4 mesh.

Proof. The lemma follows from induction on k and the recursive definitions of M^k and RC routing. Figures 6.2 and 6.3 show the routing for $k = 1$ and the first row of $k = 2$, provides the basis of the induction. The mesh M^k is subdivided recursively, the lemma is assumed to hold for $k - 1$, and new paths traversing edge $e^r_{i,j}$ are considered. The theorem results from a straightforward counting of the new paths, the details of which are omitted. A symmetric argument, with appropriate indexing, can be made for edges $e^l_{i,j}$. ∎

The vertical edges under RC routing must also be considered:

Lemma 4 *For $\alpha \in \{u, d\}$ and RC routing, the number of paths in the gossiping set that traverse $e^\alpha_{i,j}$ is given by the following formulae:*

- *For $1 \leq i \leq 2^{k-1}$,*

$$p^k_{RC}(e^\alpha_{i,j}) = p^{k-1}_{RC}(e^\alpha_{i,j}) + i(2^k)(2^{k-1}).$$

- *For $2^{k-1} < i \leq 2^k$,*

$$p^k_{RC}(e^\alpha_{i,j}) = p^{k-1}_{RC}(e^\alpha_{i,j}) + (2^{k-1})(2^k)(2^k - i).$$

Proof. An induction on k is used. Figures 6.2 and 6.3 display routing for $k = 1$ and the first row of $k = 2$, providing the basis for the induction. The mesh M^k is subdivided recursively, the lemma is assumed to hold for $p^{k-1}_{RC}(e^\alpha_{i,j})$, and new paths traversing the edge $e^d_{i,j}$ are considered. The theorem results from a straightforward counting of the new paths, the details of which are omitted. A symmetric argument, with appropriate indexing, can be made for edges $e^u_{i,j}$ ∎

Once the paths between pairs of communication requests in the communications request set have been established, a wavelength must be assigned to each path in such a way that no two paths share a wavelength and a link. Under this condition, that paths are also said to be ***properly colored.*** An algorithm that

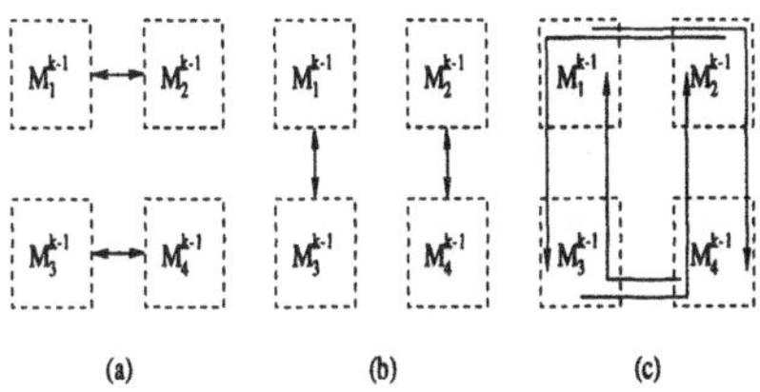

Figure 6.4 Coloring mesh M^k.

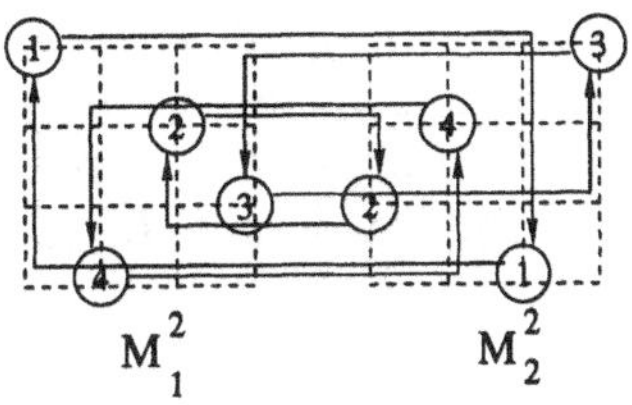

Figure 6.5 Two permutation matrices used for path coloring.

requires at most $12L_{max}/7$ wavelengths to properly color the all-to-all gossiping set is examined in the following, the recursive version of row-column routing is used.

Recall that the mesh M^k may be partitioned into four submeshes M_1^{k-1}, M_2^{k-1}, M_3^{k-1}, and M_4^{k-1}. In each of the following steps, a new set of 2^{3k-3} colors is used.

1. Paths connecting vertices in submesh M_1^{k-1} to M_2^{k-1} are colored with 2^{3k-3} colors (and vice versa). The same colors are used to color paths of points starting in M_3^{k-1} and ending in M_4^{k-1} (and vice versa). See Figure 6.4 (a).

2. Paths connecting points in submesh M_1^{k-1} to M_3^{k-1} are colored with 2^{3k-3} colors (and vice versa) . The same colors are used to color paths of points starting in M_2^{k-1} and ending in M_4^{k-1} (and vice versa). See Figure 6.4 (b).

3. Paths starting in M_1^{k-1} and ending in M_4^{k-1}, and starting in M_2^{k-1} and ending in M_3^{k-1} are colored with 2^{3k-3} colors. The same colors are used to color paths of points starting in M_3^{k-1} and ending in M_2^{k-1}, as well as paths of points starting in M_4^{k-1} and ending in M_1^{k-1}. See Figure 6.4 (c).

In each step, the coloring of paths between submeshes may be performed by using permutation matrices, one $2^{k-1} \times 2^{k-1}$ matrix for each mesh in the coloring. One color is used for each pair of permutation matrices, connecting 2^{k-1} vertices to 2^{k-1} vertices and back. Only one color is needed since, by virtue of the structure of the permutation matrices, source vertices do not share rows or columns, and likewise destination vertices. Figure 6.5 shows this technique for $2^{k-1} = 4$.

This coloring is applied iteratively to each recursive set of submeshes. For each k, $3 \times 2^{3k-3}$ new colors are used. The total number of colors used may be calculated as

$$\sum_{i=1}^{k} 3(2^{3i-3}) \quad \leq \quad \frac{3}{7}(2^{3k}).$$

One metric used to compare static wavelength assignment approximation algorithms is the approximation ratio δ [17]. It was previously shown that the optimal number of required wavelengths is bounded below by L_{max}, and for $2^k \times 2^k$ square meshes, $L_{max}^{opt} \geq 2^{3k-2}$. Hence the approximation ratio the above algorithm is:

$$\begin{aligned} \delta &= \inf\left\{\frac{3(2^{3k})}{7(2^{3k-2})}\right\} \\ &= \frac{12}{7}. \end{aligned}$$

Theorem 9 *An all-to-all gossiping on a $2^k \times 2^k$ square meshes for k being an integer can be done with at most $12L_{max}/7$ wavelengths using the row-column routing.* ∎

6.4 FILE TRANSFER IN OPTICAL NETWORKS

The scheduling of file transfers across a data network is a fundamental problem in network design and scheduling theory. Before the scheduling problem is introduced, the WAP is re-examined when the length of each path is bounded by two.

6.4.1 Arbitrary Networks with Paths of Length 2: Transformation to Graph-Edge Coloring

Arbitrary networks in which the length of any path is bounded by 2 are considered now. In Theorem'1, it was shown that the WAP for arbitrary topologies with paths of length at most 2 was NP-complete. An approximation algorithm for the WAP under this restriction is developed as follows.

The *edge-path graph* is used to transform the problem of assigning wavelengths to a path set to the problem of coloring edges in a related graph. Let P be a set of paths p_i with $|p_i| = 2$ and L_{max} as defined earlier, and G a network on which the paths are routed. Each path p_i covers exactly 2 edges of $E(G)$, say e_{i_1} and e_{i_2}, so the edges may be considered to be adjacent in terms of both a common vertex and a shared path. The *edge-path graph* is defined as the graph G' with

$$\begin{aligned} V(G') &= E(G), \\ E(G') &= \{(e_i,\ e_j) : e_i,\ e_j \in V(G') \text{ and there exists } p_k \in P \text{ that} \\ &\quad\ \text{covers } e_i \text{ and } e_j\}. \end{aligned}$$

Adjacency between vertices of $V(G')$ indicates pairs of edges that define paths. If $|p_i| = 1$, one can add one dummy edge to p_i so that $|p_i| = 2$. An example is given in Figure 6.6.

In essence, G' is the line graph of G with respect to P, which is denoted as $L(G,\ P)$. Note that $L(G,\ P)$ is a undirected graph where the maximum degree Δ of $L(G,\ P)$ is equal to L_{max}. With this transformation, the WAP then

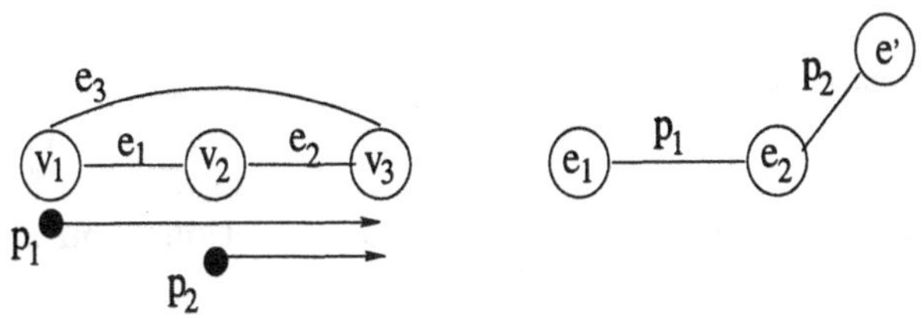

Figure 6.6 Graphs G and $L(G,\ P)$ with $P = \{p_1,\ p_2\}$.

becomes the well-known *graph edge-coloring* problem, where any multigraph can be edge-colored using at most $3\Delta/2$ colors (Shannon's theorem), and any simple graph can be edge-colored using at most $\Delta+1$ colors (Vizing's theorem). If (source, destination) pairs of the connection set are all unique (i.e. for any two paths $p_i = (s_i,\ d_i)$ and $p_j = (s_j,\ d_j)$, with $i \neq j$, either $s_i \neq s_j$ or $d_i \neq d_j$), the corresponding edge-path graph $L(G,\ P)$ is a simple graph. Therefore, the WAP can be solved using $L_{max}+1$ wavelengths for unique (source, destination) pairs and using $3L_{max}/2$ wavelengths for non-unique source-destination pairs, and the following theorem results:

Theorem 10 *For an instance of the WAP when the network topology is arbitrary and the length of any path is at most 2, then*

(i) at most $L_{max}+1$ wavelengths are required to properly color any path set in which the connection requests in the corresponding connection request set are all unique,

(ii) at most $3L_{max}/2$ wavelengths are required when the connections are not unique. ∎

In particular, if G is a ring, it can easily be verified that $L(G,\ P)$ is also a ring; therefore, the WAP can be solved using L_{max}, i.e., optimum number of wavelengths.

Theorem 11 *For an instance of the WAP when the network topology is a ring and the length of any path is at most 2, then L_{max} colors are sufficient to perform a proper coloring of any path set.* ∎

6.4.2 *File Transfer Scheduling*

The problem of scheduling file transfers is now considered. The problem is formulated in greater detail, lower bounds are deduced, and complexity results are discussed under various conditions. Using the technique developed for the WAP when the path length is boubded by two, the file transfer scheduling problem is discussed over the ring, star, and tree network topologies, and near-optimal solutions are presented.

When the scheduling of file transfers in a wavelength-routing network is considered, a distinction is made regarding the basic transfer mechanism that may be viewed as characteristic of the network and not necessarily of a schedule. In the simplest case, the network provides a constant bitrate for the transfer, so each transfer is characterized by the amount of time required to complete without interruption (a *non-preemptive* transfer). For a given set of file transfers, a feasible schedule is one that obeys the *optical constraint* (the property that no concurrent transfer of any files can share a wavelength if the files also share a directed physical link), and an optimal schedule is one that additionally minimizes the total time required to complete all of the requested transfers. A *preemptive* transfer mechanism allows a transfer, once initiated, to be interrupted for an arbitrary amount of time, and later resumed. All networks that are considered use non-preemptive transfers. Hence the schedules that are developed may be termed "non-preemptive schedules."

The problem of transferring files in a wavelength-routing network is begun by making a few definitions that model file transfers:

- G is the graph of the network, which is assumed to be of constant bitrate, i.e., each wavelength channel operates at the same rate. G represents the physical topology of the wavelength-routing network, with each edge as one of two oppositely directed fiber links.

- ω_{max} is the maximum number of wavelength channels that any link can support.

- f_i is the ith file of a set F of files. Each $f_i = (s_i, d_i, t_i, p_i)$, where s_i, d_i, t_i, and p_i are the source, destination, transmission time, and route for file f_i. Note that in a constant bitrate network, t_i may also be used to characterize the size of f_i. The network may support multiple paths for each (s_i, d_i) pair, but the routing is decided prior to the determination of a schedule. It is assumed that only one path per (s_i, d_i) is given; the issue of route selection is not addressed. It may be possible, however, to use multiple wavelength channels within a single path to speed up the file transfer.

- $start_i$ is the time at which the transfer of file f_i may begin.

- λ_i is the set of wavelengths to be used to transfer file f_i.

- a schedule $\mathcal{S}$ is a set $\{(f_i, start_i, \lambda_i) : 1 \leq i \leq |F|\}$.

- the completion time of schedule $\mathcal{S}$ is given by

$$T(\mathcal{S}) = \max_{1 \leq i \leq |F|} \left\{ start_i + \frac{t_i}{|\lambda_i|} \right\}.$$

The problem of finding optimal schedules may be stated as follows:

File Transfer Scheduling Problem (FTSP)
Given: A graph G and a file set F.

Objective: A schedule $\mathcal{S}$ is sought that satisfies the following:

- at any time during the transfers, the optical constraint is satisfied,
- the completion time $T(\mathcal{S})$ is minimized.

In the context that has been defined for the File Transfer Scheduling problem, lower bounds on optimal schedule lengths can now be deduced. For any schedule, the total data load across an edge e can be characterized as

$$\gamma(e) = \sum_{1 \leq i \leq k} \{t_i : \text{ edge } e \text{ is shared by } p_i,\ 1 \leq i \leq k\}.$$

For a given schedule, the following quantities can be calculated:

$$w(v) = \sum_{1 \leq i \leq j} \{t_i : v \text{ is the destination endpoint of } p_i,\ 1 \leq i \leq j\}.$$

$$\eta(v) = \sum_{1 \leq i \leq l} \{t_i : v \text{ is the source endpoint of } p_i,\ 1 \leq i \leq l\}.$$

In essence, $w(v)$ measures the total data reception of node v, and $\eta(v)$ measures the total data transmission of node v.

6.4.3 Some Results on FTSP

In this section, we describe some results on FTSP.

Theorem 12 *Suppose for an arbitrary instance of the problem the schedule S^* is an optimal schedule. Then, the schedule length*

$$T(S^*) \geq \max\left\{ \max_{1 \leq i \leq |F|} \{t_i\},\ \max_{e \in E}\left\{ \frac{\gamma(e)}{\omega_{max}} \right\},\ \max_{v \in V}\left\{ \max\left\{ \frac{w(v)}{\omega_{max}},\ \frac{\eta(v)}{\omega_{max}} \right\}\right\}\right\}.$$

∎

Theorem 13 *The FTSP is NP-complete for star networks.* ∎

Note that Theorem 13 implies that the FTSP remains NP-complete for tree networks even if the length of each path is bounded by two. The proofs of the above results can be found in [13].

The scheduling of file transfers in optical networks of arbitrary topology is next considered. Results from the previous section are recalled to demonstrate approximation algorithms.

Recall from the previous section that the edge-path graph $L(G,\ P)$ can be constructed from a path set P and its associated network G where the length of any path is at most 2. For each edge l_i in $L(G\ P)$, a weight $w(l_i)$ is assigned such that $w(l_i) = t_i$, where t_i is the transmission time of the file corresponding to path p_i in G. In the following, we introduce the problem of scheduling file transfers in traditional non-optical networks, called the *Link Scheduling Problem (LSP)*, studied by Coffman, Garey, Johnson, and LaPaugh [10] and Choi and Hakimi [9], and discuss how the results in [10, 9] can be applied to $L(G, P)$ to obtain a schedule for FTSP.

- *Link Scheduling Problem (LSP):* An instance of the problem consists of a weighted undirected graph $H(V, E)$, which is called a *file transfer graph.* To each edge $e \in E(H)$ there is assigned a positive number, $w(e)$, representing the amount of time needed to transfer the file corresponding to edge e between the nodes corresponding to the end vertices of e. (It is assumed that each node in the network is capable of communicating directly with every other node.) To each vertex v in $V(H)$ there is assigned a positive integer $\alpha(v)$, representing the number of communication modules at the node corresponding to v, where each communication module can be used as a transmitter and as a receiver. They also assume that once the transfer of a file begins, it continues without interruption until the transfer is complete. The problem is to find a schedule of file transfers so as to minimize the overall finishing time.

Observe that each communication module in LSP can be considered as a wavelength channel in our model; consequently, a schedule for a given network G and a path set P of FTSP can be obtained by applying a scheduling algorithm for LSP to edge-path graph $L(G, P)$, assuming that each node in $L(G, P)$ has ω_{max} communication modules. The following summarizes the results in [10, 9] related to our FTSP.

For an arbitrary topology of the file transfer graph in LSP, there exists a schedule with schedule length bounded by 2.5 times the optimum [10]. In [9], it is shown that if the file transfer graph in LSP is a tree, there exists a schedule with schedule length bounded by 2 times the optimum, and if the file transfer graph is a ring, there exists an optimum schedule. In the FTSP considered here, the topology of $L(G, P)$ is arbitrary if G is a tree, a mesh, or an arbitrary graph. If G is a star, $L(G, P)$ is a tree. So, the following theorem results:

Theorem 14 *For a given network topology G and a path set P with length of each path bounded by two, there exists a schedule with schedule length bounded by*

(i) the optimum schedule length if G is a ring,

(ii) two times the optimum schedule length if G is a star, and

(iii) 2.5 times the optimum schedule length if G is a tree, or a mesh, or an arbitrary topology. ∎

Table 6.1 Summary of Results.

topology		*path length*		
		≤ 2	$= 3$	*arbitrary*
star	*FTSP*	NPC, 2OPT	—	—
	WAP	OPT	—	—
tree	*FTSP*	NPC, 2.5OPT	NPC	NPC
	WAP	*, 5/3OPT	NPC, 5/3OPT	NPC, 5/3OPT
mesh	*FTSP*	*, 2.5OPT	*	approx-NPC
	WAP	* OPT+1 for U 3/2OPT for non-U	*	approx-NPC
ring	*FTSP*	OPT	*	NPC
	WAP	OPT	*	NPC
arbitrary	*FTSP*	NPC for U, 2.5OPT	NPC for U	approx-NPC
	WAP	NPC for U OPT+1 for U 3/2OPT for non-U	NPC for U	approx-NPC

6.4.4 Summary of Results on FTSP and WAP

Results on FTSP and WAP are summarized in Table 6.1 where unknown complexity results are denoted by *, and unique and non-unique source-destinations are respectively denoted by U and non-U. The rest of this section explains the entries.

For star topologies (i.e., with path lengths at most 2), it is shown in Theorem 13 that the FTSP is NP-complete, which implies the same result for tree topologies with path lengths of at most 2. However, an approximation algorithm that yields a solution that is at most $2OPT$ for this problem on star topologies is discussed in Theorem 14. The WAP was shown in Theorem 5 to have an algorithm that produces an optimal solution.

The complexity of the WAP for tree topologies with path lengths of at most 2 is not known, while a $2.5OPT$ approximation algorithm is discussed in Theorem 14 for the FTSP under the same conditions. For tree topologies with path lengths of at most 3, it was shown in Theorem 6 that the WAP is NP-complete, which implies the same result for the FTSP. This in turn implies that the FTSP is NP-complete for tree topologies with arbitrary path lengths. Also, the WAP result implies the NP-completeness of the WAP on tree topologies with arbitrary path lengths. Jansen, in [19], proves a $\frac{5}{3}OPT$ approximation algorithm for the WAP on trees with arbitrary path lengths, which implies the same result for the WAP on trees with path lengths of at most 3 and at most 2.

The complexities of both the FTSP and WAP, on mesh topologies with bounded path lengths are unknown. For arbitrarily long paths, the WAP was shown in Theorem 2 to be NP-complete and approximation-NP-complete in Theorem 4; that is, no approximation algorithm exists that provides a solution that is $c \cdot OPT$ for any constant c. This implies the same results for the FTSP on mesh topologies with arbitrary path lengths, and for both the WAP and FTSP on arbitrary topologies with arbitrary path lengths. Note that the approximation-NP-completeness result for the FTSP means only that no approximation algorithm exists that depends only on L_{max} (the problem is said to be "parameter independent" for L_{max}).

For ring topologies with path lengths of three, the complexities of both the FTSP and WAP are unknown, but optimal algorithms exist for both problems if the path lengths are at most 2, as was shown in Theorem 14. For arbitrary path lengths, both WAP and FTSP are known to be NP-complete.

For arbitrary topologies with path lengths of at most 2, Theorem 2 shows that the WAP is NP-complete; the proof of this result shows that all (source, destination) pairs that define the paths are unique. This implies that the FTSP is NP-complete under the same conditions. The remaining entries are based on the results discussed in Theorem 10 and 14.

6.5 SUMMARY

Three perspectives on the problem of wavelength assignment in WDM networks were developed in this chapter. The computational complexity of wavelength assignment was considered, where it was shown that the problem is approximation-hard on meshes. Next, wavelength assignment with the traffic restricted to all-to-all gossiping was examined, where lower bounds and a 12/7-approximation algorithm were produced. Finally, results on the related problem of file transfer in optical networks were given, where lower bounds and a number of approximation algorithms for wavelength assignment on several topologies were developed.

Acknowledgment

Partial funding provided by NSF Grant MIP-9628468.

References

[1] A. Aggarwal, A. Bar-Noy, D. Coppersmith, R. Ramaswami, B. Schieber, and M. Sudan, "Efficient Routing and Scheduling Algorithms for Optical Networks," *Proceedings of the ACM-SIAM Symposium on Discrete Algorithms SODA 1994*, pp. 412–423.

[2] S. Banerjee and C. Chen, "Design of Wavelength-Routed Optical Networks for Circuit Switched Traffic," *Proceedings of the IEEE Globecom '96*, London, November 1996.

[3] D. Banerjee and B. Mukherjee, "Practical Approaches for Routing and Wavelength Assignment in All-Optical Wavelength Routed Networks," *Proceedings IEEE Infocom '94*, Toronto, June 1994, pp. 110–119.

[4] D. Banerjee and B. Mukherjee, "A Practical Approach for Routing and Wavelength Assignment in Large Wavelength-Routed Optical Networks," *IEEE Journal on Selected Areas in Communications*, Vol 14, No 5, June 1996, pp. 903–908.

[5] R. A. Barry and P. A. Humblet, "On the Number of Wavelengths and Switches in All-Optical Networks," *IEEE Transactions on Communications*, Vol 42, No 2/3/4, 1994, pp. 583–591.

[6] B. Beauquier, J.C. Bermond, L. Gargano, P. Hell, S. Perennes, U. Vaccaro, "Graph Problems arising from Wavelength Routing in All Optical Networks," *Proceedings of the Second Workshop in Optics and Computer Science*, Geneva, Switzerland, April 1997.

[7] C. Chen and S. Banerjee, "A New Model for Optimal Routing and Wavelength Assignment in Wavelength Division Multiplexed Optical Networks," *Proceedings of IEEE Infocom '96*, San Francisco, April 1996, pp. 164–171.

[8] H. Choi, H.-A. Choi, and M. Azizoğlu, "Efficient Scheduling of Transmissions in Optical Broadcast Networks," *IEEE/ACM Transactions on Networking*, Vol. 14, No. 6, December 1996, pp. 913–920.

[9] H.-A. Choi and S. L. Hakimi, "Scheduling File Transfers for Trees and Odd Cycles," *SIAM Journal on Computing*, Vol. 16, No. 1, February 1987, pp. 162–168.

[10] E. G. Coffman, Jr., M. R. Garey, D. S. Johnson, and A. S. LaPaugh, "Scheduling File Transfers," *SIAM Journal on Computing*, Vol. 14, 1985, pp. 744–780.

[11] T. Erlebach and K. Jansen, "The Complexity of Call-Scheduling in Trees, Rings, and Meshes," *30th Hawaii International Conference on System Sciences*, March 1997.

[12] M.R. Garey and D.S Johnson, *Computers and Intractability, A Guide to the Theory of NP-Completeness*, W.H. Freeman and Company, 1979.

[13] E. Harder, "Routing and Wavelength Assignment in All-Optical WDM Wavelength Routing Networks," D.Sc. dissertation, The George Washington University, 1998.

[14] E. Harder. S.-K. Lee, H.-A. Choi, "On Wavelength Assignment in WDM Optical Networks," *Proceedings MPPOI '97* Montreal, Canada, June 1997.

[15] E. Harder. H.-A. Choi, "File Transfer in All-Optical Networks," submitted to *Proceedings MPPOI 98*, to be submitted to NETWORKS.

[16] M.C. Heydemann, J-C. Meyer, and D. Sotteau, "On Forwarding Indices of Networks," *Discrete Applied Mathematics*, Vol. 23, 1989, pp. 102–123.

[17] D. Hochbaum, editor, *Approximation Algorithms for NP-Hard Problems*, PWS Publishing Company, Boston, 1997.

[18] I. Holyer, "The NP-completeness of Edge-Coloring," *SIAM Journal on Computing* Vol. 10, 1981, pp. 718–720.

[19] K. Jansen, "Aproximation Results for Wavelength Routing in Directed Trees," *Proceedings of the Second Workshop in Optics and Computer Science*, Geneva, Switzerland, April 1997.

[20] R.M. Karp, "Reducibility Among Combinatorial Problems," *Complexity of Computer Computations*, R.E. Miller and J.W. Thatcher, editors, New York, Plenum Press, 1972.

[21] M. Kovačević and A. S. Acampora, "Benefits of Wavelength Translation in All-Optical Clear-Channel Networks," *IEEE Journal on Selected Areas in Communications*, Vol. 14, No. 5, June 1996, pp. 868–880.

[22] V. Kumar and E. J. Schwabe, "Improved Access to Optical Bandwidth in Trees," *Proceedings of the Fifth ACM-SIAM Symposium on Discrete Algorithms SODA*, January, 1997.

[23] F. T. Leighton, *Introduction to Parallel Algorithms and Architectures: Arrays, Trees, Hypercubes*, Morgan Kaufmann Publishers, 1992.

[24] W. Mao and R. Simha, "Routing and Scheduling File Transfers in Packet-Switched Networks," *Journal of Computing and Information*, Vol. 1, 1994, pp. 559–594.

[25] A. Mokhtar and M. Azizoğlu, "Dynamic Route Selection and Wavlength Assignment in All-Optical Networks," *LEOS*, 1995, pp. 220–221.

[26] B. Mukherjee, *Optical Communication Networks*. McGraw-Hill, 1997.

[27] B. Mukherjee, S. Ramamurthy, D. Banerjee, and A. Mukherjee, "Some Principles for Designing Wide-Area Optical Networks," *Proceedings IEEE INFOCOM '94*, Toronto, 1994, pp. 110–119.

[28] P. Raghavan and E. Upfal, "Efficient Routing in All-Optical Networks," *Proceedings of the 26th ACM Symposium on the Theory of Computing STOC 1994*, 1994, pp. 134–143.

[29] R. Ramaswami and K. N. Sivarajan, "Optimal Routing and Wavelength Assignment in All-Optical Networks," *Proceedings IEEE INFOCOM '94*, Toronto, Canada, June 1994, pp. 970–983.

[30] R. Ramaswami and K. N. Sivarajan, "Routing and Wavelength Assignment in All-Optical Networks," *IEEE/ACM Transactions on Networking,* vol. 3, No. 5, October 1995, pp. 489–500.

[31] S. Subramanian, M. Azizoğlu, and A. Somani, "Connectivity and Sparse Wavelength Conversion in Wavelength-Routing Networks," *Proceedings IEEE Infocom '96*, San Francisco, March 1996, pp. 148–155.

[32] S. Subramaniam and R. A. Barry, "Wavelength Assignment in Fixed Routing WDM Networks," *Proceedings of the IEEE International Conference on Communications '97*, May 1997, pp. 406–410.

II Models and Algorithms for Optical Interconnections

7 AN ABSTRACT MODEL FOR OPTICAL INTERCONNECTION NETWORKS

Si Qing Zheng

Department of Computer Science
Erik Jonsson School of Engineering and Computer Science
University of Texas at Dallas
Richardson, Texas 75083-0688

sizheng@utdallas.edu

Abstract: We present an abstract model, the hypernetwork model, for optical interconnection networks. A hypernetwork is a network that connects a large number of processors using multiple multiconnect components. Hypernetwork topology design can be formulated as a constrained optimization problem. Using several examples, we show how to use concepts and existing results in hypergraph theory and combinatorial block design theory to construct hypernetworks.

7.1 INTRODUCTION

The performance of most electronic digital systems today is limited by the performance of communication links between devices, rather than by devices themselves. Interconnects are often responsible for the majority of area, power, delay and cost. The interconnect bottleneck occurs at all system levels: intra-chip level, multichip module (intra-MCM or chip-to-chip) level, board level, shelf (or technically the backplane) level, and node-to-node level. As one progresses to higher and higher levels, the interconnect bottleneck tightens. Optical technologies have been developed to greatly alleviate the bottleneck problem of electrical interconnects at different levels. The main focus of this chapter is on how to design optical interconnection network topologies.

Interconnection networks have applications in high-performance computing systems, high-throughput telecommunication switching systems, local area networks, and neural network implementations. The traditional electrical interconnection networks can be divided into two major classes: router-based directed networks (also called point-to-point networks or static networks), and switch-based multistage networks (also called dynamic networks). Graph theory has been used to design and analyze these networks. Without loss of generality, we restrict our discussions to interconnections of many processors in a massively parallel computing system. Suppose the metal interconnects of a parallel system were replaced with optical interconnects in a one-to-one fashion. Then, the performance improvement as compared with the resulting cost does not worth the effort. This is because of excessive overheads of intermediate electrical-to-optical (E/O) and optical-to-electrical (O/E) conversions, and underutilization of the inherent large bandwidth and high parallelism of optical interconnecting devices.

The extremely high spatial and temporal bandwidth of optics has significant system configuration implications. An optical connecting device can be used to connect a set of more than two processing nodes. Ideally, one would prefer all-optical networks. Such a network is a single-hop network in which along any connection path there are two, but only two, conversions between electrons and photons, one at each end of the path. Due to physical limitations of an adopted technology and cost/performance constraints, a multihop network of small diameter may be considered more feasible for a massively parallel computing system. The conventional graph theoretic model appears inadequate to characterize the topologies of optical interconnection networks. The purpose of this chapter is to present an abstract model for the topologies of optical interconnection networks. We show that, by treating optical networks as hypergraphs, hypergraph theory and combinatorial block design theory can be applied to investigating many aspects of optical interconnection networks. We demonstrate this approach by presenting several hypernetwork design examples.

7.2 OPTICAL MULTICONNECT HARDWARE

We loosely define an optical multiconnect component as an integrated optical system that connects a set S, $|S| \geq 2$, of processors, and provide connection between any pair of processors in S without intermediate O/E and E/O conversions. The class of optical multiconnect components excludes those that provide fixed parallel connections in the form of image relay using lenses and mirrors. The notion of optical multiconnect components is essential for capturing the common feature of single-hop, direct interconnects among a set S of processors, and based on this notion, interesting and useful optical interconnection networks are possible.

One can utilize free-space interconnects as well as guided lightwave technologies to build optical multiconnect components. It is possible to fit a very large number of simultaneous connections within the very large available bandwidth of a single fiber. For example, the multiaccess dual-bus, folded bus, star and multilevel star shown in Figure 7.1 (a), (b), (c) and (d), respectively, can be considered as optical multiconnect components. In such arrangements, dynamically available connections between processors use different wavelengths (wavelength-division multiplexing – WDM), different time slots (time-division multiplexing – TDM), or different signal waveforms (code-division multiplexing – CDM). A fourth option would be to use one or more photonic switches to

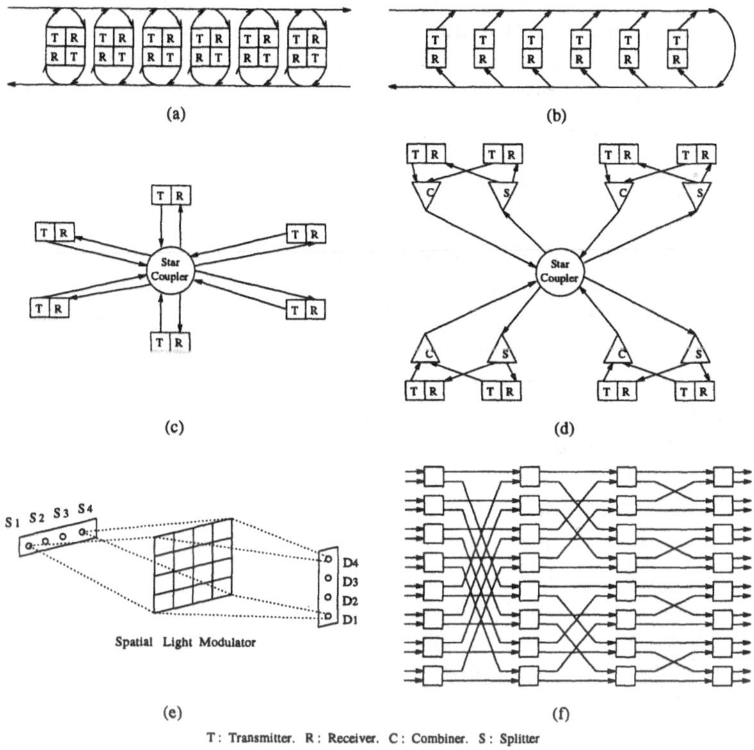

Figure 7.1 Optical multiconnect components: (a) A dual-bus. (b) A folded bus. (c) A star. (d) A multilevel star. (e) An optical crossbar system. (f) An all-optical multistage network.

form a space-division multiplexed (SDM) connecting device. For example, a free-space crossbar switch can be implemented by an optical switching array called spatial light modulator (SLM) as shown in Figure 7.1(e). An SLM steers the light beams toward their corresponding destination processors according to the interconnection patterns recorded in it, and these patterns can be dynamically changed. Various techniques for implementing SLM and light beam steering, such as matrix-vector multiplier and holograms, have been introduced. Spatial switching of light paths can be achieved by a multistage interconnection network (MIN) whose photonic switches are thrown by electrical input (with paths through the switches remaining optical). Such an all-optical MIN can also be considered as an optical multiconnect component (see Figure 7.1(f)). A common feature of all optical multiconnect components is that their multiple connections are achieved by reconfigurability: dynamic time slot allocations (for TDM), tunable transmitters and receivers (for WDM or CDM), and photonic switching (SDM). It is possible to combine two or more multiaccess schemes to construct an optical multiconnect component. For example, combining TDM or SDM with WDM, one can obtain a single-hop hybrid time-wavelength or space-wavelength division interconnect structure as a multiconnect component. For more information about multiconnect technologies, refer to [4, 7, 14].

It would be nice if a large-scale optical multiconnect component can be constructed for the purpose of interconnecting all processors in a massively parallel computing system. In reality, technology constraints may prohibit this approach. One problem is the splitting loss incurred due to the fact that the power of a transmitter is made available to all receivers in the bus and star structures. Splitting loss may be controlled by introducing photonic amplification. The performance of a TDM bus degrades as the number of processors connected by the bus increases. For a WDM multiconnect component, the number of wavelengths may not be sufficient if the number of processors is too large. An additional problem of WDM multiconnect structures is the limitation of tuning-range and tuning time of tunable receivers and transmitters. The size of an optical crossbar switch such as an SLM is limited by many factors. An MIN with photonic switches approximates a crossbar switch, but when the size of a MIN increases, the control circuitry for all the switches may become too complex. Finally, cost/performance considerations may be the decisive factor in determining the size of an optical multiconnect component.

Based on above discussions, we take the position to consider optical interconnection network design consisting of two levels: physical design of optical multiconnect components, and topological design of an all-optical (i.e. single-hop) network or a multihop network. The rest of this chapter is devoted to the high level design of optical interconnection networks.

7.3 HYPERNETWORKS

We use hypergraphs as underlying graph models of optical interconnection networks. A *hypergraph* [3] $H = (V, E)$ consists of a set $V = \{v_1, v_2, \ldots, v_N\}$ of vertices, and a set $E = \{e_1, e_2, \ldots, e_m\}$ of *hyperedges* such that each e_i is a non-empty subset of V and $\cup_{i=1}^{m} e_i = V$. If $e_i \subseteq e_j$ implies that $i = j$, then H is a *simple hypergraph*. When the cardinality of a hyperedge e, denoted as $|e|$, is 1, e is a selfloop edge. If all hyperedges have cardinality 2, then H is a graph that corresponds to a direct network. We only consider simple hypergraphs (and graphs) without selfloops. A hypergraph of n vertices and m hyperedges can also be defined by its $n \times m$ incidence matrix A, with columns representing edges and rows representing vertices such that $a_{i,j} = 0$ if $v_i \notin e_j$, and $a_{i,j} = 1$ if $v_i \in e_j$.

For a subset J of $\{1, 2, \ldots, m\}$, we call the hypergraph $H'(V', E')$ such that $E' = \{e_i | i \in J\}$ and $V' = \cup_{e_i \in E'} e_i$ the *partial hypergraph of H generated by the set J*. For a subset U of V, we call the hypergraph $H''(V'', E'')$ such that $E'' = \{e_i \cap U | 1 \leq i \leq m, e_i \cap U \neq \phi\}$ and $V'' = \cup_{e \in E''} e$ the *subhypergraph induced by the set U*.

The degree $d_H(v_i)$ of v_i in H is the number of hyperedges in V that contain v_i. A hypergraph in which all the vertices have the same degree is said to be *regular*. The *degree of hypergraph* H, denoted by $degree(H)$, is defined as $degree(H) = \max_{v_i \in V} d_H(v_i)$. A regular hypergraph of degree k is called a k*-regular hypergraph.* The rank $r_H(e_j)$ of a hyperedge e_j in H is the number of vertices it contains. A hypergraph in which all hyperedges have the same rank is said to be *uniform*. The *rank of hypergraph* H, denoted by $rank(H)$, is defined as $rank(H) = \max_{1 \leq j \leq m} |e_j|$. A uniform hypergraph of rank k is called a k*-uniform hypergraph.*

In a hypergraph H, a *path* of length q is defined as a sequence $(v_{i_1}, e_{j_1}, v_{i_2}, e_{j_2}, \ldots, e_{j_q}, v_{i_{q+1}})$ such that $v_{i_k}, v_{i_{k+1}} \in e_{j_k}$ and $i_k \neq i_{k+1}$ for $k = 1, 2, \ldots, q$. A path from v_i to v_j, $i \neq j$, is a path in H with its end vertices being v_i and v_j. A hypergraph is *connected* if for any two vertices there exists a path connecting them. We only consider connected hypergraphs. A hypergraph is *linear* if $|e_i \cap e_j| \leq 1$ for $i \neq j$, i.e., two distinct hyperedges share at most one common vertex. For any two distinct vertices v_i and v_j in a hypergraph H, the *distance* between them, denoted by $dis(v_i, v_j)$, is the length of the shortest path connecting them in H. The *diameter* of a hypergraph H, denoted by $diameter(H)$, is defined as $diameter(H) = \max_{v_i, v_j \in H} dis(v_i, v_j)$. More concepts in hypergraph theory can be found in [3].

A *hypernetwork* M is a network whose underlying structure is a hypergraph H, in which each vertex v_i corresponds to a unique processor P_i of M, and each hyperedge e_j corresponds to a multiconnect component that connects the processors represented by the vertices in e_j. We call a multiconnect component a *hyperlink*. In the rest of this chapter, the following pairs of terms are used interchangeably: (hyper)edges and (hyper)links, vertices and processors, graphs and direct networks, and hypergraphs and hypernetworks.

The set of hypernetworks contains direct networks (which have a fixed rank 2) as a subclass. Most issues in the design of hypernetworks are similar to those in the design of direct networks, but the relaxed rank requirement for hypernetworks provide considerable increased choices and flexibility in the construction of high-performance hypernetworks. A larger rank allows us to design a network with smaller degree and diameter, which imply simpler I/O interfaces and less latency caused by E/O and O/E conversions. If possible, a hypernetwork should be uniform, i.e. all hyperlinks connect the same number of processors. The uniformity implies that all hyperlinks can be implemented using optical multiconnect components of the same type. It is also desirable that a hypernetwork to be regular, i.e. all processors are connected to the same number of hyperlinks. Regularity implies simplified processor design since all processors uses the same interface circuitry. When possible, the number of hyperlinks should be minimized to reduce system costs. Enforcing a hypernetwork to be linear helps reducing the number of hyperlinks used.

Hypernetworks are capable of many simultaneous transfers. The total bandwidth of a hypernetwork is the sum of hyperlink bandwidths. It characterizes the best case bandwidth performance. The bisection bandwidth corresponds to the worst case bandwidth performance. It is obtained by partitioning the network into two parts, each has the half the processors, and summing up the bandwidths of the hyperlinks that cross the imaginary partitioning line. Larger number of hyperlinks usually results in larger total and bisection network bandwidths. Trade-offs must be considered to balance the cost and performance.

It is important to investigate the fault-tolerance aspect of hypernetworks. Hyperlink slackness is useful to enhance the reliability, and for this reason, non-linear hypernetworks can be used. Many reliability models for electrical interconnection networks have been proposed. These models include those based on graph connectivity, bandwidth availability, computation availability, performability, and graceful degradability, etc. These existing models can be adopted for hypernetworks.

Another valuable network property is the partitionability. That is, a hypernetwork should be able to be divided into several parts of equal size and similar structures, and there are enough hyperlinks to move data between these substructures rapidly. This property is useful for parallel algorithms based on the divide-and-conquer design paradigm. The inverse of partitionability is the recursive expandability of a hypernetwork, which implies that a larger hypernetwork can be constructed using smaller hypernetworks in the same class when enhancement is desired. A hypernetwork may be expected to be expanded in a incremental fashion by just adding one more processor. This is the incremental expandability discussed in the literature. In general, it is difficult to enforce both recursive and incremental expandabilities for hypernetworks with very small diameters.

A hypernetwork should support efficient parallel algorithms. Routing algorithms of a hypernetwork should be as efficient as possible. For single-hop hypernetworks and multihop hypernetworks with very small diameters, communication using routing tables may be considered acceptable. When designing a hypernetwork, one may want the network able to efficiently simulate other useful networks. The universality of a hypernetwork is its ability of simulating other directed networks and hypernetworks. The concept of graph embedding can be extended to cover hypernetworks so that the performance of simulation of one network by another is characterized by dilation, congestion, and expansion of the embedding.

A hypergraph is vertex (resp. hyperedge) symmetric if for its any two vertices (resp. hyperedges) v_i and v_j (resp. e_i and e_j) there is an automorphism of the hypergraph that maps v_i to v_j (resp. e_i to e_j). Such symmetries are desirable for a hypergraph to be used as a hypernetwork. For example, vertex (resp. hyperedge) symmetry allows for all processors (resp. hyperlinks) to be treated as identical so that simple and efficient message routing schemes can be derived. Furthermore, the uniformity and regularity of symmetric hypernetworks imply that hyperlink sharing and utilization can be expected to be balanced so that congestions are minimized for random communication patterns. Network symmetries have implications in routability, reliability, network bandwidth, partitionability, expandability and universality.

All the hypernetwork design issues mentioned above deserve investigations. Several hypernetwork topologies have been proposed and analyzed. Hypernetworks can be designed according to desired communication patterns. For example, the multiple-bus system described in [12] can be considered as a hypernetwork that supports computations based on tree communication patterns. Hypernetworks can also be obtained from a known direct network by merging its edges into a hyperedges. Based on the observation that a hypergraph can be characterized by a bipartite graph, one can transform a bipartite graph into a hypergraph as a hypernetwork. Interested readers may refer to [9, 17, 18, 19, 20, 21] for some design examples. In the next two sections, we discuss two general hypernetwork design frameworks. The first one uses the concept of the dual of an existing (hyper)network. The second is based on combinatorial block design theory.

7.4 HYPERNETWORK DESIGN BASED ON DUALS OF HYPERGRAPHS

The *dual* of a hypergraph $H = (V, E)$ with vertex set $V = \{v_1, v_2, \ldots, v_N\}$ and hyperedge set $E = \{e_1, e_2, \ldots, e_m\}$ is a hypergraph $H^* = (V^*, E^*)$ with vertex set $V^* = \{v_1^*, v_2^*, \ldots, v_m^*\}$ and hyperedge set $E^* = \{e_1^*, e_2^*, \ldots, e_N^*\}$ such that v_j^* corresponds to e_j with hyperedges $e_i^* = \{v_j^* | v_i \in e_j \text{ in } H\}$. In other words, H^* is obtained from H by interchanging the roles of vertices and hyperedges in H. The incidence matrix of H^* is the transpose of the incidence matrix of H. Thus, $(H^*)^* = H$. The following easily verifiable relations between a hypergraph and its dual are useful:

Proposition 1. *H is r-uniform if and only if H^* is r-regular.* ∎

Proposition 2. *The dual of a linear hypergraph is also linear.* ∎

Proposition 3. *A hypergraph H is vertex symmetric if and only if H^* is hyperedge symmetric.* ∎

Proposition 4. *The dual of a sub-hypergraph of H is a partial hypergraph of the dual hypergraph H^*.* ∎

Proposition 5. $diameter(H) - 1 \leq diameter(H^*) \leq diameter(H) + 1$. ∎

These facts show that some properties of the dual hypergraph H^* of a given hypergraph H can be derived from properties of H. In general, however, the structures of H and its dual H^* can be drastically different.

Let us consider using the dual K_n^* of the n-vertex complete graph K_n as a hypernetwork. Properly labeling the vertices and hyperedges in a hypergraph can greatly simplify its use as an interprocessor communication network. Vertex labels are used as processor addresses. Similarly, hyperedge labels are used as the unique identifiers of hyperlinks. There are many ways to label the vertices and hyperedges of K_n^*. Although all different labeling schemes of K_n^* are equivalent because the symmetries of K_n^* (Proposition 3), we choose to define the K_n^* hypernetwork using an interesting scheme by which the connectivity of K_n^* can be concisely derived.

Definition 1. *Let $N_n = n(n-1)/2$ for $n > 0$. The K_n^* hypernetwork, $n \geq 3$, is a hypergraph that consists of N_n vertices, $v_1, v_2, ..., v_{N_n}$, and n hyperlinks, $e_1, e_2, ..., e_n$. The connectivity of K_n^* is recursively defined as follows:*

[1] K_3^* consists of three vertices v_1, v_2, and v_3, and three hyperlinks $e_1 = \{v_1, v_2\}$, $e_2 = \{v_1, v_3\}$, and $e_3 = \{v_2, v_3\}$.

[2] K_n^* is constructed from K_{n-1}^* by adding $n-1$ more vertices $v_{N_{n-1}+1}$, $v_{N_{n-1}+2}$, $\ldots$, $v_{N_{n-1}+n-1} = v_{N_n}$, and one more hyperlink e_n such that all the newly added $n-1$ vertices are connected to e_n and $v_{N_{n-1}+i}$ is connected to hyperlink e_i, $1 \leq i \leq n-1$.

For a vertex v_i in K_n^*, we use i as its vertex label. Similarly, we use j as the label of hyperedge e_j of K_n^*. By a simple induction on n, it is easy to show that $(K_n^*)^*$ is a complete graph of n vertices. By the properties of K_n and above Propositions, we observe the following fact: K_n^* is 2-regular, $(n-1)$-uniform, linear, and vertex and hyperedge symmetric; the diameter of K_n^* is 1 if $n = 3$, and 2 if $n > 3$. Proposition 4 indicates that K_n^* can be partitioned into several smaller hypernetworks in the K_n^* family.

In the following alternative definition, the connectivity of K_n^* hypernetwork is explicitly specified.

Definition 2. *Let $N_n = n(n-1)/2$ for $n > 0$. The K_n^* hypernetwork, where $n \geq 3$, is a hypergraph that consists of N_n vertices, $v_1, v_2, \ldots, v_{N_n}$, and n hyperlinks, $e_1, e_2, \ldots, e_n$. For any two distinct vertices v_i and v_j, let $u_i = \min\{r | N_r \geq i\}$, $u_j = \min\{s | N_s \geq j\}$, $l_i = i - N_{u_i - 1}$, and $l_j = j - N_{u_j - 1}$. v_i and v_j are connected by a hyperlink if and only if one of the following conditions holds: [i] $u_i = u_j$; [ii] $u_i = l_j$; [iii] $l_i = u_j$; or [iv] $l_i = l_j$. Furthermore, if [i] or [ii] holds then $v_i, v_j \in e_{u_i}$, and if [iii] or [iv] holds then $v_i, v_j \in e_{l_i}$.*

By a simple induction on n, one can easily see that Definitions 1 and 2 use the same vertex and hyperedge labeling schemes. It is easy to verify that any vertex v_i of K_n^* is connected to exactly two hyperedges e_l and e_u, where $u = \min\{r | N_r \geq i\}$, and $l = i - N_{u-1}$. We call hyperedges e_l and e_u the *lower* and *upper hyperedge* of v, respectively. For any l and u such that $1 \leq l < u \leq n$, there is a unique vertex v_i that is connected to both hyperedges e_l and e_u, and furthermore, $i = N_{u-1} + l$. Therefore, a vertex v_i of K_n^* can be uniquely identified by an ordered pair $\langle l, u \rangle$, $1 \leq l < u \leq n$. As we will show shortly, being able to address processors by hyperlinks is a useful property of the K_n^* hypernetwork for the design and analysis of parallel algorithms.

Consider the fault tolerance aspect of the K_n^* hypernetwork. We say that a hypernetwork H is *x-processor fault-tolerant* (resp. *y-hyperlink fault-tolerant*) if it remains connected when no more than any x processors (resp. y-hyperlinks) are removed. It is easy to verify that K_n^* is $(2n-3)$-processor fault-tolerant and 1-hyperlink fault-tolerant.

Example 1. The incidence matrix A of K_6^* is

$$A = \begin{pmatrix} 1 & 1 & 0 & 0 & 0 & 0 \\ 1 & 0 & 1 & 0 & 0 & 0 \\ 0 & 1 & 1 & 0 & 0 & 0 \\ 1 & 0 & 0 & 1 & 0 & 0 \\ 0 & 1 & 0 & 1 & 0 & 0 \\ 0 & 0 & 1 & 1 & 0 & 0 \\ 1 & 0 & 0 & 0 & 1 & 0 \\ 0 & 1 & 0 & 0 & 1 & 0 \\ 0 & 0 & 1 & 0 & 1 & 0 \\ 0 & 0 & 0 & 1 & 1 & 0 \\ 1 & 0 & 0 & 0 & 0 & 1 \\ 0 & 1 & 0 & 0 & 0 & 1 \\ 0 & 0 & 1 & 0 & 0 & 1 \\ 0 & 0 & 0 & 1 & 0 & 1 \\ 0 & 0 & 0 & 0 & 1 & 1 \end{pmatrix}$$

The transpose of A is

$$A^T = \begin{pmatrix} 1 & 1 & 0 & 1 & 0 & 0 & 1 & 0 & 0 & 0 & 1 & 0 & 0 & 0 & 0 \\ 1 & 0 & 1 & 0 & 1 & 0 & 0 & 1 & 0 & 0 & 0 & 1 & 0 & 0 & 0 \\ 0 & 1 & 1 & 0 & 0 & 1 & 0 & 0 & 1 & 0 & 0 & 0 & 1 & 0 & 0 \\ 0 & 0 & 0 & 1 & 1 & 1 & 0 & 0 & 0 & 1 & 0 & 0 & 0 & 1 & 0 \\ 0 & 0 & 0 & 0 & 0 & 0 & 1 & 1 & 1 & 1 & 0 & 0 & 0 & 0 & 1 \\ 0 & 0 & 0 & 0 & 0 & 0 & 0 & 0 & 0 & 0 & 1 & 1 & 1 & 1 & 1 \end{pmatrix}$$

Clearly, A^T is the incidence matrix of a properly labeled complete graph of 6 vertices. Figure 7.2 shows the bus implementation of the K_6^* hypernetwork, where each hyperlink is represented by a horizontal line. Its incidence matrix is A given above. The vertices of K_6^* are labeled in two ways. Its corresponding complete graph, whose incidence matrix is A^T, is shown in Figure 7.3.

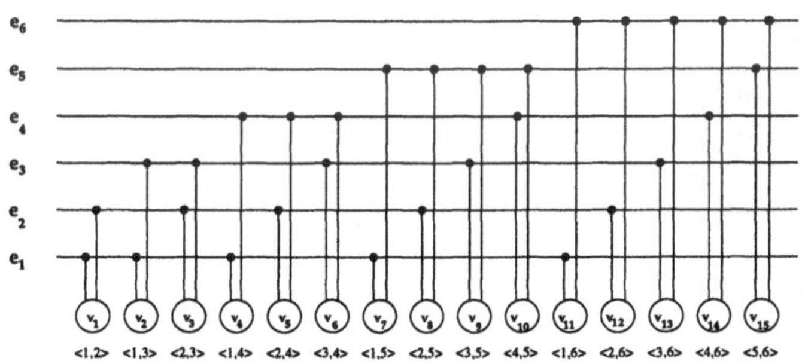

Figure 7.2 Bus implementation of K_6^*.

We observe that the gap $N_n - N_{n-1} = n-1$ between the number of processors in K_{n-1}^* and that of K_n^* is not a constant. It is desirable that hypernetworks can be expanded with incremental size increases. For any given N such that $N_{n-1} < N < N_n$, we can construct a sub-hypergraph H of K_n^* such that H has N vertices and n hyperedges. Such a subhypergraph is called an *incomplete* K_n^* hypernetwork.

Definition 3. *The incomplete K_n^* hypernetwork, where $n > 3$, of N vertices such that $N_{n-1} < N < N_n$ is the sub-hypergraph of K_n^* induced by vertex set $\{v_1, v_2, \ldots, v_N\}$.*

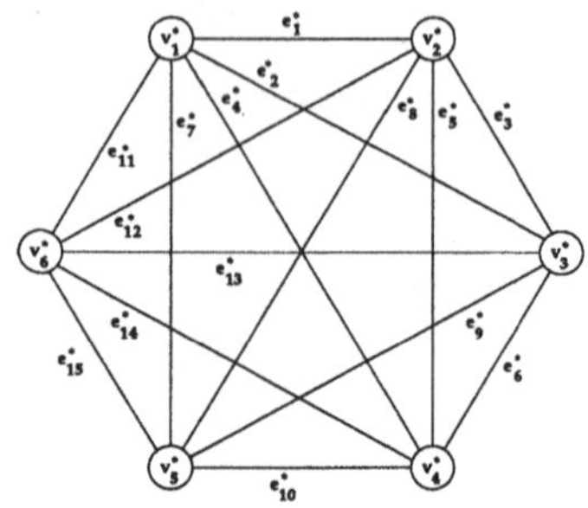

Figure 7.3 Complete graph K_6 corresponding to K_6^*.

In other words, an incomplete K_n^* hypernetwork of N vertices, $N_{n-1} < N < N_n$, corresponds to the incidence matrix obtained from the incidence matrix of K_n^* by deleting its rows corresponding vertices $v_{N+1}, v_{N+2}, \ldots, v_{N_n}$. An incomplete K_n^* is linear and 2-regular, but it is not uniform, and not (vertex and hyperedge) symmetric. It is not difficult to prove that the diameter of incomplete K_n^* hypernetwork, where $n > 3$, is 2.

We demonstrate how to use the vertex and hyperedge labels of K_n^* to design data communication algorithms. The following algorithm can be used for routing a message from $\langle a, b\rangle$ to $\langle a', b'\rangle$ in K_n^*.

procedure *ROUTE*($\langle a, b\rangle$, $\langle a', b'\rangle$)
begin
  if $b = b'$ or $b = a'$ **then**
    $\langle a, b\rangle$ sends the message to $\langle b, a'\rangle$ using hyperlink e_b
  else if $a = b'$ or $a = a'$ **then**
    $\langle a, b\rangle$ sends the message to $\langle a', b'\rangle$ using hyperlink e_a
  else /* $\langle a, b\rangle$ and $\langle a', b'\rangle$ do not share a hyperlink */
    $l = \min\{a, a'\}$;
    if $a = l$ **then**
      $\langle a, b\rangle$ sends the message to $\langle a', b'\rangle$ along the path
      $(\langle a, b\rangle, e_a, \langle a, b'\rangle, e_{b'}, \langle a', b'\rangle)$
    else $\langle a, b\rangle$ sends the message to $\langle a', b'\rangle$ along the path
      $(\langle a, b\rangle, e_b, \langle a', b\rangle, e_{a'}, \langle a', b'\rangle)$
end

By the definition of K_n^*, it is easy to verify that for any given pair of processors $\langle a, b\rangle$ and $\langle a', b'\rangle$ in the K_n^* hypernetwork, algorithm *ROUTE* routes a message from $\langle a, b\rangle$ to $\langle a', b'\rangle$ along a shortest path. Broadcasting from a processor $\langle a, b\rangle$ can be carried out using the following algorithm *BROADCAST*. One can easily prove that algorithm *BROADCAST* takes two (parallel) message steps to broadcast a message from any processor to all other processors in the K_n^* hypernetwork. Furthermore, each processor receives one and only one copy of the broadcast message. Since the diameter of K_n^* is 2, the performance of this algorithm is optimal.

procedure *BROADCAST*($\langle a, b\rangle$)
begin
  $\langle a, b\rangle$ broadcasts the message to all the processors connected by e_b;
  for all the processors $\langle a', b'\rangle$ such that $a' = b$ or $b' = b$ **do in parallel**
    if $a' = b$ **then** $\langle a', b'\rangle$ broadcasts the message to processors in
    $\{\langle i, b'\rangle | i > a'\}$ using $e_{b'}$;
    if $b' = b$ **then** $\langle a', b'\rangle$ broadcasts the message to processors in
    $\{\langle a', j\rangle | j \neq b'\}$ using $e_{a'}$
  endfor
end

Permutation is a bijection on the set of processors in K_n^*. In a permutation communication operation, each processor $\langle a, b\rangle$ sends a message to another processor $\langle a', b'\rangle$, and each processor receives a message from exactly one processor. A permutation is called a total permutation if $\langle a, b\rangle \neq \langle a', b'\rangle$ for all pairs; otherwise, it is called a partial permutation. Only total permutations are considered, since a partial permutation can be carried out using a total permutation by masking out those processors which are mapped to themselves. We use a set of N_n ordered processor pairs $(\langle a, b\rangle, \langle a', b'\rangle)$ to represent an permutation. In each pair $(\langle a, b\rangle, \langle a', b'\rangle)$, $\langle a, b\rangle$ and $\langle a', b'\rangle$ are called the source processor and destination processor of the pair, respectively. We use $A_{(\langle a,b\rangle,\langle a',b'\rangle)}$ to denote a message to be sent from $\langle a, b\rangle$ to $\langle a', b'\rangle$.

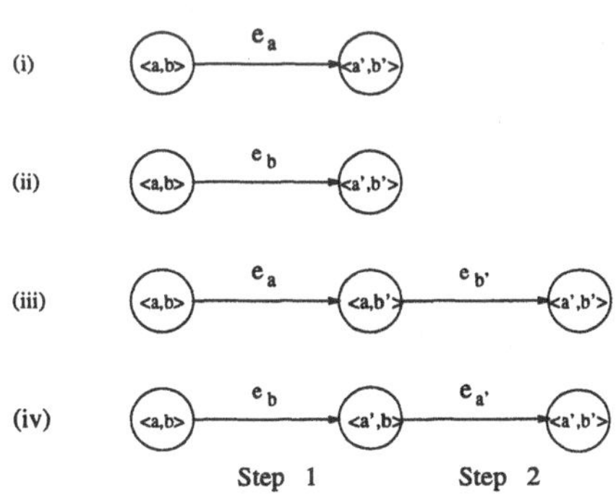

Figure 7.4 Routing paths used by algorithm PERMUTATION for messages $A_{(\langle a,b\rangle,\langle a',b'\rangle)}$: (i) $a = b'$ or $a = a'$; (ii) $a' = b$ or $b = b'$, (iii) a, a', b, and b' are distinct, and $b > b'$, and (iv) a, a', b, and b' are distinct, and $b < b'$.

We present an algorithm *PERMUTATION* which performs a permutation operation by routing messages along different paths of length at most 2. There are four cases: (i) $a = b'$ or $a = a'$, (ii) $a' = b$ or $b = b'$, (iii) a, a', b, and b' are distinct, and $b > b'$, and (iv) a, a', b, and b' are distinct, and $b < b'$.

For each of these four cases, algorithm *PERMUTATION* routes the messages strictly along paths shown in Figure 7.4. Based on these path patterns, we call a message a two-step message if it follows a path of length 2 (cases (iii) and (iv)) ; otherwise, it is called a one-step message (cases (i) and (ii)). Clearly, algorithm *PERMUTATION* transmits each message $A_{(\langle a,b\rangle,\langle a',b'\rangle)}$ along a shortest path.

procedure *PERMUTATION*
begin
  `/* Step 1 */`
  for all hyperlinks e_k **do in parallel**
    Use e_k to transmit those messages with e_k assigned for their first step
  endfor
  `/* Step 2 */`
  for all hyperlinks e_k **do in parallel**
    Use e_k to transmit the two-step messages with e_k assigned for their second step
  endfor
end

Assume that the hyperlinks of K_n^* are implemented by optical buses that do not use CDM and WDM access methods. Then, multiple messages transmitted on each bus hyperlink must be in sequence, one message at a time, and each of Step 1 and Step 2 of *PERMUTATION* requires at most $n-1$ message steps. Since there are $\frac{n(n-1)}{2}$ processors and n bus hyperlinks, a trivial lower bound for the number of parallel message steps required for an arbitrary permutation is $\frac{n-1}{2}$. Thus, the performance of algorithm *PERMUTATION* is optimal within a constant factor.

Additional algorithms for reduction (a fan-in operation), total exchange (all-to-all communication), and prefix computation on K_n^* are presented in [21]. Algorithms for shortest path routing, broadcasting, permutation, reduction, total exchange and prefix computation on the incomplete K_n^* hypernetwork can also be found in [21].

Now, consider using the dual Q_n^* of hypercube Q_n as a hypernetwork. An n-dimensional hypercube Q_n consists of 2^n vertices, each being labeled by a unique n-bit binary number. Two vertices are connected by an edge if and only if their binary labels are distinct in one bit position. Let I_n be the set of non-negative integers that can be represented by n-bit binary numbers. For $l, u \in I_n$, we use $d(l,u)$ to denote the number of different bits in the binary representations of l and u, i.e. $d(l,u)$ is the Hamming distance between the binary representations of l and u. We use a pair of integers to label a vertex in the Q_n^* hypernetwork.

Definition 4. *Let $M_n = n2^{n-1}$ for $n \geq 2$. The Q_n^* hypernetwork is a hypergraph with vertex set $\{\langle l,u\rangle | l,u \in I_n,\ l < u, \text{ and } d(l,u) = 1\}$ of M_n vertices and 2^n hyperlinks, $e_0, e_1, ..., e_{2^n-1}$. Each vertex $\langle l,u\rangle$ is connected to exactly two hyperedges e_l and e_u.*

Hypercube Q_3 and its corresponding Q_3^* are shown in Figures 7.5 and 7.6, respectively. The Q_n^* hypernetwork can also be defined in a recursive way. One can easily observe that Q_{n+1}^* (see Figure 7.6) can be constructed using two copies of Q_n^* and 2^n additional vertices. For brevity, we omit the recursive definition of Q_n^*.

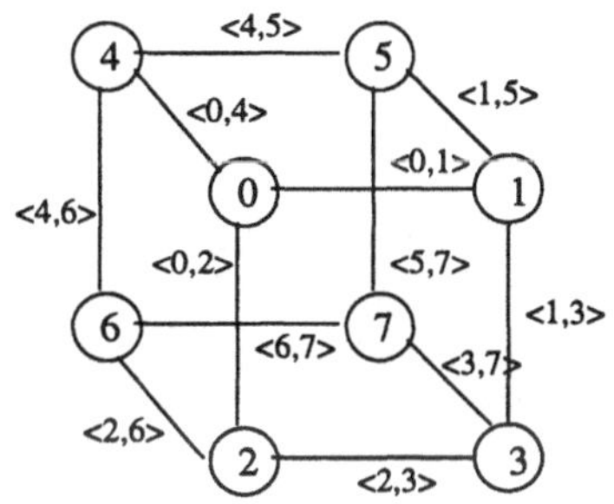

Figure 7.5 Hypercube Q_3.

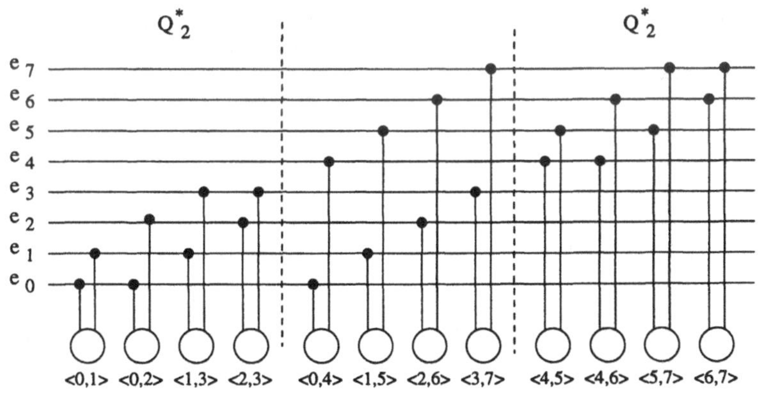

Figure 7.6 Bus implementation of Q_3^*.

Based on the properties of hypercube Q_n and Propositions 1 to 4, know that Q_n^* is 2-regular, n-uniform, linear, and vertex and hyperedge symmetric. It is not difficult to show that $diameter(Q_n^*) = n$, and Q_n^* is $(2n-3)$-processor fault-tolerant and 1-hyperlink fault-tolerant. Compared with K_n^*, the Q_n^* hypernetwork has the disadvantage that its diameter is a function of its number of vertices. However, the growing rate of the $rank(Q_n^*)$ is much smaller than that of $rank(K_n^*)$. Several data communication algorithms for the Q_n^* hypernetwork has been proposed. Interested readers may refer to [9] for details.

Using duals of direct networks one can obtain many different hypernetworks of degree 2. We say that a linear hypernetwork is ***non-trivial*** if it has at least 4 vertices, at least 2 hyperlinks, and each hyperlink contains at least 2 vertices. Let $H^+ = \{H \mid H$ is a non-trivial linear, regular hypergraph of degree 2 $\}$. We call the hypernetworks in H^+ the class of ***linear, 2-regular hypernetworks.*** For any H in H^+, its dual H^* is a conventional graph. However, the dual of a conventional graph may not be in H^+. For example, if a vertex v of a graph G has degree 1, then its corresponding hyperlink e^* in G^* contains exactly one vertex; and consequently G^* is not in H^+. Let $G^+ = \{G \mid G$ is a conventional graph such that it has at least 4 edges, and each vertex of G has degree greater than 1 $\}$. Clearly, for any network G in G^+, we can obtain a non-trivial, linear, 2-regular hypernetwork H in H^+ by applying the dual operation to G. It is easy to prove that for any hypernetwork H in H^+, $diameter(H) \geq 2$. Those degree-2 hypernetworks that are excluded from H^+ are not interesting.

Among all the hypergraphs derived from duals of conventional graphs, the dual, K_n^*, of the n-vertex complete graph K_n has the smallest m/N ratio when N is fixed and smallest diameter, where m and N is the number of hyperedges and vertices, respectively. Between the high cost/performance of fully connected network K_n and low cost/performance of linearly connected network (a ring) are a set of direct networks that constitute a wide range of trade-offs in cost and performance. For example, H can be a star graph [1], a chordal ring (including barrel shifters) [2], etc. The duals H^* of these networks also constitute a wide range of trade-offs in cost and performance.

In general, one can obtain a class of hypernetworks from a known class of hypernetworks by the dual operator. Finding hypergraph duals can be considered as a general approach to the design of new hypernetworks.

7.5 USING COMBINATORIAL BLOCK DESIGN THEORY TO DESIGN HYPERNETWORKS

Combinatorial block designs are closely related hypergraphs. For example, *balanced incomplete block designs* (BIBDs) can be treated as hypergraphs. Given a set V of v objects, a BIBD (v, b, r, k, λ) of V is a collection of b k-subsets of V such that each object appears in exactly r of b blocks, and every two objects appear in exactly λ of the b blocks. In terms of hypernetworks, BIBD (v, b, r, k, λ) is a hypernetwork of v processors, each corresponding to an object in V, and b hyperlinks, each corresponding to a block. Clearly, this hypernetwork is r-regular, k-uniform, and its diameter is 1. If $\lambda = 1$, then it is a linear hypernetwork. A hypernetwork corresponding to a BIBD with $\lambda > 1$ allows better fault tolerance.

Example 2. A BIBD $(15, 15, 7, 7, 3)$ defined on $V = \{1, 2, \ldots, 15\}$:

$$\begin{array}{ll}
B_1 = \{1,2,3,4,5,6,7\}, & B_2 = \{1,2,3,8,9,10,11\}, \\
B_3 = \{1,2,3,12,13,14,15\}, & B_4 = \{1,4,5,8,9,12,13\}, \\
B_5 = \{1,4,5,10,11,14,15\}, & B_6 = \{1,6,7,8,9,14,15\}, \\
B_7 = \{1,6,7,10,11,12,13\}, & B_8 = \{2,4,6,8,10,12,14\}, \\
B_9 = \{2,4,7,8,11,13,15\}, & B_{10} = \{2,5,6,9,11,12,15\}, \\
B_{11} = \{2,5,7,9,10,13,14\}, & B_{12} = \{3,4,6,9,11,13,14\}, \\
B_{13} = \{3,4,7,9,10,12,15\}, & B_{14} = \{3,5,6,8,10,13,15\}, \\
B_{14} = \{3,5,7,8,11,12,14\}. &
\end{array}$$

This is a hypernetwork that has 15 processors of degree 7, and 15 hyperlinks of rank 7; its diameter is 1. Every two processors are connected by exactly 3 hyperlinks.

The five parameters v, b, r, k, and λ that characterize a BIBD are not independent. There are some necessary conditions, such as, in order for (v, b, r, k, λ) to exists, it must be that

$$bk = vr, \tag{7.1}$$
$$r(k-1) = \lambda(v-1). \tag{7.2}$$

There are no known general sufficient conditions for the existences of a BIBD (v, b, r, k, λ). However, there is a rich literature on how to construct BIBDs for some combinations of the five parameters. We use several examples to demonstrate how to construct hypernetworks using some basic block design results.

Let (v, b, r, k, λ) be a BIBD defined on V, and let B_i, $1 \le i \le b$, be the blocks of this design. The *complementary design* of (v, b, r, k, λ) consists of b blocks B'_i, $1 \le i \le b$, such that $B'_i = V - B_i$. It is well-known that the complementary design of a BIBD (v, b, r, k, λ) is a BIBD $(v, b, b-r, v-k, b-2r+\lambda)$.

Example 3. A BIBD $(9, 12, 4, 3, 1)$ and its complement, which is a BIBD $(9, 12, 8, 6, 5)$:

$$\begin{array}{ll}
B_1 = \{1,2,3\}, & B'_1 = \{4,5,6,7,8,9\}, \\
B_2 = \{4,5,6\}, & B'_2 = \{1,2,3,7,8,9\}, \\
B_3 = \{7,8,9\}, & B'_3 = \{1,2,3,4,5,6\}, \\
B_4 = \{1,4,7\}, & B'_4 = \{2,3,5,6,8,9\}, \\
B_5 = \{2,5,8\}, & B'_5 = \{1,3,4,6,7,9\}, \\
B_6 = \{3,6,9\}, & B'_6 = \{1,2,4,5,7,8\}, \\
B_7 = \{1,6,8\}, & B'_7 = \{2,3,4,5,7,9\}, \\
B_8 = \{2,4,9\}, & B'_8 = \{1,3,5,6,7,8\}, \\
B_9 = \{3,5,7\}, & B'_9 = \{1,2,4,6,8,9\}, \\
B_{10} = \{1,5,9\}, & B'_{10} = \{2,3,4,6,7,8\}, \\
B_{11} = \{2,6,7\}, & B'_{11} = \{1,3,4,5,8,9\}, \\
B_{12} = \{3,4,8\}. & B'_{12} = \{1,2,5,6,7,9\}.
\end{array}$$

A BIBD (v, b, r, k, λ) is symmetric if $v = b$ (and by (7.1), $r = k$). Note that this symmetry dose not have the same meaning as vertex symmetry and hyperedge symmetry of a hypergraph. Let $B_1, B_2, \ldots, B_v$ be the blocks of a symmetric BIBD (v, v, k, k, λ). It is known that, for any i, $B_1 - B_i, B_2 - Bi, \ldots, B_{i-1} - B_i, B_{i+1} - B_i, \ldots, B_v - B_i$ form a BIBD $(v-k, v-1, k, k-\lambda, (\lambda - 1))$. Such a BIBD is called the ***residual design*** of (v, v, k, k, λ) with respect to block B_i.

Example 4. The BIBD $(15, 15, 7, 7, 3)$ of Example 2 is a symmetric BIBD. By letting $i = 1$, we obtain the following residual BIBD $(8, 14, 7, 4, 3)$:

$$\begin{array}{ll}
B_2 - B_1 = \{8, 9, 10, 11\}, & B_3 - B_1 = \{12, 13, 14, 15\}, \\
B_4 - B_1 = \{8, 9, 12, 13\}, & B_5 - B_1 = \{10, 11, 14, 15\}, \\
B_6 - B_1 = \{8, 9, 14, 15\}, & B_7 - B_1 = \{10, 11, 12, 13\}, \\
B_8 - B_1 = \{8, 10, 12, 14\}, & B_9 - B_1 = \{8, 11, 13, 15\}, \\
B_{10} - B_1 = \{9, 11, 12, 15\}, & B_{11} - B_1 = \{9, 10, 13, 14\}, \\
B_{12} - B_1 = \{9, 11, 13, 14\}, & B_{13} - B_1 = \{9, 10, 12, 15\}, \\
B_{14} - B_1 = \{8, 10, 13, 15\}, & B_{15} - B_1 = \{8, 11, 12, 14\}.
\end{array}$$

Complement and residual operations can be used to derive one BIBD from another. These operations provide trade-offs between degrees, ranks, number of hyperlinks, and redundancy for reliability (characterized by λ).

Write $B(v, k, \lambda)$ to denote a BIBD (v, b, r, k, λ) such that

$$\begin{aligned}
r &= \frac{\lambda(v-1)}{k-1}, \\
b &= \frac{\lambda v(v-1)}{k(k-1)};
\end{aligned}$$

and write $V(k, \lambda)$ for the set of all v such that a $B(v, k, \lambda)$ exists. We replace (7.1) and (7.2) by the following conditions (7.3) and (7.4) to eliminate the redundant parameters b and r.

$$\lambda(v-1) \equiv 0 \pmod{(k-1)}, \tag{7.3}$$
$$\lambda v(v-1) \equiv 0 \pmod{k(k-1)}. \tag{7.4}$$

It is shown in [13], [5] and [6] that $v \equiv 1$ or $k \pmod{k(k-1)}$ is a necessary and sufficient condition for the existence of $B(v, k, 1)$ for $k = 3, 4$ and 5, respectively. It was proved by Richard M.Wilson [16] that, given positive integers k and λ, there exists a constant $v_1 = v_1(k, \lambda)$ such that $v \in V(k, \lambda)$ for every integer $v \geq v_1$ that satisfies (7.3) and (7.4).

A BIBD (v, b, r, k, λ) is a single-hop hypernetwork. However, it may not be feasible to be implemented when v, the number of processors, is large (and consequently its number of hyperlinks b and degree r become too large). One can use BIBDs to construct multihop hypernetworks. A simple method is using BIBDs as "super-hyperlinks" to construct a *generalized hypermesh based on block designs* (GHMBD), by generalizing the k-ary n-cube network [8]. The d-dimensional generalized hypermesh based on BIBD (v, b, r, k, λ), denoted by $GHMBD(v, b, r, k, \lambda; d)$, consists of v^d processors and bdv^{d-1} hyperlinks of rank k. Each processor w is labeled by a unique d-tuple $(v_1(w), v_2(w), \ldots, v_d(w))$, where $1 \leq v_i(w) \leq v$. The connectivity of $GHMBD(v, b, r, k, \lambda; d)$ is defined as follows: k processors are connected by a hyperlink e if and only if their j-th labels form a block of (v, b, r, k, λ), and for each $i \neq j$ their i-th labels are identical. Clearly, $GHMBD(v, b, r, k, \lambda; d)$ is a k-uniform, (dr)-regular hypernetwork, and its diameter is d. If $\lambda = 1$, it is a linear hypernetwork.

We present another class of hypernetworks called *hierarchical hypernetworks based on block designs* that may have even smaller degrees and diameters. The d-level HHBD of base v_1, denoted by $HHBD(v_1, k, \lambda; d)$, is a recursively defined hypergraph.

The $HHBD(v_1, k, \lambda; 1)$ network is a BIBD $B(v_1, k, \lambda)$. It is a k-uniform, $\frac{\lambda(v_1-1)}{k-1}$-regular hypergraph with diameter equal to 1. Its each vertex w is assigned a unique label $l_1(w) = i$, $1 \leq i \leq v_1$, such that k vertices are connected by a hyperedge if and only if their labels form a block in $B(v_1, k, \lambda)$.

Let n_i denote the number of vertices in $HHBD(v_1, k, \lambda; i)$ (thus, $n_1 = v_1$). The $HHBD(v_1, k, \lambda; d)$ network, $d > 1$, is constructed using $\frac{n_{d-1}(k-1)}{\lambda} + 1$ copies, $HHBD_i(v_1, k, \lambda; d-1)$, $1 \leq i \leq \frac{n_{d-1}(k-1)}{\lambda} + 1$, of $HHBD(v_1, k, \lambda; d-1)$. Each of these component $HHBD(v_1, k, \lambda; d-1)$'s is called a $(d-1)$-*level sub-HHBD*. Each vertex w of $HHBD(v_1, k, \lambda; d)$ is labeled by a d-tuple $(l_1(w), l_2(w), \ldots, l_d(w))$. For any vertex w in $HHBD_i(v_1, k, \lambda; d-1)$, $l_d(w) = i$, and $l_s(w)$, $1 \leq s \leq d-1$, remain the same as in $HHBD(v_1, k, \lambda; d-1)$. All hyperedges connecting vertices in the same $HHBD_i(v_1, k, \lambda; d-1)$ follow the connectivity conditions specified by their $(d-1)$-tuples formed by digits $l_1, l_2, \ldots, l_{d-1}$ in $HHBD_i(v_1, k, \lambda; d-1)$. That is, the level-$s$ hyperedges in each $HHBD_i(v_1, k, \lambda; d-1)$ are the level-s hyperedges of $HHBD_i(v_1, k, \lambda; d)$. The additional hyperedges, which are called *level-d hyperedges*, are those connecting vertices from different $(d-1)$-level sub-HHBDs. More specifically, a hyperedge $\{w_{i_1}, w_{i_2}, \ldots, w_{i_k}\}$ is a level-d hyperedge of $HHBD(v_1, k, \lambda; d)$ if and only if

(a) All w_{i_j}, $1 \leq j \leq k$, are from k different $(d-1)$-level sub-HHBDs;

(b) $l_s(w_{i_1}) = l_s(w_{i_2}) = \cdots = l_s(w_{i_k})$ for $1 \leq s \leq d-1$;

(c) $\{l_d(w_{i_1}), l_d(w_{i_2}), \ldots, l_d(w_{i_k})\}$ is a block of BIBD $B(n_{d-1}(k-1)/\lambda+1, k, \lambda))$, which is

$$(v_d, b_d, r_d, k, \lambda) = \left(\frac{n_{d-1}(k-1)}{\lambda} + 1, \left[\frac{n_{d-1}(k-1)}{\lambda} + 1\right] \frac{n_{d-1}}{k}, n_{d-1}, k, \lambda\right);$$

(d) each vertex of $HHBD(v_1, k, \lambda; d)$ is included in exactly one level-d hyperedge.

Clearly, v_d, b_d, r_d, k and λ satisfy (7.3) and (7.4), and by Wilson's proof, a BIBD $(v_d, b_d, r_d, k, \lambda)$ can be constructed. Since $r_d = n_{d-1}$, all conditions (a) to (d) can be satisfied.

The $HHBD(v_1, k, \lambda; d)$ network is $(\frac{\lambda(v-1)}{k-1} + d - 1)$-regular and k-uniform. Its diameter is d. If $\lambda > 1$, then any two distinct vertices in each level-1 sub-HHBD are connected by exactly λ hyperedges, and any two distinct vertices in different level-i, $i > 1$, sub-HHBDs are connected by at most one hyperedge. Thus, $HHBD(v_1, k, \lambda; d)$ is linear only if $\lambda = 1$.

The BIBDs $B(v, 3, 1)$ are called *Steiner triple systems.* This is an extensively studied class of block designs [11]. We list $B(3,3,1)$, $B(7,3,1)$ and $B(15,3,1)$ below. Note that for admissible $v > 3$, $B(v, 3, 1)$ is not unique.

$$
\begin{aligned}
B(3,3,1) &= \{\{1,2,3\}\}.\\
B(7,3,1) &= \{\{1,2,4\},\{1,3,7\},\{1,5,6\},\{2,3,5\},\{2,6,7\},\\
&\quad \{3,4,6\},\{4,5,7\}\}.\\
B(15,3,1) &= \{\{1,2,3\},\{1,4,5\},\{1,6,7\},\{1,8,9\},\{1,10,11\},\\
&\quad \{1,12,13\},\{1,14,15\},\{2,4,6\},\{2,5,7\},\{2,8,10\},\\
&\quad \{2,9,11\},\{2,12,14\},\{2,13,15\},\{3,4,7\},\{3,5,6\},\\
&\quad \{3,8,11\},\{3,9,10\},\{3,12,15\},\{3,13,14\},\{4,8,12\},\\
&\quad \{4,9,13\},\{4,10,14\},\{4,11,15\},\{5,8,13\},\{5,9,12\},\\
&\quad \{5,10,15\},\{5,11,14\},\{6,8,14\},\{6,9,15\},\{6,10,12\},\\
&\quad \{6,11,13\},\{7,8,15\},\{7,9,14\},\{7,10,13\},\{7,11,12\}\}.
\end{aligned}
$$

For example, $HHBD(3,3,1;2)$ constructed using $B(3,3,1)$ and $B(7,3,1)$ is shown in Figure 7.7.

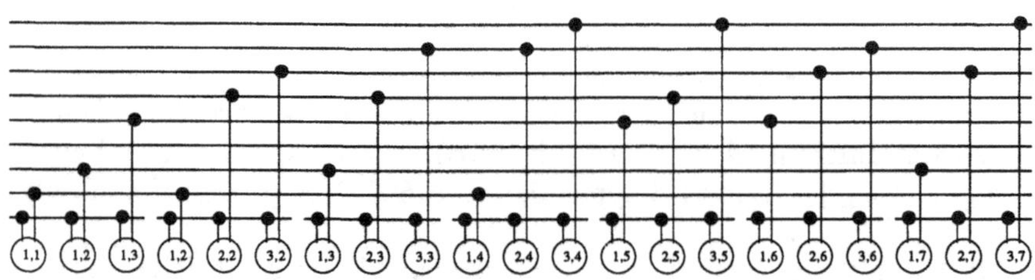

Figure 7.7 Bus implementation of $HHBD(3,3,1;2)$.

Consider routing a message from one processor x to another processor y in $HHBD(v_1, k, \lambda; d)$. Based on the hierarchical structure of $HHBD(v_1, k, \lambda; d)$, we have the following routing algorithm *HHBD_ROUTE*. There are two cases.

Case 1: If x and y are in the same $(d-1)$-level sub-HHBD, say $HHBD_i(v_1, k, \lambda; d-1)$, the problem of finding a path from x to y in $HHBD_i(v_1, k, \lambda; d)$ is reduced to finding a path from x to y in $HHBD_i(v_1, k, \lambda; d-1)$.

Case 2: If x and y are in two different $(d-1)$-level sub-HHBDs, say $HHBD_i(v_1, k, \lambda; d-1)$ and $HHBD_j(v_1, k, \lambda; d-1)$, respectively, then the problem of finding a path from x to y in $HSN(b,d)$ is also reduced to finding paths between two processors in the same $(d-1)$-level sub-HHBD. There is a level-d hyperlink e that connects a processor a of $HHBD_i(v_1,k,\lambda;d-1)$ and a processor b of $HHBD_j(v_1,k,\lambda;d-1)$. Algorithm *HHBD_ROUTE* finds a path $path(x,a)$ from x to a in $HHBD_i(v_1,k,\lambda;d-1)$ and a path $path(b,y)$ from b to y in $HHBD_j(v_1,k,\lambda;d-1)$, and then connects the two paths by e.

Let $length_d(x,y)$ denote the length of the path from x to y determined by *HHBD_ROUTE* in a d-level HHBD. Then,

$$\begin{aligned} length_d(x,y) &= length_{d-1}(x,y) \text{ for Case 1, and} \\ length_d(x,y) &= length_{d-1}(x,a) + length_{d-1}(y,b) + 1 \text{ for Case 2.} \end{aligned}$$

We claim that $length_d(x,y) \le 2^d - 1$. For $d=1$, $length_1(x,y)=1$ obviously holds. Suppose $length_{d-1}(x,a) \le 2^{d-1}-1$ and $length_{d-1}(y,b) \le 2^{d-1}-1$ for Case 2. Then, $length_d(x,y) \le 2(2^{d-1}-1)+1 = 2^d-1$. It is not difficult to prove that there exists a pair of vertices x and y in $HHBD(v_1,k,1;d)$ such that $dis(x,y) = 2^d - 1$. For brevity, we omit this proof, and claim that $diameter(HHBD(v_1,k,1;d)) = 2^d - 1$. The diameter of $HHBD(v_1,k,\lambda;d)$ for $\lambda < 1$ can be smaller than $2^d - 1$.

The operation of broadcasting from any processor x in $HHBD(v_1,k,\lambda;d)$ can also be carried out recursively. Suppose that x is in $HHBD_i(v_1,k,\lambda;d-1)$. Broadcasting from x consists of three major operations: Broadcast from x to all processors in $HHBD_i(v_1,k,\lambda;d-1)$. Then, all processors in $HHBD_i(v_1,k,\lambda;d-1)$. transmit the broadcast message using the level-d hyperlinks in parallel. After this step, one and only one processor x_j in each $(d-1)$-level sub-HHBD would have received the broadcast message. Finally, in parallel, each x_j broadcast in the $(d-1)$-level HHBD that contains it recursively. By a simple induction, we know that the total number of parallel communication steps taken by this algorithm is exactly $2^d - 1$.

In Table 7.1, we show n (number of processors), m (number of hyper links), degree, diameter and m/n ratio of several $HHBD(v_1,3,1;d)$ hypernetworks. From this table, we observe that the number of processors in an $HHBD(v_1,3,1;d)$ grows rapidly and the number of hyperlinks, degree and diameter grow very slowly as d increases. Such growing rates of processors, hyperlinks, degree and diameter are achieved by only relaxing the rank of 2 (of conventional graphs) to 3. For a larger k, the growing rate of the number of processors of $HHBD(v_1,k,\lambda;d)$ is even faster, and, in contrast, the growing rates of the number of hyperlinks, degree and diameter are much slower. Thus, for $k \ge 5$, one may assume that d is a very small constant for an $HHBD(v_1,k,\lambda;d)$ of any practical size.

A major problem of the HHBD hypernetworks is that there is a big gap between the sizes of $HHBD(v_1,k,\lambda;d-1)$ and $HHBD(v_1,k,\lambda;d)$. Let V_d denote the set of vertices in $HHBD(v_1,k,\lambda;d)$, and let $n_d = |V_d|$. For any

Table 7.1 Comparison of $HHBD(3,3,1;d)$, $HHBD(7,3,1;d)$ and $HHBD(15,3,1;d)$ Networks.

	n, number of vertices	m, number of (hyper)edges	degree	diameter	$\frac{m}{n}$ ratio
$HHBD(3,3,1;1)$	3	1	1	1	0.33
$HHBD(7,3,1;1)$	7	7	3	1	1
$HHBD(15,3,1;1)$	15	35	7	1	2.33
$HHBD(3,3,1;2)$	21	14	2	3	0.67
$HHBD(7,3,1;2)$	105	140	4	3	1.33
$HHBD(15,3,1;2)$	465	1244	8	3	2.67
$HHBD(3,3,1;3)$	903	903	3	7	1
$HHBD(7,3,1;3)$	22155	36925	5	7	1.67
$HHBD(15,3,1;3)$	432,915	1,298,745	9	7	3
$HHBD(3,3,1;4)$	1,631,721	2,175,628	4	15	1.33
$HHBD(7,3,1;4)$	981,710,205	1,963,420,410	6	15	2
$HHBD(15,3,1;4)$	374,831,227,365	1,249,437,424,550	10	15	3.33

given N such that $n_{d-1} < N < n_d$, we want to identify a subset U of V_d such that $|U| = N$ and the subhypergraph of $HHBD(v_1, k, \lambda; d)$ induced by U has a small diameter. We call this subhypergraph of $HHBD(v_1, k, \lambda; d)$ an *incomplete $HHBD(v_1, k, \lambda; d)$ hypernetwork*, and denote it by $HHBD^-(v_1, k, \lambda; d)$. Clearly, $rank(HHBD^-(v_1, k, \lambda; d+1)) = k$ and $degree(HHBD^-(v_1, k, \lambda; d)) = d$.

For convenience, assume that $HHBD(v_1, k, \lambda; 0)$ is a hypergraph that has one vertex. Let v_i be the number of copies of $HHBD(v_1, k, \lambda; i)$ used in the construction of $HHBD(v_1, k, \lambda; i+1)$. Let $N_d = N$, $v'_i = \lfloor \frac{N_{i+1}}{n_i} \rfloor$, and $N_i = N_{i+1} - v'_i \cdot n_i$, $0 \leq i \leq d-1$. Clearly, $1 \leq v'_{d-1} < v_{d-1}$, $0 \leq v'_j \leq v_j$ for $0 \leq j < d-1$, and $N = \sum_{i=0}^{d-1}(v'_i \cdot n_i)$. Suppose that all non-zero v'_i are $v'_{i_1}, v'_{i_2}, \ldots, v'_{i_p}$, $i_1 < i_2 < \cdots < i_p = d-1$. With v'_i, $0 \leq i \leq d-1$, available, we use the following procedure *CONSTRUCT* to identify a collection $U = \{U_0, U_1, \ldots, U_{d-1}\}$ of vertex subsets of $HHBD(v_1, k, \lambda; d)$ that satisfy the following conditions:

(i) U_i, $0 \leq i \leq d-1$, are disjoint, and U_i contains vertices of v'_i i-level sub-HHBDs that are in the same $(i+1)$-level sub-HHBD;

(ii) there exists at least one pair of vertices u and v such that u is in U_{i_j}, v is in $U_{i_{j+1}}$ and u and v are connected by a level-(i_j+1) hyperedge in $HHBD(v_1, k, \lambda; d)$.

Let $HHBD^-(v_1, k, \lambda; d)$ be the subhypergraph induced by the vertices in the union of all member subsets of U. Condition (i) ensure that $HHBD^-(v_1, k, \lambda; d)$ has exactly N vertices, and condition (ii) further ensures that $HHBD^-(v_1, k, \lambda; d)$ is a connected subhypergraph of $HHBD(v_1, k, \lambda; d)$.

procedure *CONSTRUCT*($v'_0, v'_1, \ldots, v'_{d-1}$)
begin
  `/* Assume that initially` $U_i := \phi$, $0 \le i \le d-1$. `*/`
  for $i = d-1$ **down to** 0 by -1 **do**
    if $i = d-1$ **then**
      $U_i :=$ vertex set of any v'_i $(d-1)$-level sub-HHBDs of
          $HHBD(v_1, k, \lambda; d)$;
      $V_i :=$ a $(d-1)$-level sub-HHBD of $HHBD(v_1, k, \lambda; d)$ that
          is not in U_i;
      $u :=$ a vertex in V_i such that it is connected to a $(d-1)$-level
          sub-HHBD of U_i by a level-d hyperedge;
    else
      if $v'_i = 0$ **then**
        $V_i =$ the i-level sub-HHBD of $HHBD(v_1, k, \lambda; d)$ in V_{i+1}
          that contains u;
      else
        $U_i :=$ vertex set of any v'_i i-level sub-HHBDs of V_{i+1}
          that contains vertex u;
        if there does not exists an i-level sub-HHBD that is in V_{i+1}
          but not in U_i **then stop**;
        $V_i :=$ an i-level sub-HHBD in V_{i+1} but not in U_i;
        $u :=$ a vertex in V_i such that it is connected to an i-level
          sub-HHBD of U_i by a level-$(i+1)$ hyperedge
        endif
      endif
    endif
  endfor
end

It is easy to see that vertex subsets U_i produced by this procedure satisfy conditions (i) and (ii). We sketch a proof for an upper bound for the distance between any two vertices in the subhypergraph $HHBD^-(v_1, k, \lambda; d)$ induced by the vertices of non-empty U_{i_j}'s generated by the procedure *CONSTRUCT*.

Let $H(U_{i_j})$ denote the subhypergraph of $HHBD^-(v_1, k, \lambda; d)$ induced by vertices in U_{i_j}. Condition (i) implies that for any two i_j-level sub-HHBDs $HHBD'(v_1, k, \lambda; i_j)$ and $HHBD''(v_1, k, \lambda; i_j)$ of $H(U_{i_j})$, there exists a level-(i_j+1) hyperedge that connects a vertex a in $HHBD'(v_1, k, \lambda; i_j)$ and a vertex b in $HHBD''(v_1, k, \lambda; i_j)$. By Case 2 of algorithm *HHBD_ROUTE*, $diameter(H(U_{i_j})) \le 2(2^{i_j}-1)+1 = 2^{i_j+1}-1 < 2^d-1$. Then, an upper bound for maximum distance between any two vertices x and y in $HHBD^-(v_1, k, \lambda; d)$ can be derived by assuming that x is in U_{i_1} and y is in U_{i_p}, and the path P from x to y contains a subpath in each $H(U_{i_j})$. Let $len(P)$ denote the length of path P. From (ii), we have

$$\begin{aligned} len(P) &\leq diameter(H(U_{i_1})) + \sum_{j=2}^{p}(1 + diameter(H(U_{i_j})) \\ &\leq (2^{i_1+1} - 1) + \sum_{j=2}^{p} 2^{i_j+1} \leq \left(\sum_{j=1}^{p} 2^{i_j+1}\right) - 1 \\ &\leq \sum_{j=0}^{d} 2^j = 2^{d+1} - 1. \end{aligned}$$

Hence, for any two vertices x and y in $HHBD^-(v_1, k, \lambda; d))$, $dis(x, y) \leq 2^{d+1} - 1$, which leads to $diameter(HHBD^-(v_1, k, \lambda; d)) \leq 2^{d+1} - 1$.

Summarizing above discussions, we know that for any given N such that $n_{d-1} < N < n_d$, one can construct subnetwork $HHBD^-(v_1, k, \lambda; d)$ of $HHBD$ $(v_1, k, \lambda; d)$ such that $HHBD^-(v_1, k, 1; d)$ has exactly N vertices, and the diameter of $HHBD^-(v_1, k, \lambda; d)$ is no lager than $2^{d+1} - 1$, which is the diameter of $HHBD^-(v_1, k, 1; d+1)$.

7.6 SUMMARY

Based on the observation that optical multiconnect components have limited scalability, we presented the hypernetwork abstract model for optical interconnection networks. A hypernetwork is a network that connects a large number of processors using hyperlinks, which are an abstraction of multiconnect components, as building blocks. The relaxation in the number of processors connected by a link in an interconnection network has important implications in network design and parallel computing. Since an optical multiconnect component may connect tens and even hundreds of processors, it is possible to construct all-optical networks (those without intermediate O/E and E/O conversions) or almost all-optical networks (those with very small diameters).

Hypernetwork topology design can be formulated as a constrained optimization problem. For instance, one may fix or set a selection range for one or more parameters and optimize others. Trade-offs may be sought among contradicting parameters such as degree and diameter. We showed how to use the concepts and existing results in hypergraph theory and combinatorial block design theory to construct hypernetworks by giving several examples. Better hypernetworks are yet to be discovered.

The efficiency of parallel algorithms on a massively parallel computer system using a hypernetwork for interprocessor communications depends on the topology of the hypernetwork and the implementation of the hyperlinks. Designing efficient parallel algorithms involves taking the advantages of the topological structure of the hypernetwork and the capacities of hyperlinks. Using massively parallel architectures based on hypernetworks, one can achieve significant improvement in algorithm time complexities, compared with using parallel architectures based on directed and multistage interconnection networks.

We would like to point out that, according to current VLSI technology, it is possible to build an electrical crossbar switch that connects tens of processors. Shared medium electrical interconnection such as a bus and a ring operating in pipelined mode (token ring) also provide full connections for a set of processors. Hence, our hypernetwork model also has applications in designing electrical interconnection networks using such electrical crossbar switch, bus and ring structures as building blocks. It is possible to generalize our hypernetwork model to obtain a unified model for all interconnection networks.

Acknowledgments

The author's work in developing the hypernetwork model has been supported by National Science Foundation under Grant NSF ECS-9626215 and Louisiana Board of Regents of Louisiana under grant LEQSF (1996-99)-RD-A-16.

References

[1] S. Akers, D. Harel, and B. Krishnamurthy. The star graph: an attractive alternative to the n-cube. *Proceedings of 1987 International Conference on Parallel Processing*, 393-400, 1987.

[2] B.W. Arden, and H. Lee. Analysis of chordal ring network. *IEEE Transactions on Computers*, 30:291-295, 1981.

[3] C. Berge. *Hypergraphs.* North-Holland, 1989.

[4] P. E. Green, Jr. *Fiber Optical Networks.* Prentice Hall, 1993.

[5] H. Hanani. The existence and construction of balanced incomplete block designs, *Ann. Math. Statist.*, 32:361-386, 1961.

[6] H. Hanani. On balanced incomplete block designs with blocks having five elements, *Journal of Combinatorial Theory* 12A:184-201, 1972.

[7] J. Jahns and S. H. Lee. *Optical Computing Hardware.* Academic Press, 1994.

[8] F. T. Leighton. *Introduction to Parallel Algorithms and Architectures: Arrays · Trees · Hypercubes.* Morgan Kaufmann Publishers, 1992.

[9] Y. Li, J. Wu and S. Q. Zheng. An optical interconnection structure based on the dual of a hypercube. To appear in *Informatica.*

[10] C. C. Lindner. A survey of embedding theorems for Steiner systems. *Topics on Steiner Systems*, edited by C. C. Lindner and A. Rosa, North-Holland, 1980.

[11] C.C Lindner and A. Rosa (editors). *Topics on Steiner Systems.* North-Holland, 1980.

[12] O. M. Dighe, R. Vaidyanathan and S. Q. Zheng. The bus-connected ringed tree: a versatile interconnection network. *Journal of Parallel and Distributed Computing*, 33:189-196, 1996.

[13] D. K. Ray-Chaudhui and R. M. Wilson, Solution of Kirkman's School-girl Problem, *Proc. Symp. Pure Math. 19*, 1971.

[14] C. Tocci and H. J. Caulfield (editors). *Optical Interconnection: Foundations and Applications.* Artech House, 1994.

[15] W. D. Wallis. *Combinatorial Designs.* Marcel Dekker, 1988.

[16] R. M. Wilson. An existence theory for pairwise balanced designs III. Proof of the existence conjectures. *Journal of Combinatorial Theory*, 18A:71-79, 1975.

[17] S. Q. Zheng. Hypernetworks - a class of interconnection networks with increased wire sharing: Part I - Part IV. Technical Report, #94-008, Department of Compute Science, Louisiana State University, Baton Rouge, LA 70803, 1994.

[18] S. Q. Zheng. Hypernetworks: a class of interconnection networks for new parallel computers. *Proceedings of 8th ISCA International Conference on Parallel and Distributed Computing Systems*, 189-194, 1995.

[19] S. Q. Zheng. Hypercube hypernetworks: implementations of hypercube with increased wire sharing. *Proceedings of 8th ISCA International Conference on Parallel and Distributed Computing Systems*, 452-457, 1995.

[20] S. Q. Zheng. Sparse hypernetworks based on Steiner triple systems. *Proceedings of 1995 International Conference on Parallel Processing* (ICPP), I:92-95, 1995.

[21] S. Q. Zheng and J. Wu. Dual of a complete graph as an interconnection network. *Proceedings of the 8th IEEE Symposium on Parallel and Distributed Processing* (SPDP), 433- 442, 1996.

8 A UNIQUE DESIGN OF FIBER-OPTIC INTERCONNECTION NETWORKS AND ALGORITHMS

Chunming Qiao

Department of Electrical and Computer Engineering
State University of New York at Buffalo
Buffalo, New York 14260

qiao@computer.org

Abstract: A unique 2D mesh called reconfigurable array with spanning optical buses (RASOB) is described, which supports a high degree of transparency (e.g. bit-rate) and reconfigurability with a low control and hardware (e.g. optical switches) complexity. Efficient algorithms that can take advantage of the unique features of the architecture are also described.

8.1 INTRODUCTION

Imagine a "future-proof" interconnection network in which any transmission rate can be supported without having to replace any of its links or switches. In order to plug in future generations of increasingly powerful microprocessors in such an interconnection network, only appropriate I/O interfaces with a higher transmitting and receiving speed are needed. An interconnection network with wires and electronic switches will have to be replaced when the bit rate to be supported exceeds its design threshold. On the other hand, fiber-optic interconnects enable such a vision because of their virtually unlimited bandwidth, and format (e.g. addressing, coding) transparency.

An important step towards realizing the vision of building such a future-proof fiber-optic interconnection network is to keep data in the optical form during transition by sending the data over an all-optical data path. In other words, conversions between optical and electronic signals can only take place at the source and destination, but not at any intermediate switches. In addition, the network has to provide strong connectivities via flexible reconfiguration, but should involve little or no electronic processing to establish or release all-optical data paths. Unfortunately, current photonic switches (e.g. Lithium Niobate couplers [2, 3]) introduce crosstalk, power loss and high implementation costs. Due to the superiority of the electronic processing logic, it is likely that switch control will be done electronically. Accordingly, it is also important to keep the hardware complexity (e.g. in terms of the switches and I/O interfaces) of the network low while ensuring its reconfigurability.

Note that some of these design goals are in conflict, which necessitates trade-offs and makes an ideal future-proof model difficult, if not impossible, to be implemented today. Specifically, single-hop, broadcasting networks such as linear optical buses or passive star couplers have many desirable features but are not scalable in terms of limited power budget and bandwidth available per node. For switched (or non-broadcasting) networks such as photonic Multistage Interconnection Networks (MINs), and multi-hop networks using switches (e.g. meshes), a concept called *RTDM* (Reconfiguration with Time Division Multiplexing) may be applied whose basic idea is to let the network go through a sequence of configurations to emulate complete connectivity or to realize certain permutations [22]. Such a sequence of configurations, or schedule, is made known to the processors (and switches), which then transmit/receive (and set their states, respectively) at appropriate time. This is efficient for static or compiled communications. For dynamic communications, applying RTDM is a way to trade some of the vastly available bandwidth for the elimination of run-time control while achieving all-optical data communications.

Other trade-offs in the design of mesh-like networks using switches, including multi-dimensional buses [21, 20], are related to the number of photonic switches used along a typical path and in the entire network. While increasing the number of switches strengthens the network's connectivity, it also increases the power loss and crosstalk suffered by the signals going through the switches, and thereby limiting the network's scalability. On the other hand, having a large number of transmitting and receiving interfaces per processor may effectively reduce the number of switches needed, but costly O/E and E/O conversion circuits will be required at these interfaces.

In this chapter, we first illustrate how to exploit the unique characteristics of fiber-optic interconnects in designing interconnection networks through an example. Although the interconnection network to be discussed is not ideally future-proof as a result of trade-offs, it has many desirable features mentioned earlier. We then illustrate how to fully take advantage of the novel interconnect network in the designs of application algorithms.

The chapter is organized as follows. In Section 8.2, the architecture of the RASOB (Reconfigurable Array with Spanning Optical Buses) is presented. A *unique* feature of the *RASOB* architecture which distinguishes it from other two dimensional arrays with either optical or electronic buses [10, 14, 25] is that in a RASOB one can set up a direct (all-optical) path between *any* two processors, not just between those at the same row or the same column. A RASOB provides rich connectivities by reconfiguring itself through two phases, namely Row and Column Phases, during which a processor can communicate with another processor at the same row and at a different row, respectively. In addition, a RASOB only uses one 2 × 2 switch (and one transmitter interface and two receiving interfaces) per processor and all the switches are set to the same state (straight or cross) at the same time. The price paid for such a low control and hardware complexity is that during a Column Phase, only one processor per row can transmit to any given column, making it interesting and yet challenging to develop applications using minimal number of phases.

In Section 8.3, a parallel matrix transposition algorithm on a RASOB is described, which utilizes the unique feature of the RASOB mentioned above and achieves an optimal time complexity. Based on such an algorithm, other matrix operations such as multiplications can also be performed efficiently [19]. In addition, we also describe two embedding algorithms with which binary trees of $N^2 - 1$ and $2^k \cdot N^2 - 1$ nodes (for any $k > 1$), respectively, can be emulated efficiently by a RASOB. The first algorithm lets a non-square $N \times (N + 1)$ RASOB emulate small (or large) binary trees with an optimal dilation cost and a near-optimal expansion cost (or load-balance). The second algorithm improves on the first by emulating small and large binary trees in a square $N \times N$ RASOB, thereby achieving optical expansion cost and load-balance, respectively, while maintaining the optimal dilation cost. We summarize the chapter in Section 8.4.

8.2 THE RASOB MODEL

The *RASOB* architecture is similar to the array structure described in [21]. A main difference is that the former is suitable for SIMD applications, and the latter is suitable for MIMD applications but has higher hardware and control complexities since it employs an all-optical addressing mechanism [6, 21, 23].

Figure 8.1 illustrates the architecture of an $N \times N$ RASOB, which consists of N folded row buses and N folded column buses. Each processor transmits to the upper segment of a row bus (only), and receives from the lower segment of the row bus and the right segment of a column bus. Note that each processor may be replaced with a *module* of a number of processors that are electronically interconnected in some desirable way. The processors in a module would then share the same transmitting and receiving interfaces of the module via either *wavelength* or *time division multiplexing.*

There is a 2 × 2 switch at the intersection of each row bus (to be more precise, its lower segment) and each column bus (its left segment) so that a packet transmitted to a row bus may continue to propagate on the row bus, or be *switched* to a column bus. Once switched, a packet continues to propagate on the column bus and will not be switched back to a row bus as to be discussed later in this section.

In the RASOB, the processors communicate with each other by letting a source send a packet at a specific point in time and by letting its destination receive the packet at another specific point in time later. In particular, the RASOB may alternate between two phases, namely, a row phase and a column phase. In a row phase, all the switches stay at *straight* to allow each processor to

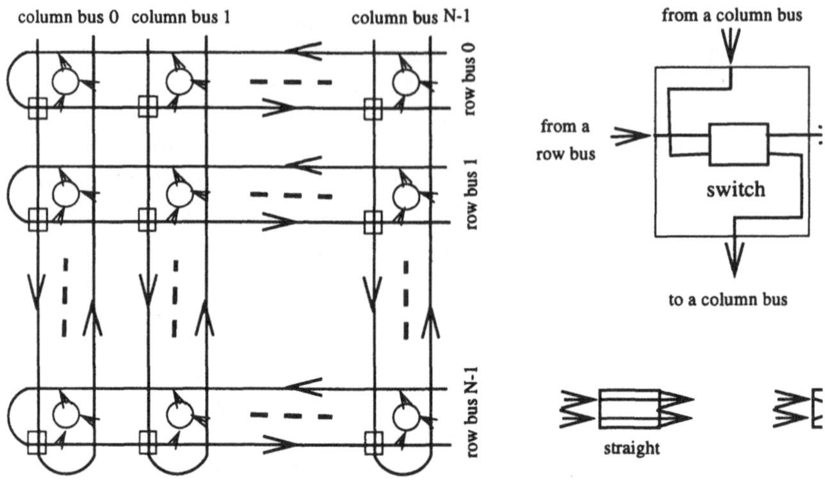

Figure 8.1 An $N \times N$ RASOB.

communicate with others at the same row. In a column phase, all the switches will be set simultaneously to *cross* at a certain point in time to allow each processor to communicate with others at different rows, which may or may not be at the same column. Note that, a processor may also communicate with others at the same row in a column phase if needed. If the switching speed is high (e.g. tens to hundreds of picoseconds when Lithium Niobate switches are used), the switch control overhead is almost negligible.

Like in other optical bus based architectures employing Time Division Multiplexing (TDM) communications, synchronization is a crucial issue. Such an issue, however, is beyond the scope of this chapter. Suffice it to say that the larger fan-out, less power loss, better noise immunity and other desirable features of optics make it possible to distribute a high-frequency global clock to hundreds of receiver modules with little clock skew [26]. Additionally, the jitters that optical data signals encounter in single-mode fibers of a length less than one kilometer is small and can be tolerated by using narrow (e.g. 10 bits) guard bands at each end of a time slot (up to a few hundreds of bits) [24].

A bus cycle (or simply *cycle* hereafter) in the RASOB is defined to be $4N$ units, where a unit (on the order of nanoseconds) is the propagation delay between two adjacent nodes (or their transmitting/receiving interfaces) at a row or column bus. The advantage of RASOB (as any optical bus system) is that multiple packets can be transmitted (received) within a bus cycle concurrently by multiple transmitters (receivers), unlike in an electronic bus that requires exclusive access. However, as described in [19, 21], the duration of a unit limits the size of each packet in a RASOB since when nodes transmit packets concurrently, collision would occur if packets are too long. Nevertheless, at a transmission speed of 10 Gb/s, for example, a 128-bit packet can fit into a car with a duration of $13ns$. Accordingly, in a RASOB of a reasonable size, e.g. a few hundreds of nodes (say $N < 20$), a cycle is less than a microsecond, which would be comparable to a communication cycle in a conventional system.

Since optical signals propagate unidirectionally, it will take a signal at most $2N$ units to traverse a row bus (first the upper segment and then the lower segment), and at most another $2N$ units to traverse a column bus (first the left segment and then the right segment). Note that, alternatively, one may use two buses per row (or per column), one for downstream and another for upstream transmissions in order to reduce the propagation time on each row or column bus to N, and the cycle to $2N$ as in [9, 10]. However, each node will have to have four transmitters and receivers instead of one transmitters and two receivers, which is an expensive proposition.

Communication Phases

As described in [19, 21], each row or column phase in the RASOB takes at most $4N$ units and thus can be completed in one cycle. In the following, we concentrate on a row bus and use the train loading/unloading model to illustrate how row and column phases can be ***pipelined*** in the RASOB so that a throughput of four phases per cycle can be achieved.

Figure 8.2 shows how a row phase and a column phase may be pipelined (or co-exist). Specifically, let us imagine that at the beginning of a cycle, $t = 0$, two trains (each having N cars) are formed on the upper segment of a row bus, as shown in Figure 8.2 (a). The first train is called a "row" train, and the second a "column" train. In either train, the cars are numbered 0 through $N - 1$ from left to right, and each car can be regarded as an empty packet slot with a duration of a unit time.

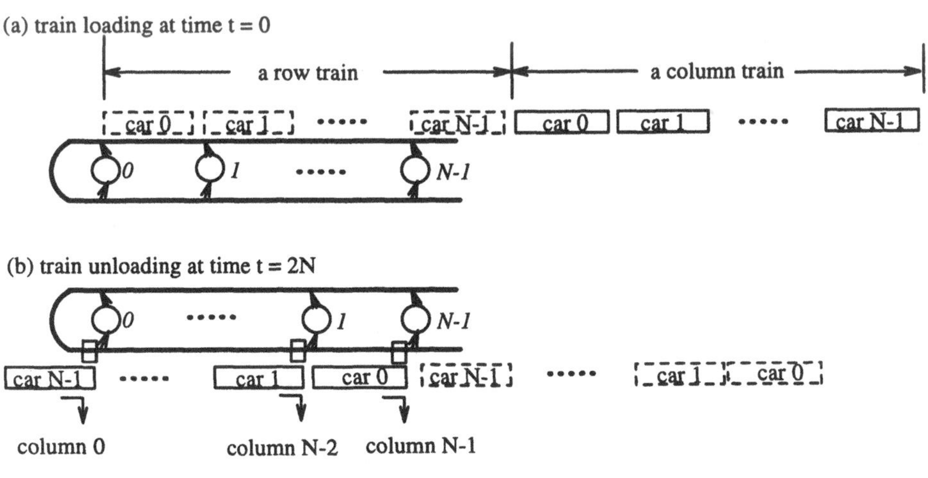

Figure 8.2 Train load/unloading in a RASOB.

Row Phase: One method of allocating the cars in a train is called *Time Division Source-oriented Multiplexing* (TDSM) in [21], which allows all the processors to transmit their packets simultaneously. Specifically, let the processors on each row bus be numbered according to their column positions, that is, 0 through $N-1$ from left to right. Processor i may load car i of a row train at $t=0$, where $0 \le i \le N-1$. The loaded row train then continues on the lower segment of the row bus, enabling each packet to be unloaded (or copied) by one or more destination processors at that row. The row train will leave the row bus at $t = 2N$ units (see Figure 8.2 (b)).

The above discussions imply that any one-to-one or one-to-many (multicasting) communication patterns among the processors at the same row may be accomplished in a row phase with TDSM. Specifically, let $SrcCol$ and $DestCol$ be the column positions of the source and destination processors, respectively, at the row bus. We can program the source processor to transmit a packet at time (hereafter, all timings are in units relative to the beginning of the corresponding cycle)

$$T_{RowSend} = 0. \tag{8.1}$$

Similarly, we can program the destination processor(or processors in a one-to-many situation) to receive the packet from the source at

$$T_{RowRec} = T_{RowSend} = SrcCol + DestCol + 1. \tag{8.2}$$

Note that any many-to-one communication patterns among the processors at the same row can also be accomplished in a row phase with TDSM by programming the sources (to transmit simultaneously) and their common destination (to receive at multiple points in time) as above.

A different method of allocating the cars, called *Time Division Destination-oriented Multiplexing* (or TDDM) in [21], is also possible. In TDDM, processor i will *receive* a packet from car $(N-1)-i$, and all the processors will receive their packets simultaneously. To describe TDDM, we have to imagine a row train which originates from the right end of the upper segment of a row bus at $t = 0$, much like the column train depicted in Figure 8.2 (a). We will omit the detailed descriptions of a row phase with TDDM, but the descriptions of column phases provided below are similar. It is worth noting, however, that TDDM would allow a processor to send two or more different packets to other processors (also known as ***simulcasting***) at the same row. In particular, any simulcasting pattern among the processors at the same row, in which no two (or more) processors try to send to the same processor, can be accomplished in one row phase with TDDM.

Column Phase: Communications in a column phase is slightly more complicated since cars (or packets) are to turn (or be switched) from a row bus to column buses. In particular, the simple switch control in the RASOB dictates that all the cars in a column train will turn onto different column buses at the *same* time since all the switches will be set to cross at the same time. Specifically, as shown in Figure 8.2 (b), car 0 of the column train will turn onto column bus $N-1$, car 1 onto column bus $N-2$ and so on, and car $N-1$ will turn onto column bus 0. Assume that the switches are placed at (or closely to) the receiving interfaces on the lower segment of each row bus, every car of a column train will arrive at its turning point at $t = 2N$ units, as illustrated in

Figure 8.2(b). Note that at this moment, a third and a fourth trains, which can be either row or column trains, may form on the upper segment of the row bus as in Figure 8.2(a) in a pipelined fashion. In this way, four trains are loaded $4N$ units, achieving a pipelined throughput of 4 phases in one cycle.

Once the cars of the above column train finish the turning (at $t = 2N + 1$ units), all the switches are *reset* to straight. Each column bus now looks quite similar to a row bus at the start of a row phase, which is rotated 90° degree counter-clockwise. More specifically, a train of N cars, one from each row bus, is formed on the left segment of each column bus.

Since the cars of the next (third) train will not arrive at their respective turning points until $2N$ units later, the switches will remain in the straight state until each train formed on a column bus leaves the left segment. In other words, each train form on a column bus will continue onto the right segment without possibly being switched back to any row bus, enabling each packet carried in a car to be unloaded (or copied) by one or more destination processors at that column. Note that since no packets will need to be switched back from a column bus to a row bus, a simpler switching device than a 2×2 switch may be sufficient.

Based on the above discussions, we can now derive the timings for transmitting and receiving packets in a column phase as follows. Let *SrcRow* and *DestRow* be the row numbers of the source and destination processors, respectively, and let *SrcCol* and *DestCol* be defined as before. In a column phase, a processor may send a packet to a processor at any column *DestCol* by loading the $[(N-1) - DestCol]$-th car of the column train (as in TDDM). Hence, the transmitting time in a column phase (relative to $t = 0$) is

$$\begin{aligned} T_{ColSend} &= (N - 1 - SrcCol) \\ &\quad +(N - 1 - DestCol) + 1 \\ &= 2N - SrcCol - DestCol - 1. \end{aligned} \tag{8.3}$$

As mentioned before, all the switches are set to cross at $T_{SwitchSet} = 2N$, and accordingly the receiving time in the column phase is

$$\begin{aligned} T_{ColRec} &= T_{SwitchSet} + (N - 1 - SrcRow) \\ &\quad +(N - 1 - DestRow) + 1 \\ &= 4N - SrcRow - DestRow - 1. \end{aligned} \tag{8.4}$$

Note that a processor may send to, in addition to receiving from, processors at different columns by loading multiple cars of a column train as long as there is no conflicts. Specifically, we have

Condition for Conflict: A conflict occurs if and only if two or more packets from the same row need to be transferred to the same processor at a different row, or to two or more processors at the same column in a column phase (since these packets will contend for the same car).

Obviously, there is also a limitation imposed by the bandwidth, which requires that no more than N packets be transmitted in one (row or column)

phases. A key to the development of efficient embedding (and other algorithms) in a RASOB is to be able to fully take advantage of the RASOB's strong ability to support many other conflict-free one-to-one, one-to-many and many-to-one communication patterns so as a minimal number of phases will be needed for the embedding (and execution of the algorithms). Finally, data dependency between two phases may prevent them from being pipelined.

Connectivity and Complexity

The control complexity of a RASOB is low because each of the above equations (8.1 to 8.3) involves simple arithmetic calculations. In addition, the hardware complexity of a RASOB is also low because each processor uses only one two-state 2×2 switch and has only one transmitter. Although two receiver interfaces are needed by each processor, a single high-speed receiving circuit may be shared among these two interfaces. As a comparison, most mesh-based reconfigurable architectures would require at least an equal number of switches having four or more states and four I/O interfaces per each processor [15, 17].

Despite of the low control and hardware complexities, the *RASOB* provides strong connectivities due to the following characteristics. First, a direct connection between any two processors can be established, as mentioned before. The existence of such a direct "***all-optical***" path is important, because conversions between optical and electronic signals required for buffering and address decoding at intermediate nodes are costly. Secondly, reconfiguration is flexible as one may ***interleave*** Row and Column phases in many ways to provide communication bandwidth required by an application. Finally, because only a portion of optical power is tapped off at each receiver interface, ***multicasting*** can be supported simply by programming multiple receivers to receive at different points of time during the same phase. Noting that both one-to-one and broadcast are special cases of multicasting, we may summarize the communication capability of the *RASOB* below:

- all processors at row i can multicast to the processors at the same row at the same time; and such a row-to-row multicasting can be performed at all N rows simultaneously.
- all processors at column j can multicast to the processors at the same column at the same time; and such an column-to-column multicasting can be performed at all N columns simultaneously.
- A processor at row i and column j, hereafter denoted by $p(i,j)$, can multicast to several processors at any column k and such a processor-to-column multicasting can be performed by all the N^2 processors at the same time, with a ***restriction*** being that two or more processors at the same row i cannot multicast to the same column k at the same time.

Note that while the first two items on the list alone mean that the RASOB is at least as powerful as any mesh with row and column buses, the third item clearly shows that the RASOB has a stronger connectivity than other meshes with row and column buses. In addition, we note that the *RASOB* has many other capabilities that are not included in this list but may be deducted from its basic features described earlier. For example, *simulcasting*, in which a processor sends different messages to different processors simultaneously can be supported by assigning more than one car in a train to a source processor. However, a high-speed transmitter interface that can transmit one packet every packet slot time is required at each processor. Similarly, many-to-one communications can be supported with high-speed receiver interfaces that can receive one packet every packet slot time.

8.3 ALGORITHM DEVELOPMENT

Although the *RASOB* has a strong connectivity, it, like many practically scalable interconnection networks, has a weaker connectivity than a completely-connected one. The capabilities as well as restrictions of the RASOB makes it an interesting yet challenging task to design efficient algorithms.

In designing algorithms on a RASOB, one may use the idea proposed in [22] to partition the set of connections required by an application into subsets such that the connections in each subset can be established in a Row or Column phase. However, such a partition may not result in optimal (i.e. minimal) number of communication phases and therefore a customized design may be necessary. As an example of how one can take advantage of the capabilities while overcoming the restrictions of the *RASOB* architecture, we first describe a parallel matrix transposition, which requires only two phases (a row and a column, which are not pipelined) and thus has an optimal time complexity.

8.3.1 *Matrix Transposition*

In the following discussions, we assume, for simplicity, that each matrix is of $N \times N$ and processor $p(i,j)$ stores the element of the matrix at row i and column j. For a larger matrix of $\hat{n} \times \hat{n}$ where $\hat{n} > N$, we may first treat the $\hat{n} \times \hat{n}$ matrix as a $kN \times kN$ matrix where $kN \geq \hat{n}$ for some integer k. We then divide this matrix into blocks, each being a $k \times k$ submatrix and assign the submatrix to a processor in the *RASOB*. This would be natural if each processor is in fact a module consisting of a number of processors, as mentioned earlier.

Consider matrix B, whose elements are denoted by $b(i,j)$. To transpose matrix B, one needs to route element $b(i,j)$ from processor $p(i,j)$ to processor $p(j,i)$, where $0 \leq i,j < N$ and $i \neq j$. In the following discussion, we will use the notation $[x]_N$ to represent operation x *mod* N.

Our matrix transposition algorithm accomplishes data movement from $p(i,j)$ to $p(j,i)$ in two steps: (1). route an element from $p(i,j)$ (where $j \neq i$) to an intermediate location, which is $p(j,[i+j-1]_N)$ and (2). route the element from that intermediate location to its final destination $p(j,i)$.

Figure 8.3 (a) and (b) illustrate the step one routing of the elements from the *upper* and *lower* triangle of a 5×5 matrix B to their intermediate locations, respectively. In both (a) and (b), virtual locations marked by dotted circles are resulted from the calculation of the column index $i+j-1$ without performing the *mod* N operation. They are used to show the *skewed* pattern of the step one routing and each of them corresponds to a real location at the same row and at column $[i+j-1]_N$.

In general, let f_1 be the routing function used in the first step, that is, $f_1(i,j) = (j, [i+j-1]_n)$. One can easily prove that [19]

Lemma 1. *(a) f_1 is a one-to-one mapping. (b) Let $f_1(i,j) = (j,k)$ and $f_1(i,j') = (j',k')$ where $j \neq j'$, then $k \neq k'$.* ∎

Part (b) of the Lemma means that *no two* elements originated from the same row will end up at the same column, or in other words, the condition for conflict in RASOB is NOT met. As a result, step one routing requires only one Column phase in the *RASOB*. Note that after the first step is completed, element $b(i,j)$ is at $p(j,[i+j-1]_N)$, which is at the same row as the final destination $p(j,i)$.

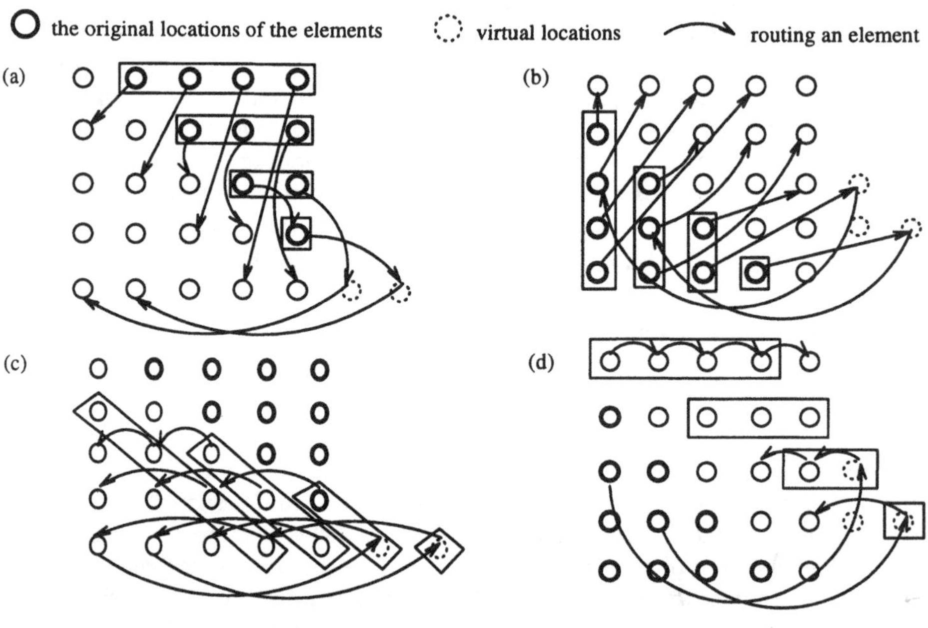

Figure 8.3 Transpose a matrix in two steps: (a) & (c): move elements from the upper triangle. (b) & (d): move elements from the lower triangle.

To formalize the routing function used in the second step, we define $f_2(j,[i+j-1]_N) = (j,i)$, or equivalently, $f_2(i,j) = (i,[j-i+1]_N)$. It is easy to see that f_2 is also a one-to-one mapping function. In addition, since f_2 routes an element to the same row, the second step routing can be accomplished with *Row communications* in an *RASOB*. Specifically, we have

Lemma 2. *Step two routing can be completed in one* Row phase. ∎

Figure 8.3 (c) and (d) shows the step two routing of the elements of the *upper* and *lower* triangle of matrix B, respectively, from their intermediate locations to their final destinations. Again, virtual locations marked by dotted circles are used to show the routing pattern. We note that in (c), processor $p(2,1)$ already holds $b(2,1)$ and hence does not have to participate in the step two routing. Similarly, in (d), processors $p(2,j)$ already hold $b(j,2)$, for $j = 3,4,5$ respectively, and hence do not have to participate in the step two.

Based on Lemmas 1 and 2, we have

Theorem 1. *Matrix transposition can be done in 2 communication phases.* ∎

We note that in the *RASOB*, it is impossible to relocate $b(0,1)$ and $b(0,2)$, both from the same row 0, to $p(1,0)$ and $p(2,0)$, both of which are at the same column 0, respectively, in one phase without causing a conflict. Therefore, it is clear that the matrix transposition algorithm which requires only two phases is optimal. It is also noted that during each of the two phases, the algorithm requires that a processor send and receive at most one packet and thus only a subset of the architectural capabilities has been utilized. We also note that matrix elements have to arrive at their intermediate locations before they can be forwarded to their destinations in step two, the second phase cannot start until the first finishes, and hence the two phases cannot be pipelined.

A detailed description of the two phases including the time at which each processor should send and receive can be found in [19].

8.3.2 Binary Tree Embedding

Having demonstrated how to take advantage of the unique features of the RASOB architectures in developing applications such as matrix transpositions, we now demonstrate the same in emulating other interconnection networks or communication structures such as binary trees.

A binary tree is an important communication structure in multiprocessor systems, based on which many fundamental algorithms have been developed [1, 7, 8, 13]. Efficient embeddings of binary trees into other common topologies including meshes have been studied extensively in the past [4, 5, 11, 12, 18, 27, 28]. Of particular relevance is the embedding of binary trees in arrays with optical buses described in [9, 10]. In these embeddings, the size of the binary trees does not exceed that of the target machine.

In this subsection, we study the embeddings of a small binary tree (having $N^2 - 1$ nodes) as well as a large binary tree (having $2^k \cdot N^2 - 1$ nodes for some integer k). We first consider their embeddings in a non-square $N \times (N+1)$ RASOB such that any one-step communication operations in a tree can be completed in one RASOB cycle. This means that such embeddings achieve a minimum dilation cost of 1. We then improve the embeddings by considering a square $N \times N$ RASOB such that an optimal expansion cost of nearly 1 and an optimal load balance of about 2^k tree nodes per RASOB node are achieved for the small and large binary trees, respectively, while maintaining the optimal dilation cost of 1.

Preliminary

Let $N = 2^l$ for some l. Accordingly, the large binary tree of $2^k \cdot N^2 - 1$ has $2l+k$ levels, while the small binary tree of $N^2 - 1$ nodes has $2l$ levels (or $k = 0$). We number the levels from 0 to $2l + k - 1$ from top to bottom as shown in Fig. 8.4. In addition, the nodes are number in a breadth-first fashion, starting at 1 for the root. In this way, given a non-leaf node X, where $X < 2^{2l+k-1}$, its left child and right child (for are $2X$ and $2X + 1$, respectively, and its parent (for $X > 1$) is $\lfloor \frac{X}{2} \rfloor$.

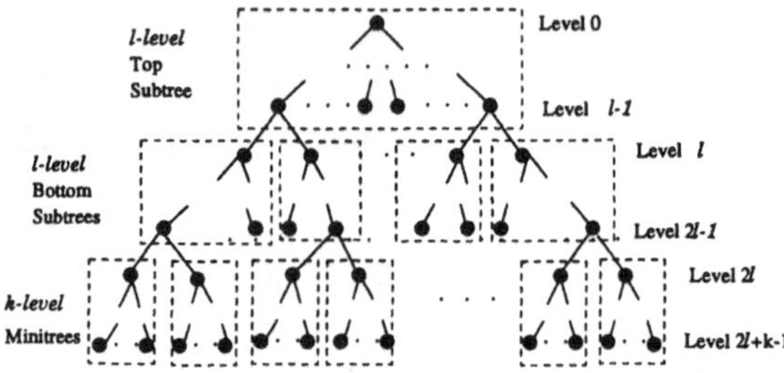

Figure 8.4 A partitioned $2l + k$-level tree.

The two basic packet transfer patterns in a binary tree that we will embed in a RASOB are *parent-to-children* (P-to-C) and *children-to-parent* (C-to-P). In P-to-C, every node (except the leaves) sends a packet to both its left child and right child. In C-to-P, every node (except the root) sends a packet to its parent. Our goal is to embed *both* packet transfer patterns in a RASOB in one cycle so that any communication operation requiring one step in a tree can be emulated in one cycle in the RASOB. Specifically, let $R()$ be the mapping function used in the embedding. Then, every RASOB node emulating a tree node X, denoted by $R(X)$, will be able to send a packet to $R(2X)$ and $R(2X + 1)$ (except if X is a leaf), and send a packet to $R(\lfloor \frac{X}{2} \rfloor)$ (except if X is the root) within a cycle *concurrently* with all other RASOB nodes.

To facilitate the embedding, we cut the $2l + k$-level tree between levels $l - 1$ and l, and if $k > 0$, between levels $2l - 1$ and $2l$ as well, as shown in Fig. 8.4. For the small tree having only $2l$ levels, this creates $N + 1$ *subtrees* each having l-levels and $2^l - 1$ (or $N - 1$) nodes: the top l levels of the tree becomes what we call the *top subtree* and the rest *subtrees* are called *bottom subtrees*, which are assigned indices 0 to $2^l - 1$ from left to right. For the large tree, the additional $k > 0$ levels at bottom will contain N^2 *minitrees*, each having $2^k - 1$ nodes.

Embedding in a Non-Square RASOB

We first embed the small binary tree in a non-square RASOB having $N + 1$ rows and N columns. Such an embedding, whose detailed descriptions can be found in [16], serves as the basis of the improved embedding to be described later.

The partition of the tree into one top-subtree and N bottom-subtrees leads to a straight-forward embedding of one subtree per row, as illustrated in Fig. 8.5. Specifically, bottom-subtree s, where $0 \leq s \leq N - 1$, is embedded at row s, while the top-subtree is embedded at row N. In this way, intra-subtree P-to-C (or C-to-P) packet transfers can be completed in one row phase. At least one additional column phase is needed to complete inter-subtree P-to-C and C-to-P packet transfers.

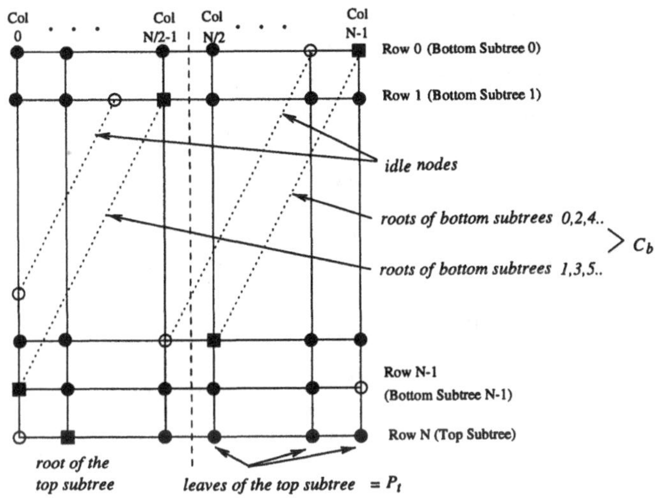

Figure 8.5 Embedding a $2l$-level tree in a $(N+1)$-row and N-column RASOB.

Denote the leaves of the top subtree by $\mathcal{P}_t$ and the roots of the bottom subtrees by $\mathcal{C}_b$ (note that they are the parents and children, respectively). Accordingly, *inter-subtree* P-to-C contains packet transfers from $\mathcal{P}_t$ to $\mathcal{C}_b$), while *inter-subtree* C-to-P contains packet transfers from $\mathcal{C}_b$ to $\mathcal{P}_t$.

Given that no conflict is possible in intra-subtree packet transfers in a row phase, node X in the top-subtree, where $1 \le X \le N-1$, can be simply embedded at column X (and row N). However, to avoid conflicts in inter-subtree P-to-C packet transfers (in a column phase), each bottom-subtree root has to be embedded at a distinct column. Since there are N nodes in $\mathcal{C}_b$, N columns are needed[1], although this leads to an idle node at each row as there are only $N-1$ nodes in each subtree. From the figure, it can also be seen that the nodes in $\mathcal{C}_b$ exhibit a skewed pattern (and so do the idle nodes). Such a skewed pattern makes it easy to complete the inter-subtree P-to-C packet transfer as the nodes in $\mathcal{P}_t$ can send to their left children at the same time, and then to their right children at the same time as described in [16].

To formally describe the embedding of any node X in the bottom-subtrees, where $N \le X \le N^2-1$, we let $m = \lfloor \log X \rfloor$ be the level at which node X is (note that $l \le m \le 2l-1$). In addition, to facilitate our discussions, let $1X_{m-1}...X_0$ be the binary representation of X (note that its most significant bit has to be 1 due to the way m is defined). In this way, the index of the

[1]This is why the tree is not embedded in a $(N+1) \times (N-1)$ RASOB.

bottom subtree in which X belongs to, or in other words, the row at which X is embedded, can be determined as

$$Row(X) = X_{m-1} \dots X_{m-l}. \tag{8.5}$$

To determine the column at which X is embedded, we first *rank* each node in a subtree according to the node number. Since the nodes are numbered in a breadth-first fashion, their ranks in a subtree are also assigned in a breadth-first fashion. Specifically, the root of a subtree is ranked as 1, its left child as 2, and its right child as 3, and so on. That is, we can determine its rank as follows:

$$Rank(X) = \begin{cases} 1 & \text{if } N \leq X \leq 2N-1 \text{ (or } m = l\text{)}; \\ 1X_{m-l-1}\dots X_0 & \text{if } X \geq 2N \text{ (or } m > l\text{)}. \end{cases} \tag{8.6}$$

Accordingly, we can determine the column number as follows:

$$Col(X) = \begin{cases} N + Rank(X) - \frac{Row(X)}{2} - 2 & \text{if } Row(X) \text{ is even;} \\ \frac{N}{2} + Rank(X) - \frac{Row(X)-1}{2} - 2 & \text{if } Row(X) \text{ is odd.} \end{cases} \tag{8.7}$$

As an example, a 4-level binary tree (where $N = 4$) depicted in Fig. 8.6 (a) is embedded into a 5×4 RASOB as shown in (b).

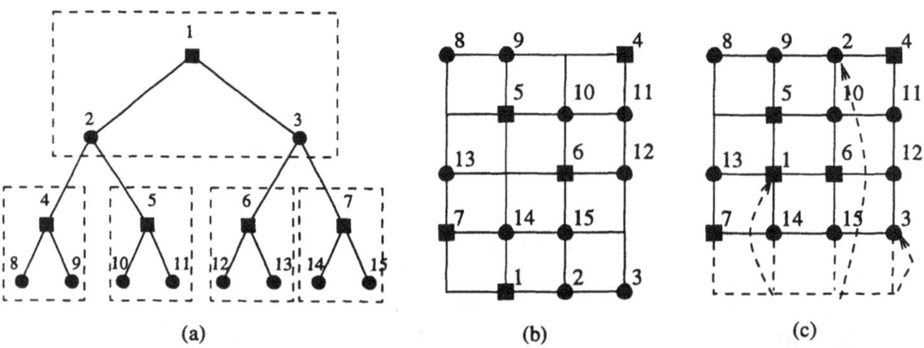

Figure 8.6 Examples (a) A 4-level tree. (b) Embedding in a 5×4 RASOB. (c) Embedding in a 4×4 RASOB.

As a result of the embedding, the P-to-C and C-to-P packet transfers can be completed in four phases as follows, whose detailed descriptions are in [16].

- Phase 1: this is a row phase for intra-subtree P-to-C packet transfers.
- Phase 2: this is a column phase for inter-subtree P-to-C packet transfers.
- Phase 3: this is a row phase for intra-subtree C-to-P packet transfers.
- Phase 4: this is a column phase for inter-subtree C-to-P packet transfers.

Note that, in either Phase 1 or Phase 3, $N-2$ packets are sent and received at each row, which approaches the bandwidth limit (of N packets per row). On the other hand, inter-subtree P-to-C and C-to-P can share the bandwidth available in a column phase. That is, Phases 2 and 4 can be consolidated in one since in these two phases, either the nodes in $\mathcal{P}_t$ at row N transmit and those in $\mathcal{C}_b$ at rows 0 through $N-1$ receive, or vice versa. This means that P-to-C and C-to-P packet transfers among the nodes in a $2l$-level binary tree can be completed in three phases, which is minimal.

Based on this embedding of the small tree in three phases, the large tree can be embedded in four phases as follows. Given that there are $\frac{N}{2}$ leaves in any bottom subtree, there are N minitrees as its children. These N minitrees are embedded at the *same row* as their "parent" bottom subtree, one minitree per RASOB node, as illustrated in Figure 8.7. As a result, intra-minitree P-to-C and C-to-P communications do not require any packets be sent between RASOB nodes. In addition, at each row s, all but one node have a load of 2^k (and the node which is left idle when embedding the bottom subtree has a load of 2^k-1). This results in a good load balance except at row N, where all but one node have a load of 1 since no minitrees are embedded there.

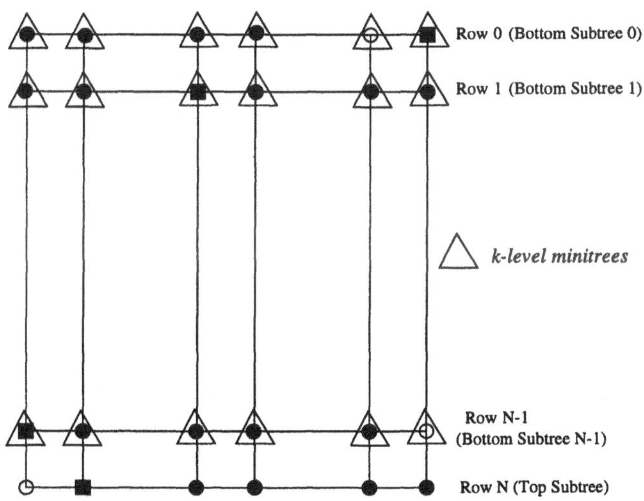

Figure 8.7 Embedding a $(2l+k)$-level tree.

Specifically, let $\mathcal{P}_b^s$ denote the $\frac{N}{2}$ leaves of a bottom subtree having index s, and let $\mathcal{C}_m^s$ denote the N roots of the minitrees under this bottom subtree. For any node $X \in \mathcal{P}_b^s$, we embed its left child, node $2X \in \mathcal{C}_m^s$ (and the entire minitree rooted at $2X$) at the same location as node X, that is, at column $Col(X)$ and row $Row(X)$. On the other hand, we embed the right child of X, node $2X+1 \in \mathcal{C}_m^s$ (and the entire minitree rooted at $2X+1$) at the same row as node X but $\frac{N}{2}$ columns away, that is, at column $Col(X) + \frac{N}{2}$ $(mod\ N)$.

As a result of this embedding of the minitrees, communications between the nodes in $\mathcal{P}_b^s$ and their left children do not require any packets to be sent between RASOB nodes. However, communications between the nodes in $\mathcal{P}_b^s$ and their right children do. In particular, the P-to-C and C-to-P communications between the nodes in $\mathcal{P}_b^s$ and their right children will require $\frac{N}{2}$ packets be sent from $\mathcal{P}_b^s$ to their right children and *vice versa*. Accordingly, such communications can be accomplished in one row phase [16].

Embedding in a Square RASOB

We now describe how the small binary tree can be embedded in a square $N \times N$ RASOB. A square RASOB is more practical to assume and such an embedding has a better (in fact, an optimal) expansion cost. As to be discussed later in this section, this also leads to the embedding of the large binary tree (shown in Figure 8.4) in the square RASOB with an improved load balance.

The key observation leading to our proposed embedding in the square RASOB is that previously when embedding the small tree in a non-square RASOB, there was one *idle* RASOB node at each row, as illustrated in Fig. 8.5. In addition, the idle RASOB nodes at rows 0 through $N-1$ were at distinct columns. Accordingly, every node X in the top subtree, which was previously embedded at row N and column X, can now be moved up to a row at which the RASOB node at column X was previously idle. This eliminates the need to have row N, as shown in the example in Fig. 8.6 (c).

Formally, given a node X in the top-subtree, where $1 \leq X \leq N-1$, we can determine the row at which the RASOB node at column X was idle in the previous embedding, or equivalently, the row at which X will be embedded in the square RASOB as follows:

$$IdleRow(X) = \begin{cases} N-3-2X & 1 \leq X \leq \frac{N}{2}-2; \\ 2N-4-2X & \frac{N}{2}-1 \leq X \leq N-2; \\ N-1 & X = N-1. \end{cases} \tag{8.8}$$

Transfer Packets in Four Phases: Now that the nodes in the top-subtree are scattered among rows 0 through $N-1$, *intra-top-subtree* P-to-C and C-to-P packet transfers can no longer be completed in two row phases, namely 1 and 3, respectively, as before. Rather, they will have to be completed in two column phases.

Our idea is to complete intra-top-subtree P-to-C packet transfers from in Phase 2, or in other words, along with inter-subtree P-to-C packet transfers. Similarly, we will complete intra-top-subtree C-to-P packet transfers in Phase 4 along with inter-subtree C-to-P packet transfers. The reasons that no conflicts is possible in the modified Phases 2 and 4 are illustrated in Fig. 8.8 (a) and (b).

Specifically, let the set of nodes in the top subtree that do not belong to $\mathcal{P}_t$ be denoted by $\bar{\mathcal{P}}_t$. An important observation is that the nodes in the top subtree (i.e those in $\mathcal{P}_t \cup \bar{\mathcal{P}}_t$) are at different rows and different columns. To clarify this observation, a diagonal line (having two segments) in Fig. 8.8 may be used to indicate their locations. Similarly, another diagonal line may be used to illustrate that the nodes in $\mathcal{C}_b$ are also at different rows and different columns. Note that since the sources of the packet transfers for both intra-top-subtree

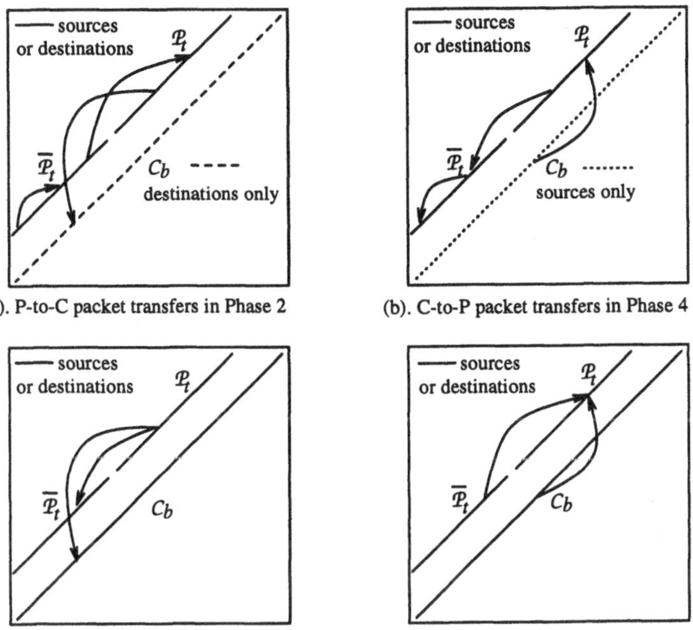

(a). P-to-C packet transfers in Phase 2

(b). C-to-P packet transfers in Phase 4

(c). a type 1 conflict

(d). a type 2 conflict

Figure 8.8 Two column phases are both sufficient (see (a) and (b)) and necessary (see (c) and (d)) for all intra-top-subtree and inter-subtree packet transfers.

and inter-subtree P-to-C are all in the top-subtree and thus at distinct rows as shown in Fig. 8.8 (a), no conflicts are possible if a column phase is dedicated to them. Similarly, no conflicts are possible in a column phase dedicated to intra-top-subtree and inter-subtree C-to-P packet transfers communications since all the destination nodes of these packet transfers are in the top-subtree and thus at distinct columns, as shown in Fig. 8.8 (b). In short, the following four phases can be used to complete the packet transfers:

- Phase 1 (modified): this is a row phase for *intra-bottom-subtree* P-to-C packet transfers. Note that, this phase can be replaced by a *column phase* (see the discussions in the last paragraph in Sec. 2.).

- Phase 2 (modified): this is a column phase for both inter-subtree and *intra-top-subtree* P-to-C packet transfers.

- Phase 3 (modified): this is a row phase for *intra-bottom-subtree* C-to-P packet transfers. Note that, this phase cannot be replaced by a column phase since a node needs to receive two packets from its children that are embedded at the same row.

- Phase 4 (modified): this is a column phase for both inter-subtree and *intra-top-subtree* C-to-P packet transfers.

Note that, we can no longer consolidate Phases 2 and 4 into one column phase because such a consolidation will result in potential conflicts that are illustrated in Fig. 8.8 (c) and (d). Specifically, if all intra-top-subtree and inter-subtree P-to-C and C-to-P packet transfers are carried out in one consolidated column phase, two types of conflicts may occur. A type 1 conflict occurs when a node in $\mathcal{P}_t$ (a parent) needs to send two packets to two nodes, one in $\bar{\mathcal{P}}_t$ (a grand-parent) and the other in $\mathcal{C}_b$ (a child) that are at the same column, as shown in Fig. 8.8 (c). In the embedding example shown in Fig. 8.6 (c) for instance, both nodes 1 and 5 are at column 1, and hence node 2 cannot send packets to them in one column phase. Similarly, a type 2 conflict occurs when a grand-parent and a child are at the same row, and both need to send a packet to a parent, as shown in Fig. 8.8 (d). In Fig. 8.6 (c) for instance, both nodes 1 and 6 are at row 3, and hence they cannot send packets to node 3 in one column phase.

Transfer Packets in Three Phases: Note that it is possible to consolidate the four phases into three by distributing the conflicting packets transfers among different phases. Specifically, Figure 8.9 (a) shows the only possible conflicts when combining Phases 2 and 4, which can be avoided by the following two Lemmas established in [20]:

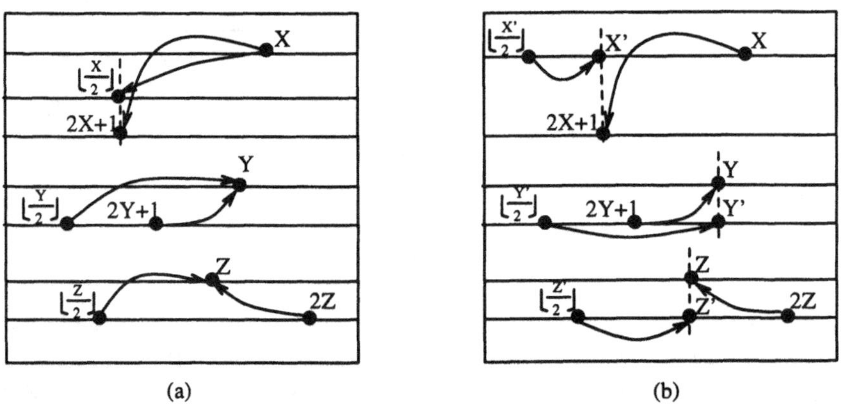

Figure 8.9 (a) Conflicts after combining Phases 2 and 4. (b) Conflicts in Phase 1 after moving three packet transfers from Phases 2 and 4 into Phase 1.

Lemma 3. *No type 1 conflicts will occur when Phases 2 and 4 are consolidated into one column phase* as long as *we remove the packet transfer from* X $(\in \mathcal{P}_t)$ *to* $2X+1$ $(\in \mathcal{C}_b)$ *from the consolidated phase (and put it into a separate phase), where* $X = N/2+(N-4)/6 = 2(N-1)/3$ *if* $N = 2^{2n}$ *or* $X = N/2+(N-2)/6 = (2N-1)/3$ *if* $N = 2^{2n+1}$. ∎

Lemma 4. *Let* $Z = N-1 \in \mathcal{P}_t$. *No type 2 conflicts will occur when Phases 2 and 4 are consolidated into one column phase* as long as *we remove the packet transfer from* $2Z$ *to* Z, *and either (1).* $N = 4^n$, *or (2). if* $N = 2^{2n+1}$, *we also remove the packet transfer from* $2Y+1$ *to* Y, *where* $Y = N/2+(N-8)/6 = 2(N-2)/3 \in \mathcal{P}_t$. ∎

Based on these Lemmas, we can combine the Phases 2 and 4 by excluding at most two packet transfers from Phase 4 namely, $2Y+1 \rightarrow Y$ and $2Z \rightarrow Z$, and one packet transfer from Phase 2 namely, $X \rightarrow 2X+1$, as illustrated in Fig. 8.10. Note that from Eqs. 8.8 and 8.5, one may verify that the sources of the excluded packet transfers, namely X, $2Z$ and/or $2Y+1$, are at different rows, and hence these packet transfers can be accommodated in a column phase, e.g. Phase 1, without conflicting with each other.

Phase (modified)		1 (column)	2 (column)	3 (row)	4 (column)
Basic Function		intra-bottom-subtree P-to-C	intra-top-subtree and inter-subtree P-to-C	intra-bottom-subtree C-to-P	intra-top-subtree and inter-subtree C-to-P
CHANGE	+	$X \rightarrow 2X+1$ $2Y+1 \rightarrow Y$* $2Z \rightarrow Z$	All other inter-subtree C-to-P All intra-top-subtree C-to-P	$\lfloor \frac{X'}{2} \rfloor \rightarrow X'$ $\lfloor \frac{Y'}{2} \rfloor \rightarrow Y'$* $\lfloor \frac{Z'}{2} \rfloor \rightarrow Z'$	
	—	$\lfloor \frac{X'}{2} \rfloor \rightarrow X'$ $\lfloor \frac{Y'}{2} \rfloor \rightarrow Y'$* $\lfloor \frac{Z'}{2} \rfloor \rightarrow Z'$	$X \rightarrow 2X+1$		EVERYTHING: $2Y+1 \rightarrow Y$* $2Z \rightarrow Z$ All other inter-subtree C-to-P All intra-top-subtree C-to-P

* only if $N = 2^{2n+1}$

Figure 8.10 Rearrangement of the packet transfers to eliminate Phase 4.

However, as illustrated in Fig. 8.9 (b), each of the excluded packet transfers would conflict with one existing intra-bottom-subtree P-to-C packet transfer in Phase 1. In order to accommodate these excluded packet transfers in Phase 1, existing packet transfers, namely, $\lfloor \frac{X'}{2} \rfloor \rightarrow X'$ and $\lfloor \frac{Z'}{2} \rfloor \rightarrow Z'$, and possibly packet transfer denoted by $\lfloor \frac{Y'}{2} \rfloor \rightarrow Y'$, need to be *replaced* and moved to a different phase (e.g. Phase 3). Note that, the value of X' can be determined by solving Eqs. 8.5 and 8.7 given that $Row(X') = Row(X)$ and $Col(X') = Col(2X+1)$. Similarly, the values of other variables in this paragraph can be determined as well.

Given that in Phase 3, there are only $N-2$ packet transfers sent at every row as discussed earlier, including $Row(X')$, $Row(Y')$ and $Row(Z')$, it is easy to derive (see [20]):

Lemma 5. *These replaced packet transfers involving nodes X', Y' and Z' can be completed in Phase 3 along with all intra-bottom-subtree C-to-P packet transfers already in Phase 3.* ∎

In short, we have shown that, by rearranging the packet transfers in the modified Phases 1 through 4, Phase 4 can be eliminated. Specifically, as shown in Fig. 8.10, Phase 2 will be used for all intra-top-subtree packet transfers (both P-to-C and C-to-P), as well as inter-subtree packet transfers except the three (at the most) packet transfers involving nodes X, $2Y+1$ and $2Z$. These inter-subtree packet transfers are moved to Phase 1 and replace the three (at the most) intra-bottom-subtree packet transfers involving nodes X', Y' and Z'. Finally, these replaced packet transfers are moved to Phase 3. hence, we have

Theorem 2. *The packet transfers in a 2l-level tree can be completed in two column phases and one row phase in the square RASOB.* ∎

Note that since the additional nodes in the large binary tree were embedded at the first N rows in a non-square $(N+1) \times N$ RASOB as described earlier, they can be embedded in the same way in the square RASOB. Moreover, additional packet transfers can also be completed in an extra phase (e.g. Phase 4) as before. As a result, the large binary tree can be embedded in four phases in the square RASOB. In addition, an optimal load balance is achieved since all RASOB nodes in the square RASOB now emulate 2^k tree nodes, except one which emulates only $2^k - 1$ tree nodes.

8.4 SUMMARY

In designing fiber-optic interconnection networks, it is important to consider exploiting the virtually unlimited bandwidth and achieving a high degree of transparency of fiber-optic interconnects, while simplifying electronic control (including avoiding unnecessary O/E/ and E/O conversions), and reducing undesirable effects of the electro-optical switches (e.g. crosstalk and power-loss) associated with reconfiguration. In this chapter, we have demonstrated how to take these into consideration in designing a unique spanning optical bus based array called RASOB. In addition, we have developed algorithms that can take advantage of the unique features of RASOB. Specifically, an efficient matrix transposition algorithm, which is essential to high-performance matrix operations in computer geometry and image processing, has been described. The algorithm takes two (2) RASOB cycles while as a comparison, we note that matrix transposition on a hypercube and a mesh would require $O(\log n)$ and $O(n)$ steps respectively. We have also presented optimal (or near optimal) embeddings of binary trees of $2^k \cdot N^2 - 1$ nodes (for any integer k) in a square $N \times N$ (or a non-square $(N+1) \times N$) RASOB. As a result, any one-step communication operation such as packet transfers among parents and children in the tree can be completed in one (1) cycle. As an added benefit of the embedding, not only can any communication operation requiring T steps in the tree be completed in T cycles in the RASOB, the embedding allows the root to broadcast to all the leaves (in fact, to all the nodes) in just one cycle in the RASOB, whereas such an operation would take $O(k + \log N)$ steps in the tree. Since in general, it is difficult to express the communication time of an algorithms in optically interconnected networks in terms of the electronic processing speed, care must be taken in interpreting such comparison results. Nevertheless, the advantage of fiber-optic interconnects should be clear.

Acknowledgment

Partial funding provided by a grant from NSF under contract MIP-9409864.

References

[1] S. Akl, D. Barnard, and R. Doran. Design, analysis, and implementation of a parallel tree search algorithm. *IEEE Trans. on Pattern Analysis and Machine Intelligence*, PAMI-4(2):192–203, March 1982.

[2] R. Alferness, L. Buhl, S. Korotky, and R. Tucker. High-speed $\Delta\ \beta$-reversal directional coupler switch. In *Photonic Switching, OSA Technical Digest*, volume 13, pages 77–78, 1987.

[3] A. Benner, H. Jordan, and V. Heuring. Optically switched lithium niobate directional couplers for digital optical computing. In *SPIE Proceedings, Digital Optical Computing II*, volume 1215, pages 343–352, January 1990.

[4] F. Bermon and L. Snyder. On mapping parallel algorithms into parallel architectures. *Journal of Parallel and Distributed Computing*, 4:439–458, 1987.

[5] S. Bhatt, F. Chung, T. Leighton, and A. Rosenberg. Optimal simulations of tree machines. In *Symp. on Foundations of Computer Sciences*, pages 274–282, October 1986.

[6] D. Chiarulli, R. Ditmore, R. Melhem, and S. Levitan. An all optical addressing circuit : Experimental results and scalability analysis. *IEEE/OSA Journal of Lightwave Technology*, 9(12):1717–1725, 1991.

[7] J. Cooper and S. Akl. Efficient selection on a binary tree. *Information Processing Letters*, 23(3):123–126, 1986.

[8] E. Dekel and S. Sahni. Binary trees and parallel scheduling algorithms. *IEEE Transactions on Computers*, 32(3):307–315, 1983.

[9] Z. Guo. Optically interconnected processor arrays with switching capability. *Journal of Parallel and Distributed Computing*, 23:314–329, 1994.

[10] Z. Guo et al. Array processors with pipelined optical busses. *Journal of Parallel and Distributed Computing*, 12(3):269–282, 1991.

[11] A. Gupta and S. Hambrusch. Embedding complete binary trees into butterfly networks. *IEEE Transactions on Computers*, 40(7):853–863, 1991.

[12] A. Gupta and S. Hambrusch. Load balanced tree embeddings. *Parallel computing*, 18:595–614, 1992.

[13] E. Horowitz and A. Zorat. The binary tree as an interconnection network: applications to multiprocessor systems and VLSI. *IEEE Transactions on Computers*, 30:247–253, April 1981.

[14] V.P. Kumar and C.S. Raghavendra. Array processors with multiple broadcasting. *Journal of Parallel and Distributed Computing*, 2:173–190, 1987.

[15] R. Lin and S. Olariu. Reconfigurable buses with shift switching - architecture and applications. In *Int'l Phoenix Conf. on Computers and Communications*, pages 23–29, 1993.

[16] Y. Mei and C. Qiao. Embedding binary trees in arrays with optical buses. In *International Conference on Massively Parallel Processing Using Optical Interconnections (MPPOI)*, pages 136–141, October 1997.

[17] R. Miller, V.K.P.Kumar, D.I. Reisis, and Q.F. Stout. Parallel computations on reconfigurable meshes. *IEEE Transactions on Computers*, 42:678–692, 1993.

[18] B. Monien and H. Sudborough. Simulating binary trees on hypercubes. In *3rd Aegean Workshop on Computing: VLSI algorithms and architectures (also in Vol. 319 of Lecture Notes in Computer Science, Springer-Verlag)*, pages 170–180, July 1988.

[19] C. Qiao. On designing communication-intensive algorithms for a spanning optical bus based array. *Parallel Processing Letters*, 5(5):499–511, 1995.

[20] C. Qiao and Y. Mei. An optimal embedding of binary trees in a square reconfigurable arrays with spanning optical buses. *Parallel Processing Letters*, 1998. to appear.

[21] C. Qiao and R. Melhem. Time-division optical communications in multiprocessor arrays. *IEEE Transactions on Computers*, 42(5):577–590, May 1993.

[22] C. Qiao and R. Melhem. Reconfiguration with time-division multiplexed MINs for multiprocessor communications. *IEEE Transactions on Parallel and Distributed Systems*, 5(4):337–352, 1994.

[23] C. Qiao, R. Melhem, D. Chiarulli, and S. Levitan. Optical multicasting in linear arrays. *International Journal of Optical Computing*, 2(1):31–48, 1991.

[24] D. Sarrazin, H. Jordan, and V. Heuring. Fiber optic delay line memory. *Applied Optics*, 29(5):627–637, February 1990.

[25] Q.F. Stout. Mesh connected computers with broadcasting. *IEEE Transactions on Computers*, 32:826–830, 1983.

[26] S. Tang et al. 1-GHz clock signal distribution for multiprocessor supercomputer. In *International Conference on Massively Parallel Processing Using Optical Interconnections (MPPOI)*, pages 186–191, 1996.

[27] A. Wagner and D. Corneil. Embedding trees in a hypercube is NP-complete. *SIAM Journal on Computing*, 15(1):570–590, 1990.

[28] A. Wu. Embedding of tree networks into hypercubes. *Journal of Parallel and Distributed Computing*, 2:238–249, 1985.

9 FUNDAMENTAL ALGORITHMS FOR THE ARRAY WITH RECONFIGURABLE OPTICAL BUSES

Sanguthevar Rajasekaran and Sartaj Sahni

Department of Computer and Information Science and Engineering
University of Florida
Gainesville, Florida 32611
{raj,sahni}@cise.ufl.edu

Abstract: In this chapter we present a survey of algorithms that have been developed for packet routing, sorting, and selection on the AROB (Array with Reconfigurable Optical Buses) model. The algorithms to be considered are both randomized and deterministic.

9.1 INTRODUCTION

An Array with Reconfigurable Optical Buses (AROB) [22, 16] is an $m \times n$ reconfigurable mesh [11] where the buses are implemented using optical technology. AROBs have been studied by many researchers in the recent past owing to their promise in superior practical performance.

Figure 9.1 shows a 4×4 reconfigurable mesh. Each processor can use its local switches to connect together subsets of the four bus segments connected to it. Reconfigurable meshes that use electronic buses have been studied widely. Variants such as the RN [3], RMESH [9], PARBUS [11], M_r [24], RMBM [31], REBSIS [14], and DMBC [30] have been proposed and investigated.

Reconfigurable meshes with optical buses have been less explored. In the AROB model proposed in [22, 16], the switch settings of the processors permitted are the same as those in the RN model of [3]. These settings are shown in Figure 9.2. A bus link connects two adjacent processors x and y and has

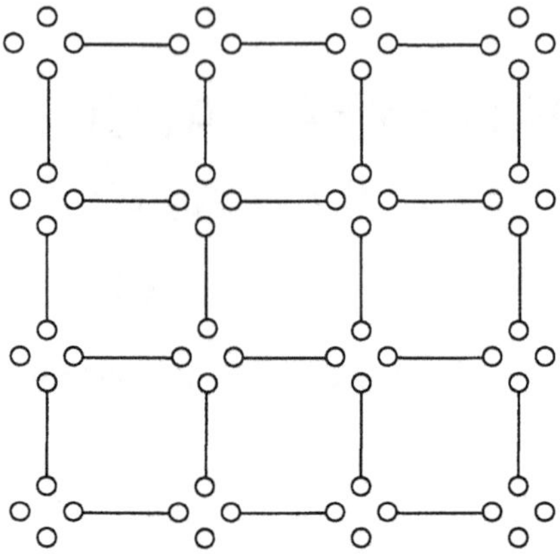

Figure 9.1 A 4 × 4 reconfigurable mesh.

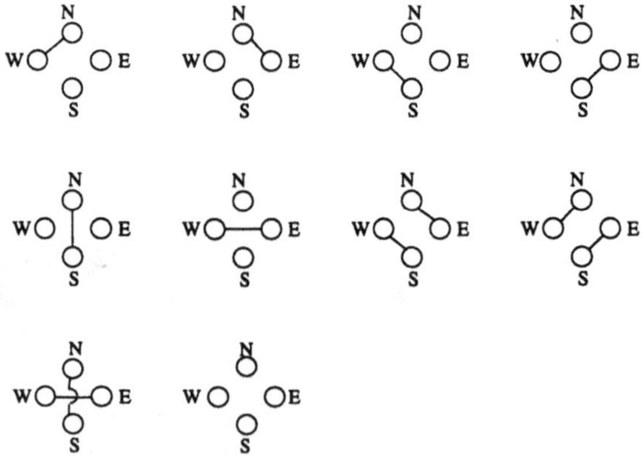

Figure 9.2 Possible switch connections.

two associated wave guides. One of the wave guides is used for transmission of signals from x to y and the other wave guide is used for transmission from y to x. Bus links can be connected together to form disjoint buses by appropriately setting the local switches. On each bus, we have to specify which orientation of the waveguide on each link of the bus is to be used. The resulting directed graph that represents the bus should be a directed chain. The root of this chain is called the bus 'leader'. The number of links on the chain representing a bus is referred to as the length of the bus. The position of a processor on any bus is its distance from the bus leader. By a *cycle* we mean the time it takes to transmit a message on a bus. A cycle consists of *slots* of duration τ and each slot can carry a different optical signal. τ is the time needed for an optical pulse to move down one bus link and τ encompasses the time to send a b-bit message. Here a bit is a light pulse with a w second duration (for appropriate values of $b \geq 1$ and w). It has been argued [22] that for reasonable size meshes (say up to 1000×1000), the number of slots in a cycle may be assumed to be n for an $n \times n$ mesh. Further, the duration of a cycle may be assumed constant and comparable to the time for a CPU operation.

Each processor has a slot counter that is used to determine which slot of a cycle has to be used. These counters may be started at the beginning of a cycle. The bus leader initiates a light pulse at the beginning of a cycle (i.e., it writes a one to the bus). The slot counter of any processor stops when the light pulse reaches that processor. Note that this timing mechanism does not require any bus read operation. Clearly, the terminal counter value is the distance of the processor from the bus leader. In cases of write conflicts, the last written message remains in the slot. Realize that since a processor can read/write from/to its bus during only one slot of a cycle, it cannot poll the up to n light pulses moving through it in one cycle. An additional AROB feature that facilitates the development of algorithms is the delay unit at each processor. A processor can introduce a one time slot delay in the light pulses passing through it.

The AROB model has been defined in [22]. Similar models have been proposed before (see e.g., [21]). A related model known as the ***Optical Communication Parallel Computer (OCPC)*** has also been considered in the literature (see e.g., [1], [7], [32], [8]). In an OCPC any processor can communicate with any other processor in one unit of time, provided there are no conflicts. If two or more processors try to send a message to the same processor, then no message reaches the intended destination.

By a one-dimensional (1D) AROB we mean a $1 \times n$ AROB.

In Section 9.2 we provide some preliminaries related to randomized algorithms. and survey known results in the area of AROBs. Sections 9.3, 9.4, and 9.5 are devoted to the problems of packet routing, sorting, and selection, respectively. In Section 9.6 we provide our conclusions and list some open problems.

9.2 PRELIMINARIES

This section is devoted to providing some preliminary facts and results.

9.2.1 Randomized Algorithms

Asymptotic functions such as $O(.)$, $\Omega(.)$, etc. are used in characterizing the resource bounds of deterministic algorithms. Similar functions can be defined for randomized algorithms as well. A randomized algorithm is said to use $\widetilde{O}(f(n))$ amount of any resource (like time, space, etc.) if the amount of resource used is no more than $c\alpha f(n)$ with probability $\geq (1-n^{-\alpha})$ for any α, $c > 0$ being a constant. Functions such as $\widetilde{\Theta}(.)$, $\widetilde{o}(.)$, etc. can be defined in a similar fashion.

By high probability we mean a probability of $\geq (1-n^{-\alpha})$ for any constant $\alpha \geq 1$.

Chernoff Bounds [5]. These bounds are useful in closely approximating the tail ends of a binomial distribution.

A Bernoulli trial has two outcomes namely *success* and *failure*, the probability of success being p. A binomial distribution with parameters n and p, denoted as $B(n,p)$, is the number of successes in n independent Bernoulli trials.

If X has a distribution of $B(n,p)$ and m is any integer $> np$, then the following are true:

$$Prob.[X > m] \leq \left(\frac{np}{m}\right)^m e^{m-np};$$

$$Prob.[X > (1+\delta)np] \leq e^{-\delta^2 np/3}; \text{ and}$$

$$Prob.[X < (1-\delta)np] \leq e^{-\delta^2 np/2};$$

for any $0 < \delta < 1$.

9.2.2 Problems Definition

Let $x_1, x_2, \ldots, x_n$ be a sequence of elements from the domain Σ. Also let $\oplus$ be a binary, associative, and unit-time computable operator defined on Σ. The problem of prefix computation is to compute $x_1, x_1 \oplus x_2, x_1 \oplus x_2 \oplus x_3, \ldots, x_1 \oplus x_2 \oplus \cdots \oplus x_n$. This is a fundamental operation essential for any parallel machine.

Sorting is the problem of rearranging a given sequence of numbers in non-decreasing order.

The problem of selection is to identify the ith smallest number from out of n given numbers (where i is an input and $1 \leq i \leq n$).

A popular class of parallel computing models are the interconnection networks. Packet routing plays a vital role in the design of efficient interprocessor communication mechanisms in interconnection networks. In particular, a single step of interprocessor communication in any network can be thought of as a packet routing task. We define the routing problem as follows: There is a packet of information at each node that is destined for some other node. Send all the packets to their destinations as quickly as possible in such a way that at most one packet crosses any edge at any time. It is noteworthy that the packet routing problem is equivalent to the random access write operation first defined by Nassimi and Sahni [18]. The *run time* of any packet routing algorithm is the time taken by the last packet to reach its destination. The *queue size* is the maximum number of packets that any processor has to store during the algorithm.

Partial permutation routing is the task of routing where at most one packet originates from any node and no more than one packet is destined for any node. If at most h packets originate from any node and at most h packets are destined for any node then we have a $h-h$ *routing* or h-*relations routing* problem [32].

9.2.3 Prefix Computation

In this section we provide a summary of algorithms developed for the problem of prefix computation on a 1D AROB.

Lemma 1 *Consider a* $1 \times n$ *processor AROB. If each processor has a bit, then the prefix sums of these bits can be computed in* $O(1)$ *cycles [22].*

Proof. This Lemma is proved with the following algorithm. Processor 1 initiates a light pulse in time slot one of a cycle if its bit is zero and in slot two otherwise. Slot counters of all the processors are started at the start of the cycle. Processors set their delay units to introduce a one slot delay in case the processor's bit is one. The slot counter of any processor is turned off when the light pulse initiated by processor 1 reaches it. Each processor now computes its prefix sum value making use of the terminal counter value, its data bit, and its distance from processor 1. ∎

The above Lemma can be strengthened as follows [22].

Lemma 2 *We can add* n $\log n$*-bit numbers in* $O(1)$ *cycles on a* $\log n \times n$ *AROB.* ∎

Similar results have been obtained for computing prefix sums on a 2D AROB [22].

Lemma 3 *If there is a bit at each node of a* $\sqrt{n} \times \sqrt{n}$ *AROB, the prefix sums of these bits can be computed in* $O(1)$ *cycles.* ∎

The algorithm for Lemma 3 uses the algorithm of [20] for computing the prefix sums of an integer sequence. The maximum bus length employed by this algorithm is $3\sqrt{n}$.

The bus length for the prefix sums problem can be reduced to to $\sqrt{n}$ (which is a factor of 3 improvement over [22]'s algorithm) as has been shown by Rajasekaran and Sahni [28].

Lemma 4 *Prefix sums of bits on a* $\sqrt{n} \times \sqrt{n}$ *AROB can be computed in* $O(1)$ *cycles, keeping the maximum bus length as* $\sqrt{n}$.

Proof. Assume that the prefix sums have to be computed in row major order. Use the algorithm of Lemma 1 to compute the prefix sums along each row in $O(1)$ cycles. Now, each processor in the last column has the sum of all 1's in the corresponding row. The original problem of computing prefix sums thus has reduced to computing prefix sums of $\sqrt{n}$ numbers where each number is at most $\sqrt{n}$ (i.e. each number is an $O(\log n)$-bit number). The prefix sums of $\sqrt{n}$ $O(\log n)$-bit numbers can be computed in $O(1)$ cycles on a $\sqrt{n} \times \sqrt{n}$ AROB using the following algorithm.

1. Group the numbers with $\log n$ numbers in each group. Allocating a subarray of size $\log n \times \log^2 n$, compute the sum of numbers in each group as done in [22]. This step takes $O(1)$ cycles and the maximum bus length is $O(\log^2 n)$.

2. Compute prefix sums of the $\frac{\sqrt{n}}{\log n}$ group sums. Lemma 2 can be used to do this as follows. For each prefix sum (there are $\frac{\sqrt{n}}{\log n}$ of them) allocate a subarray of size $\log n \times \frac{\sqrt{n}}{\log n}$. Broadcast the appropriate numbers to each subarray. Within each subarray add the numbers using Lemma 2. The maximum bus length for this step is $\sqrt{n}$.

3. Finally, compute the prefix sums local to each group of $\log n$ numbers. Note that this step is similar to step 2. Each group will get a subarray of size $\log n \times \log^2 n$. The maximum bus length is $O(\log^2 n)$.

The run time of this algorithm is $O(1)$ cycles. ∎

9.3 PACKET ROUTING

In this section we consider routing problems on 1D as well as 2D AROBs.

9.3.1 1D Packet Routing

We start with the simple problem of routing any partial permutation on a 1D AROB.

Lemma 5 *In an AROB of size* $1 \times n$ *any permutation can be routed in* $O(1)$ *cycles [22].*

Proof. We assume that it takes τ time for a packet to move from one processor to the next. Since there are two wave guides (one for left-to-right transmission and the other for right-to-left transmission), the flow of packets from left to right is not affected by the flow of packets from right to left. Consider any permutation to be routed. Number the processors $1, 2, \ldots, n$ starting from left. We consider only the packets that have to travel from left to right (since the other packets can be analyzed in the same way). In any cycle, each processor has a time slot assigned to it for reading from (and writing into) the bus. Let the reading time slot for processor i be $2i$. Processor 1 creates a 'time slot' for each packet that moves one edge per τ time. The time slots created will be in the order of the processors, i.e., the first time slot is meant for processor 1, time slot 2 is meant for processor 2, and so on. If processor p has a message for processor q, p will write this message at time $t + (p+q)\tau$, where t is the start time. Clearly, this algorithm terminates in 2 cycles (or $2n\tau$ time). ∎

We can strengthen the above Lemma as follows.

Lemma 6 *Consider a routing problem on a $1 \times n$ AROB where $O(1)$ packets originate from any node and $O(1)$ packets are destined for any node. This problem can also be solved in $O(1)$ cycles.*

Proof. Let c (resp. d) be an upper bound on the number of packets destined for (originating from) any node. It suffices to consider the case $d = 1$, since that algorithm can be repeated d times to take care of the general case. There will be $2c$ runs of the algorithm given above for Lemma 5. The above algorithm has the property that the message read by any processor q is the last message written in time slot q. If more than one processors wrote in time slot q, they can determine if their message was read by q or not by reversing the routing process. This way, in every two executions of the above routing algorithm, a processor receives one message destined for it. ∎

A further extension of the above ideas leads to the following Lemma [22].

Lemma 7 *Let there be k elements arbitrarily distributed (at most one per processor) in a $\sqrt{n} \times \sqrt{n}$ AROB. 'Compact' the elements in the first $\lceil \frac{k}{\sqrt{n}} \rceil$ rows. This problem can be solved in $O(1)$ cycles.*

Proof. The algorithm for the above problem figures out a unique address for each element (using the prefix algorithm of Lemma 3) and then routes the elements using greedy paths. There is no possibility of a collision. ∎

Next we consider the problem of h-relations routing. Several h-relations routing algorithms have been proposed for the OCPC model. 1) Anderson and Miller's algorithm can route $\log n$-relations on an n-node OCPC in $\widetilde{O}(\log n)$ time [1]; 2) Valiant's algorithm can route any h-relation in time $\widetilde{O}(h + \log n)$ [32]; 3) Geréb-Graus and Tsantilas have given a simple algorithm that can route any h-relation in time $\widetilde{O}(h + \log n \log\log n)$ [7]; 4) Goldberg, Jerrum, Leighton, and Rao [8] have given a more complex algorithm with a run time of $\widetilde{O}(h + \log\log n)$.

The OCPC and the AROB models differ in two critical aspects. If more than one messages are sent to some processor π at the same time in an OCPC, none of them reaches π. In contrast, under the same scenario, one of the messages will reach π on the AROB model. In addition, operations such as prefix sums (limited to integers of certain magnitude) and compaction can be performed in $O(1)$ time on the AROB model and not on the OCPC model.

Since a single step of an n-node OCPC can be simulated on a $1 \times n$ AROB in $O(1)$ cycles, all the OCPC packet routing algorithms mentioned above can be used on a $1 \times n$ AROB without any change in the asymptotic run times.

Lemma 8 *Any h-relation can be routed on a $1 \times n$ AROB in $\widetilde{O}(h + \log\log n)$ cycles.* ∎

An interesting question is if there exists a better algorithm for the one-dimensional AROB. Consider the simple problem where there are two processors (on an n-node machine) that want to send a message to the same processor. If both of them attempt to transmit at the same time, then no message will ever reach the destination on the OCPC. But a randomized approach can break the deadlock. For instance, each processor can attempt a transmission with some probability less than one (say 1/2). Such an algorithm can be seen to terminate in $\widetilde{O}(\log n)$ time. The same problem can be solved in three cycles on a one dimensional AROB.

Rajasekaran and Sahni [28] have given a better algorithm for the AROB that runs in $\widetilde{O}(h)$ cycles. Consider a $1 \times n$ AROB $\mathcal{L}$. A special case where $h = O(1)$ has already been considered in Lemma 6. It is beneficial to look at some more special cases of routing before dealing with the general case.

Problem 1 [$\mathsf{Load\_Balancing}(n, k, N)$]: In an n-node network there are no more than k packets at any node such that N is the total number of packets. The problem is to do a load balancing, i.e., to rearrange the packets such that each node has $\leq \lceil \frac{N}{n} \rceil$ packets.

Lemma 9 $\mathsf{Load\_Balancing}(n, k, N)$ *can be solved in $O(k)$ cycles on a $1 \times n$ AROB. We can also solve the same problem in $O(k)$ cycles on a $\sqrt{n} \times \sqrt{n}$ AROB.*

Proof. The proofs for 1D and 2D cases are similar. Consider a 1D AROB. The processors order their packets from 1 to k. Perform a prefix sums computation for the first packets of all the processors (using Lemma 1) and as a result obtain a unique address for each such packet. Route these first packets. Since prefix takes $O(1)$ cycles, the routing takes the same time also. Similarly process the second packets, the third packets, and so on. If q is the number of first packets, then, the second packets will be routed to nodes starting from $q + 1$, and so on. The whole algorithm takes $O(k)$ cycles. ∎

Problem 2 [$\mathsf{Route}(\ell, k, \mathcal{N})$]: There are no more than ℓ packets originating from any node of a network $\mathcal{N}$ and at most k packets are destined for any node. Route the packets.

Lemma 10 $\mathsf{Route}(\ell, k, \mathcal{N})$ *can be done on a $1 \times n$ AROB in $O(k\ell)$ cycles.*

Proof. We route the packets in *stages*. In any stage, a processor chooses one of its remaining packets (if any) and tries to send it. In cases of conflicts, the packet may not reach its destination. If the packet succeeds, the processor takes up the next packet; otherwise the processor tries to send the same packet again. A packet can meet with failure in at most $k-1$ cycles. Therefore, every processor will be able to transmit all of its ℓ packets in ℓk cycles or less. ∎

Theorem 1 *Any h-relation can be routed in $\widetilde{O}(h)$ time on a $1 \times n$ AROB.*

Proof. A basic step in [28]'s algorithm consists of each processor randomly choosing one of its packets and sending it. There is some constant probability of a processor succeeding in one basic step. The algorithm runs in *stages*. A load balancing is performed at the end of every stage. Let k_i be the maximum number of packets in any node at the beginning of stage i, with $k_1 = h$. Also let N_i be the number of packets that have not yet been routed at the beginning of stage i, where $N_1 \leq nh$. The value of k_i decreases by a constant factor with high probability at the end of the stage. Thus there will be $\widetilde{O}(\log h)$ stages of routing. After $\widetilde{O}(\log h)$ stages of routing, there will be $\widetilde{O}(n)$ packets left. These packets can be routed by load balancing followed by an application of Lemma 6. A detailed description of the algorithm follows.

Algorithm **Route**

$i = 1$; $k_i = h$; $N_i = nh$;
repeat

for $j = 1$ *to* $1.5k_i$ *do*

Step 1. Let π be any processor. Processor π picks one of the remaining packets uniformly at random and sends it in the bus. There is a probability of $\frac{1}{q}$ that any packet from π will be sent, q being the number of remaining packets. Depending on conflicts, the packet sent may or may not successfully reach its destination.

Step 2. Reverse the direction of the above routing to inform the processors as to whether or not their packets have been successfully sent.

Perform a load balancing operation to ensure that each processor gets very nearly the same number of remaining packets; compute k_{i+1} and N_{i+1}; $i = i + 1$;

until $N_i \leq cn$ (c being a constant).

Analysis. The analysis consists of a proof by induction that at the beginning of stage $i + 1$, for almost all the processors, the number of remaining packets destined for any processor is $\leq \frac{h}{2^i}$ and also that k_{i+1} is no more than $\frac{h}{2^i}$. The base case is easy.

Let the hypothesis be true for all the stages up to i. We'll prove it for stage $i + 1$. Let π be any processor. If at any time during stage i the number (call it m) of packets yet to be routed to π falls below $\frac{k_i}{2}$, we are fine. Therefore assume that m is at most k_i and at least $\frac{k_i}{2}$.

At any time during stage i, the probability that at least one of the packets destined for π will be sent is $\geq 1 - \frac{1}{\sqrt{e}}$. This is proved as follows. Let the number of packets left at processor q and destined for π be x_q, $q = 1, 2, \ldots, n$. Clearly, $\sum_{q=1}^{n} x_q = m$. Probability that one of the x_q packets at q will be sent is $\geq \frac{x_q}{k_i}$. Thus, probability that none of these x_q packets will be sent is $\leq (1 - \frac{x_q}{k_i}) \leq e^{-x_q/k_i}$, using the fact that $(1 - \frac{1}{y})^y \leq \frac{1}{e}$ for any y. Therefore, the probability that none of the m packets destined for π will be sent is $\leq e^{-(x_1+x_2+\ldots+x_n)/k_i} = e^{-m/k_i}$. As a result, the probability that at least one packet destined for π will be sent is $\geq 1 - e^{-m/k_i} \geq 1 - \frac{1}{\sqrt{e}}$.

Since at least one of these m packets has been sent, it follows that one packet will successfully reach π. Call this event (of sending at least one packet destined for π) a *success*.

The expected number of successes in $1.5k_i$ steps of routing is $= 1.5k_i(1-\frac{1}{\sqrt{e}})$. Using Chernoff bounds, this number is $\geq \frac{k_i}{2}$ with probability $\geq (1 - \frac{1}{k_i^4})$. In other words, the expected number of processors that have not received $\geq \frac{k_i}{2}$ packets in this stage is $\leq \frac{n}{k_i^4}$. The expected total number of such packets is $\leq \frac{n}{k_i^3}$. It can be shown that the number of such packets is $\widetilde{O}(\frac{n}{k_i^3})$.

In summary, for almost all the processors, the number of remaining packets destined for any processor at the end of stage i is $\leq \frac{k_i}{2}$. Since the total number of packets that can not be delivered (as expected) is $\widetilde{O}(\frac{n}{k_i^3})$, after load balancing, k_{i+1} will be $\leq \frac{k_i}{2}$.

Since the value of k_i decreases by a factor of 2 in every stage, there will be $\widetilde{O}(\log h)$ stages. The number of packets that cannot be routed in stage i is $\widetilde{O}(\frac{n}{k_i^3})$. The sum of this over all the stages is $\sum_{i=1}^{d \log h} \widetilde{O}(\frac{n}{k_i^3}) = \sum_{i=1}^{d \log h} \widetilde{O}\left(\frac{n}{(h/2^i)^3}\right) = \widetilde{O}(n)$.

Therefore, after $\widetilde{O}(\log h)$ stages, the number of packets remaining to be routed is $\widetilde{O}(n)$. These packets can be routed using a load balancing (in $O(1)$ cycles) followed by an application of Lemma 10. The additional time needed is $O(h)$ cycles.

The time spent in each stage is the sum of routing time and load balancing time. Routing takes $O(k_i)$ cycles and load balancing also takes $O(k_i)$ cycles (c.f. Lemma 9). Thus, the total time spent in all the stages is $\sum_{i=1}^{d \log h} O(k_i) = \sum_{i=1}^{d \log h} \widetilde{O}(\frac{h}{2^i}) = \widetilde{O}(h)$. ∎

9.3.2 2D Packet Routing

In this section we consider the problem of routing h-relations on a 2D AROB. Rao and Tsantilas [29] have given an OCPC algorithm that will route any h-relation in $\widetilde{O}(h + \frac{\log n}{\log \log n})$ time. It is not clear if the algorithm of [8] can be extended to get a run time of $\widetilde{O}(h + \log \log n)$ for the 2D OCPC.

The algorithm of [28] can route an arbitrary h-relation on a $\sqrt{n} \times \sqrt{n}$ AROB in time $\widetilde{O}(h + \log\log n)$. The special case of $h = O(1)$ is easy to solve using an analogue of [29]'s algorithm.

Lemma 11 *$O(1)$-relations can be routed in $\widetilde{O}(\log\log n)$ cycles on a 2D AROB.* ∎

Theorem 2 *Any h-relation can be routed in $\widetilde{O}(h + \log\log n)$ cycles on a $\sqrt{n} \times \sqrt{n}$ AROB.*

Proof. The algorithm is similar to that of **Route** given for the 1D AROB. The changes are:

(a) In step 1, 'sends it in the bus' is replaced by:

Step 1.1. Each processor selects a random row index. The processors in each column try to route their selected packets to the randomly selected rows using the algorithm of Lemma 5. Since the routes in each column do not necessarily define a permutation, only some of the packets will get through.

Step 1.2. Packets that are successful in step 1.1 are routed along rows to their destination columns using the permutation routing scheme of Lemma 5. Here also since the destination columns in a row may not form a partial permutation, some packets may not get through.

Step 1.3. The destination rows of the packets in each column that got through in step 1.2 form a partial permutation. The partial permutation in each column may be realized using the method of Lemma 5.

(b) The load balancing step is done using Lemma 9 as applied to a 2D AROB.

Note that steps 1.1, 1.2 and 1.3 are similar to the three phase algorithm of [33].

This algorithm can be analyzed along the same lines as that of **Route**. The value of k_i reduces by a factor of 2 per iteration of the *for* loop with high probability. The total number of packets that remain following the adapted algorithm **Route** is $\widetilde{O}(n)$. These packets can be balanced at the end (in $O(h)$ cycles) and routed in $\widetilde{O}(h+\log\log n)$ cycles (c.f. Lemma 11). The total number of cycles spent in routing in the adapted algorithm **Route** is $\widetilde{O}(h + \log\log n)$. So the overall number of cycles is $\widetilde{O}(h + \log\log n)$. ∎

Deterministic Routing. A deterministic routing algorithm for the 2D AROB has been given in [27].

Lemma 12 *Any partial permutation can be routed in $O(\log n)$ cycles on a $\sqrt{n} \times \sqrt{n}$ AROB.*

Proof. Start by sorting the packets into ascending order of their destinations. To do this, the destinations are mapped into a single number using the row major mapping scheme. The sorted packets are in processors $1, 2, \ldots$ (in row major order). We can perform this sorting in $O(\log n)$ cycles using a binary radix sort. Lemma 9 is used to accomplish the sort on each bit. Note that the load balancing scheme of Lemma 9 is equivalent to a stable sort of bits. After the sort, no two packets in the same column have the same row as their destination since there are $\sqrt{n}-1$ packets between them. Now use Lemma 5 to route packets in each column to their destination rows. Following this, no two packets in the same row have the same column as their destination. So Lemma 5 may be used again to route packets in the same row to their destination columns. The work done following the sort takes $O(1)$ cycles. So, the overall number of cycles is $O(\log n)$. ∎

Lemma 13 *Any h-relation can be routed in $O(h \log n)$ cycles on a 2D AROB of size $\sqrt{n} \times \sqrt{n}$.*

Proof. Here also we sort the packets into nondescending order of their destinations. Following the sort, the packets are in the first few processors (in row major order), h packets to a processor. This is accomplished using a binary radix sort on the row major index of the packet's destination processor.

In order to sort on bit k of this index, we first concentrate the packets with bit k equal to 0, h packets to a processor and then concentrate those with bit k equal to 1. The process for each bit value is similar. Consider the case of packets with bit k equal to 0. Call these packets ***selected packets.*** A processor may have up to h selected packets.

We combine the selected packets in each processor to form a 'superpacket' of size at most h. The superpackets are compacted into processors $1, 2, \ldots$ (in row major order) using the 2D compaction algorithm of [22]. Since the superpacket size is $O(h)$, this takes $O(h)$ cycles. The superpackets are now decomposed into the original packets. The original packets are to be further compacted so that we have h packets to a processor. Each packet in a processor is assigned a level number corresponding to its order in the processor. Level numbers are in the range 1 to h.

Prefix sums for the level i packets, $1 \le i \le h$ can be computed using the 2D prefix sum algorithm given in section 9.2. The rank $r(i,j)$ of a level i packet in processor j is $\sum_{k=1}^{h} ps(k, j-1) + (i-1)$, where $ps(k, j-1)$ is the prefix sum of the level k packet in processor $j-1$. The processor $P(i,j)$ to which this packet is to be routed is $\lfloor r(i,j)/h \rfloor$. Furthermore, this packet will be the $round(i,j) = r(i,j) \bmod h + 1$-th packet in this processor. Since the number of packets in each row is at most $h\sqrt{n}$, no two packets (i,j) and (k,l), where j and l are processors in the same row, have $column(P(i,j)) = column(P(k,l))$ and $(row(P(i,j)) \neq row(P(k,l))$ or $round(i,j) = round(k,l))$. As a result, the compaction may be completed as follows.

1. Perform h rounds of row permutation routing on each row. In round k, route the packets (i,j) with $round(i,j) = k$ to the processor in column $column(P(i,j))$.

2. Perform h rounds of column permutation routing. In round k, route the packets (i, j) with $round(i, j) = k$ to the processor in row $row(P(i, j))$.

The radix sort described above takes $O(h \log n)$ cycles. To complete the h-relation routing we perform h rounds of column and row permutations. In round i, the level i packets in each column are first routed to the correct row using a column permutation. There can be no collision as for a collision the number of packets destined to the same row needs to be $> h\sqrt{n}$. Next, the level i packets are routed to the correct column using row permutations. Again, collisions are not possible. ∎

In [27] algorithms have also been given for routing various classes of 1-relations in $O(1)$ cycles.

9.4 SORTING ON THE AROB

In this section we consider both general and integer sorting.

9.4.1 General Sorting

Several $O(1)$ time algorithms have been discovered for sorting on the reconfigurable mesh (e.g., [11], [19], [15]). The same algorithm runs on the AROB preserving the run time.

Lemma 14 *Sorting of n numbers can be performed in $O(1)$ time on a reconfigurable mesh of size $n \times n$.* ∎

Rajasekaran and Sahni [28] have presented a better general sorting algorithm. This algorithm sorts n numbers in an AROB of size $n^\epsilon \times n$ for any constant $\epsilon > 0$. The run time is $O(1)$ cycles and hence the algorithm is optimal in view of the following lower bound that was originally proved for the parallel comparison tree model.

Lemma 15 *Sorting of n numbers needs $\Omega\left(\frac{\log n}{\log(1+\frac{p}{n})}\right)$ time using p parallel comparison tree processors [4].* ∎

This lower bound indicates that if sorting of n numbers has to be done in $O(1)$ time, then $\Omega(n^{1+\epsilon})$ processors will have to be used, for some constant $\epsilon > 0$. Since a parallel comparison tree can simulate an AROB step per step, the same lower bound applies to the AROB as well.

We now give a summary of the general sorting algorithm of [28]. Let $k_1, k_2, \ldots, k_n$ be the n given numbers. Think of these numbers as forming a matrix M with $r = n^{2/3}$ rows and $s = n^{1/3}$ columns. They employ the column sort algorithm of Leighton [13]. The same result can be obtained using the recent algorithm ((l, m)-merge sort) of Rajasekaran [23]. There are 7 steps in the algorithm.

1. Sort the columns of M in increasing order;

2. Transpose the matrix preserving the dimension as $r \times s$. In other words, pick the elements in column major order and fill the rows in row major order;
3. Sort the columns in increasing order;
4. Apply to the numbers the reverse of the permutation used in step 2;
5. Sort the columns such that adjacent columns are sorted in reverse order;
6. Apply two steps of odd-even transposition sort to the rows. In particular, perform a comparison-exchange between processors $2i + 1$ and $2i$, for $i = 0, 1, \ldots$ followed by a comparison-exchange between processors $2i$ and $2i + 1$, for $i = 1, 2, \ldots$; and
7. Sort each column in increasing order. It can be shown that the numbers will now be sorted in column major order.

AROB Implementation. Store the n given numbers in the first row of the $n^\epsilon \times n$ AROB one key per processor. Each key will always know which row and which column of the matrix M it belongs to. Whenever we need to sort the columns, we will make sure that the numbers belonging to the same column will be found in successive processors.

Note that on a $1 \times n$ AROB any permutation can be routed in $O(1)$ cycles. As a result, steps 2 and 4 can be completed in $O(1)$ cycles. We can complete step 6 in $O(1)$ cycles as follows. Rearrange the numbers such that elements in the same row appear in successive processors followed by an application of two steps of the odd-even transposition sort. After this, move the keys to where they came from.

Details on implementing steps 1, 3, 5, and 7 follow. Assume that we have an AROB of size $n^{2/3} \times n$. We will subsequently indicate how to reduce the size to $n^\epsilon \times n$ for any $\epsilon > 0$.

Partition the AROB into $n^{1/3}$ parts each of size $n^{2/3} \times n^{2/3}$. Each part corresponds to a column of M. Distribute the n given numbers such that the first column of M is in the first $n^{2/3}$ processors of row 1; the second column is in the next $n^{2/3}$ processors of the first row; and so on. Sort the numbers in each part (i.e., each column of M) using Lemma 14 in $O(1)$ cycles. Clearly, steps 1, 3, 5, and 7 of column-sort can be performed in $O(1)$ cycles. Thus it follows that n numbers can be sorted in $O(1)$ cycles on an AROB of size $n^{2/3} \times n$.

The AROB size can be reduced to $n^{4/9} \times n$ as follows: Use the column sort with $r = n^{2/3}$ and $s = n^{1/3}$. In steps 1, 3, 5, and 7, each part of $n^{2/3}$ numbers will be sorted using an AROB of size $n^{4/9} \times n^{2/3}$. This is accomplished using the AROB algorithm above.

We can reduce the size to $n^{8/27} \times n$, $n^{16/81} \times n$, and so on in a similar fashion. Thus we get the following Theorem:

Theorem 3 *We can sort n numbers in $O(1)$ cycles using an AROB of size $n^\epsilon \times n$, where ϵ is any constant > 0.* ∎

9.4.2 Integer Sorting

An integer sorting algorithm that runs on a $1 \times n$ AROB and can sort n k-bit numbers in $O(k)$ cycles has also been given in [28]. Even on the CRCW PRAM model, such an algorithm is not possible in view of the lower bound result of Beame and Hastad [2].

We now give a summary of this algorithm. This algorithm makes use of the idea of *radix sorting* and Lemmas 1 and 5. The idea of radix sorting can be summarized as follows. If n numbers in the range $[0, R]$ can be stable sorted using P processors in time T, then we can also stable sort n numbers in the range $[0, R^c]$ in $O(T)$ time using P processors, c being any constant. A sorting algorithm is *stable* if equal keys remain in the same relative order in the output as they were in the input.

The algorithm of [28] has k *stages* where stage i involves sorting the numbers with respect to their ith LSBs. In the first stage the numbers are sorted with respect to their LSBs, and so on.

Each stage can be completed in $O(1)$ cycles as follows. Realize that a stage is nothing but sorting n 1-bit numbers. We can perform two prefix sums operation one for the zeros and one for the ones in the input. Each processor can determine the position of its data in the sorted list using the prefix sums. Now, permute the data according to sorted order. Since the prefix sums as well as the permutation operations take $O(1)$ cycles each, every stage takes $O(1)$ cycles as well (c.f. Lemmas 1 and 5).

We arrive at the following Theorem.

Theorem 4 *We can sort n k-bit numbers in $O(k)$ cycles on a $1 \times n$ AROB.* ∎

9.5 SELECTION

Given a sequence of n numbers $k_1, k_2, \ldots, k_n$ and an $i \leq n$, the problem of selection is to identify the ith smallest of the n numbers. Many optimal sequential algorithms are known for selection (see e.g., [10]). Elegant randomized algorithms are also known for sequential selection [6]. Parallel algorithms have been developed on various models of computing as well (see e.g., [25]). Most of the parallel selection algorithms (be they deterministic or randomized) make use of sampling.

Pan has given a selection algorithm for the 1D AROB [21].

Lemma 16 *Selection problem on an input of n elements can be solved in an expected $O(\frac{n}{p} \log n)$ cycles on an AROB of size $1 \times p$. The worst case run time is $O(n)$ cycles.*

Proof. Pan's algorithm picks a random element as the 'pivot' element and partitions the input into two around the pivot. It decides which partition the ith element is in. The algorithm throws away the irrelevant partition and performs an appropriate selection in the relevant partition.

Each partitioning step can be expected to reduce the number of surviving keys by a constant factor. Thus the expected run time is $O(\frac{n}{p}\log n)$. ∎

Rajasekaran and Sahni's algorithm [28] runs in $\widetilde{O}(1)$ cycles on an AROB of size $\sqrt{n} \times \sqrt{n}$. To begin with there is a number input at each processor. The basic idea behind this algorithm is the following. 1) Pick a random sample S of size $q = o(n)$ from the input; 2) Identify two elements ℓ_1 and ℓ_2 from the sample whose ranks in S are $i\frac{q}{n} - \delta$ and $i\frac{q}{n} + \delta$ for some suitable δ. It can be shown that these elements 'bracket' the element to be selected with high probability; 3) Eliminate input keys whose values are outside the range $[\ell_1, \ell_2]$; 4) Perform an appropriate selection from out of the remaining keys.

A Sampling Lemma. Let X be a sequence of n numbers from a linear order and let $S = \{k_1, k_2, \ldots, k_s\}$ be a random sample from X. Let the sorted order of the sample be $k'_1, k'_2, \ldots, k'_s$. If r_i is the rank of k'_i in X, the the following Lemma can be used to obtain a high probability confidence interval for r_i. (We define the rank of any element k in X as one plus the number of elements $< k$ in X.)

Lemma 17 *For every* $\alpha > 0$, *Prob.* $\left(|r_i - i\frac{n}{s}| > \sqrt{3\alpha}\frac{n}{\sqrt{s}}\sqrt{\log n}\right) < n^{-\alpha}$. ∎

For a proof of this Lemma the reader is referred to [26]. The selection algorithm of [28] makes use of this sampling Lemma.

The algorithm of [28] selects the ith smallest key from the input. Let N_j denote the number of *alive* keys at the beginning of the jth iteration of the *repeat* loop. Each key is alive at the beginning, i.e., $N_1 = n$. Let s_j denote the number of alive keys that do not get deleted in the jth iteration. Also let S_j denote the set of sample keys employed in the jth iteration.

Algorithm **Select**

$j = 1$; $N_j = n$; $i_j = i$;
repeat

1. An alive key gets included in the sample S_j with probability $\frac{1}{N_j^{0.6}}$. $|S_j|$ will be $\widetilde{\Theta}(N_j^{0.4})$.
2. Collect the sample keys in the first row. The algorithm of Lemma 7 is used to do this. With high probability only the first row will be nonempty.
3. Sort the numbers in the first row using Theorem 3. Processors which do not get a sample key will have a key valued ∞. Let q_j be the number of keys in the sample. Choose keys ℓ_j^1 and ℓ_j^2 from S_j with ranks $\left\lceil \frac{i_j q_j}{N_j} \right\rceil - d\sqrt{q_j \log N_j}$ and $\left\lceil \frac{i_j q_j}{N_j} \right\rceil + d\sqrt{q_j \log N_j}$, respectively, d being a constant $> \sqrt{3\alpha}$. (α is as in Lemma 17.) This takes $O(1)$ time.

4. Keys that fall outside the range $[\ell_j^1, \ell_j^2]$ are killed. Count the number, s_j, of surviving keys. It can be shown that this number is $\widetilde{O}(N_j^{0.8}\sqrt{\log N_j})$.

5. Count the number del_j of keys $< \ell_j^1$ that got killed in step 4. If $del_j \geq i_j$ or $i_j > del_j + s_j$ (i.e., if the key to be selected is not one of the remaining keys), start all over again (i.e., go to step 1 with $j = 1$, $N_j = n$, and $i_j = i$).

   Set $i_{j+1} = i_j - del_j$; $N_{j+1} = s_j$; and $j = j + 1$.

until $N_j \leq \sqrt{n}$

6) As in step 2, compact the surviving keys in the first row Sort the first row and output the i_jth smallest key from out of the remaining keys.

Theorem 5 *Algorithm* **Select** *runs in* $\widetilde{O}(1)$ *cycles on a* $\sqrt{n} \times \sqrt{n}$ *AROB.*

Proof. Step 1 takes $O(1)$ cycles. The number of keys in the sample S_j has the distribution $B(N_j, \frac{1}{N_j^{0.6}})$. Applying Chernoff bounds, $|S_j|$ is $\widetilde{\Theta}(N_j^{0.4})$. Step 2 takes $O(1)$ cycles (c.f. Lemma 7). Using Theorem 3, Step 3 runs in $O(1)$ cycles. In steps 4 and 5 counting can be done in $O(1)$ cycles as well (cf. Lemma 1). In step 6, compaction and sorting take $O(1)$ cycles each.

The Theorem will be proven if we can show that the number of times the *repeat* loop will be executed is $\widetilde{O}(1)$. The number of keys alive at the end of jth iteration is $\widetilde{O}(N_j^{0.8}\sqrt{\log N_j})$ in accordance with Lemma 17. Therefore, the number of alive keys at the end of the first k iterations is $\widetilde{O}(n^{0.8^k}(\log n)^3)$. As a result, with high probability, there will be ≤ 4 iterations of the *repeat* loop. ∎

9.6 CONCLUSIONS

In this chapter we have presented efficient algorithms for packet routing, sorting, and selection on the AROB. We have considered both integer sorting and general sorting problems. Our general sorting and selection algorithms are optimal. So is the h-relations algorithm for the one-dimensional AROB. An interesting open problem is if there exists an $O(h)$ routing algorithm (deterministic or randomized) for the 2D AROB. Also, can our integer sorting algorithm be improved?

Acknowledgments

Research of S. Rajasekaran was partially supported by an NSF Grant CCR-9596065 and an EPA Grant R-825-293-01-0. Research of S. Sahni was supported in part by the Army Research Office under Grant DAA H04-95-1-0111.

References

[1] R.J. Anderson and G.L. Miller, Optical Communication for Pointer Based Algorithms, Technical Report CRI-88-14, Computer Science Department, University of Southern California, 1988.

[2] P. Beame and J. Hastad, Optimal Bounds for Decision Problems on the CRCW PRAM, Journal of the ACM, 36(3), 1989, pp. 643-670.

[3] Y. Ben-Asher, D. Peleg, R. Ramaswami, and A. Schuster, The Power of Reconfiguration, Journal of Parallel and Distributed Computing, 1991, pp. 139-153.

[4] R. Bopanna, A Lower Bound for Sorting on the Parallel Comparison Tree, Information Processing Letters, 1989.

[5] H. Chernoff, A Measure of Asymptotic Efficiency for Tests of a Hypothesis Based on the Sum of Observations, Annals of Mathematical Statistics 2, 1952, pp. 241-256.

[6] R.W. Floyd, and R.L. Rivest, Expected Time Bounds for Selection, Communications of the ACM, Vol. 18, No.3, 1975, pp. 165-172.

[7] M. Geréb-Graus and T. Tsantilas, Efficient Optical Communication in Parallel Computers, Symposium on Parallel Algorithms and Architectures, 1992, pp. 41-48.

[8] L. Goldberg, M. Jerrum, T. Leighton, and S. Rao, A Doubly-Logarithmic Communication Algorithm for the Completely Connected Optical Communication Parallel Computer, Proc. Symposium on Parallel Algorithms and Architectures, 1993, pp.300-309.

[9] E. Hao, P.D. McKenzie and Q.F. Stout, Selection on the Reconfigurable Mesh, Proc. Frontiers of Massively Parallel Computation, 1992, pp. 38-45.

[10] E. Horowitz, S. Sahni, and S. Rajasekaran, *Computer Algorithms*, W.H. Freeman Press, 1998.

[11] J. Jang and V.K. Prasanna, An Optimal Sorting Algorithm on Reconfigurable Mesh, Proc. International Parallel Processing Symposium, 1992, pp. 130-137.

[12] J. Jenq and S. Sahni, Reconfigurable Mesh Algorithms for Image Shrinking, Expanding, Clustering, and Template Matching, Proc. International Parallel Processing Symposium, 1991, pp. 208-215.

[13] T. Leighton, Tight Bounds on the Complexity of Parallel Sorting, IEEE Transactions on Computers, C-34(4), 1985, pp. 344-354.

[14] R. Lin, S. Olariu, Reconfigurable Buses with Shift Switching: Concepts and Applications, IEEE Transactions on Parallel and Distributed Systems, Volume 6(1), 1995, pp. 93-102.

[15] R. Lin, S. Olariu, J. L. Schwine, and J. Zhang, Sorting in $O(1)$ Time on a Reconfigurable Mesh of Size $N \times N$, Proc. European Workshop on Parallel Computing, 1992, pp. 16-27.

[16] R. G. Melhem, D. M. Chiarulli, and S. P. Levitan, Space Multiplexing of Waveguides in Optically Interconnected Multiprocessor Systems, Computer Journal, 32(4), 1989, pp. 362-369.

[17] R. Miller, V.K. Prasanna-Kumar, D. Reisis and Q.F. Stout, Meshes with Reconfigurable Buses, IEEE Transactions on Computers, vol. 42, 1993, pp. 678-692.

[18] D. Nassimi and S. Sahni, A Self-Routing Benes Network and Parallel Permutation Algorithms, IEEE Transactions on Computers, C-30(5), 1981, pp. 332-340.

[19] M. Nigam and S. Sahni, Sorting n Numbers on $n \times n$ Reconfigurable Meshes with Buses, Proc. International Parallel Processing Symposium, 1993, pp. 174-181.

[20] S. Olariu, J.L. Schwing and J. Zhang, Integer Problems on Reconfigurable Meshes, with Applications, Proc. 1991 Allerton Conference, 4, 1991, pp. 821-830.

[21] Y. Pan, Order Statistics on Optically Interconnected Multiprocessor Systems, Proc. First International Workshop on Massively Parallel Processing Using Optical Interconnections, 1994, pp. 162-169.

[22] S. Pavel and S.G. Akl, Matrix Operations using Arrays with Reconfigurable Optical Buses, manuscript, 1995.

[23] S. Rajasekaran, A Framework For Simple Sorting Algorithms On Parallel Disk Systems, *Proc. 10th Annual ACM Symposium on Parallel Algorithms and Architectures*, 1998.

[24] S. Rajasekaran, Meshes with Fixed and Reconfigurable Buses: Packet Routing, Sorting and Selection, Proc. First Annual European Symposium on Algorithms, Springer-Verlag Lecture Notes in Computer Science 726, 1993, pp. 309-320.

[25] S. Rajasekaran, Sorting and Selection on Interconnection Networks, to appear in Proc. *DIMACS Workshop on Interconnection Networks and Mapping and Scheduling Parallel Computation*, 1995.

[26] S. Rajasekaran and J.H. Reif, Derivation of Randomized Sorting and Selection Algorithms, in *Parallel Algorithm Derivation and Program Transformation*, Edited by R. Paige, J.H. Reif, and R. Wachter, Kluwer Academic Publishers, 1993, pp. 187-205.

[27] S. Rajasekaran and S. Sahni, Deterministic Routing on the Array with Reconfigurable Optical Buses, *Parallel Processing Letters* 7(3), 1997, pp. 219-224.

[28] S. Rajasekaran and S. Sahni, Sorting, Selection, and Routing on the Array with Reconfigurable Optical Buses, *IEEE Transactions on Parallel and Distributed Systems* 8(11), pp. 1123-1131, 1997.

[29] S. Rao and T. Tsantilas, Optical Interprocessor Communication Protocols, Proc. Workshop on Massively Parallel Processing Using Optical Interconnections, 1994, pp. 266-274.

[30] S. Sahni, Data Manipulation on the Distributed Memory Bus Computer, to appear in Parallel Processing Letters, 1995.

[31] R.K. Thiruchelvan, J.L. Trahan, and R. Vaidyanathan, On the Power of Segmenting and Fusing Buses, Proc. International Parallel Processing Symposium, 1993, pp. 79-83.

[32] L.G. Valiant, General Purpose Parallel Architectures, in Handbook of Theoretical Computer Science: Vol. A (J. van Leeuwen, ed.), North Holland, 1990.

[33] L.G. Valiant and G.J. Brebner, Universal Schemes for Parallel Communication, Proc. 13th Annual ACM Symposium on Theory of Computing, 1981, pp. 263-277.

10 COMPUTING THE HOUGH TRANSFORM ON ARRAYS WITH RECONFIGURABLE OPTICAL BUSES

Sandy D. Pavel and Selim G. Akl

Department of Computing and Information Science
Queen's University
Kingston, Ontario, K7L 3N6
Canada

akl@qucis.queensu.ca

Abstract: This chapter examines the possibility of implementing the Hough transform for line and circle detection on the array with reconfigurable optical buses (AROB). The AROB combines some of the advantages and characteristics of reconfigurable meshes and meshes with optical pipelined buses. It is shown that the Hough transform for line and circle detection in an $N \times N$ image can be implemented optimally in a constant number of steps. The costs of the two algorithms are $O(N^2p)$ and $O(N^2p^2)$, respectively, where p is the magnitude of one dimension in the parameter space. These values are optimal with respect to the corresponding time complexities of the best known sequential algorithms.

10.1 INTRODUCTION

The Hough Transform (HT) was introduced as a method of detecting the shape of object boundaries in binary images [14]. In order to achieve this, a number of parameters are associated to the patterns of interest. The main objective of the HT is to transform the spatially-extended patterns, represented by the edge pixels in the image, so that they produce spatially-compact features in the associated parameter space. The global shape detection problem is converted into a local, less difficult, peak detection problem.

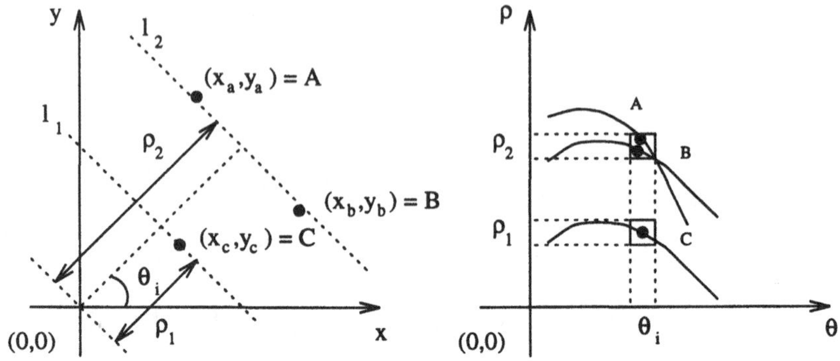

Figure 10.1 The Hough transform. (a) The (ρ, θ) parameterization of the image space. (b) The (ρ, θ) parameter space.

Suppose that we have an $N \times N$ digitized image which contains a set of n *edge pixels*, $n \leq N^2$, obtained after some preprocessing (edge detection) step. A pixel can be specified by its coordinates (x, y), $0 \leq x, y \leq N-1$, in the image plane. Herein, the origin of the system of coordinates $(0,0)$ is considered to be the bottom left corner of the image plane. Thus, the n edge pixels are represented by the set $\{(x_i, y_i) \mid 0 \leq i \leq n-1\}$ for which pixel (x_i, y_i) is "1". For *line detection*, the edge pixel (x_i, y_i), $0 \leq i \leq n-1$, is transformed into a sinusoidal curve in the (ρ, θ)-parameter plane which is described by:

$$\rho = \lfloor x_i \cos\theta + y_i \sin\theta \rfloor. \tag{10.1}$$

The pair (ρ, θ) represents a parameterization of a line in the image space where ρ is the perpendicular distance from the origin to that line and θ is the slope of the perpendicular from the origin to that line. Figure 10.1.(a) shows two *digitally collinear* pixels, A and B, that is, pixels which are within some ρ_{res} distance from some line l_2. Parameter ρ_{res} is called the *resolution* of the problem. In the case presented in Fig. 10.1.(a) the two supporting lines l_1 and l_2 of the three pixels in the image space have the same angle θ_i and two different values ρ_1 and ρ_2 for the normal distances. Their possible contribution in the parameter space is represented in Fig. 10.1.(b).

In the image space, the normal distance ρ satisfies $\rho \leq \sqrt{2}N$. If p values are used to represent ρ, then $p \leq \sqrt{2}N$. The slope θ of the perpendicular from the origin is in the interval $[0, \pi]$. Due to the limited resolution and for simplicity, the number of distinct values for the θ parameter is also considered to be p, with $p \leq \sqrt{2}N$. Thus, the size of the (ρ, θ)-space is $p \times p$. Assume that the values of parameters θ and ρ are $\theta_0, \theta_1, \ldots, \theta_{p-1}$ and $\rho_0, \rho_1, \ldots, \rho_{p-1}$, respectively. Associated with the parameter space is a $p \times p$ array, *count*$[\rho, \theta]$. For each edge pixel (x_i, y_i) and each θ_k the value of $\rho_{i,k}$ is computed using Equation 10.1 and the value corresponding to the particular entry of *count*$[\rho_{i,k}, \theta_k]$ is incremented. Finally, a threshold function can be applied to all the elements of *count*. The time complexity of the sequential algorithm for computing the *count* array is $O(pn)$. This is proportional to the number of edge pixels and the number of quantized values along the θ direction. It is easy to verify that the *circle detection* algorithm has $O(p^2n)$ time complexity, due to an increased parameter space (one more parameter), and in the case of *ellipse detection* the time complexity is $O(p^4n)$. (A detailed algorithm for circle detection is given in Section 10.4.)

So far we have assumed that the edge pixels are available and they are the only input data manipulated by these algorithms. Another possibility which occurs in practice is to have the entire $N \times N$ image as input to the algorithm. Since p is $O(N)$, the three time complexities for line, circle and ellipse detection become $O(N^3)$, $O(N^4)$ and $O(N^6)$, respectively. The same orders of magnitude are obtained when n, the number of edge pixels, is $O(N^2)$. The detection of non-analytic curves of arbitrary shape, scale and orientation could take even longer. Thus, the amount of computation required by the HT is proportional to the number of edge pixels n (or to the number of image pixels N^2) and also to the dimension of the parameter space. In general, the resolution of the parameter space is determined by the accuracy sought and the resources available. As the amount of computation can become very large, it is essential to be able to speed up the computation of the HT. Parallel processing was proposed as a possible solution. Parallel algorithms for line detection using the HT have been designed for different parallel architectures including the tree [13], the linear array [7], the pyramid [17], the mesh [5, 8, 18, 30, 39, 40] and the hypercube [29].

A number of algorithms for line detection using the HT of an $N \times N$ image on reconfigurable networks have also been proposed in the literature. The algorithms given in [16, 26] have $O(p \log(N/p))$ time complexity and use an $N \times N$ reconfigurable mesh. In [16] the time complexity is improved to $O((p/N) \log N)$ and $O((p/N^{1/2}) \log N)$ by increasing the two-dimensional array to $N \times N^2$ and $N^{1.5} \times N^{1.5}$, respectively. An $O(p \log p + \log N)$ time algorithm for the HT is given in [2] where an $O(N^2)$-processor architecture, called Reconfigurable Multi-Ring Network, is used. When the input is reduced to the n edge pixels, constant time algorithms are obtained in [19] and [28] using reconfigurable arrays of dimensions $p \times n \times n \times n$ and $n \times \log^2 n \times p \times p$, respectively. As n can be $O(N^2)$, the two architectures could use $O(pN^6)$ and $O(p^2N^2 \log^2 N)$ processors, respectively. In [20] three parallel HT algorithms are given for circle detection. The best of these algorithms runs in $O(N^2)$ time and uses an $N \times N$ mesh architecture.

We propose two algorithms for line and circle detection using the Hough transform. Both algorithms take a constant number of steps and are implemented on a computational model called the array with reconfigurable optical buses. The next section presents the model of computation together with some of the basic techniques implied by the use of optics for communication. In Section 10.3 the line detection algorithm is described. The circle detection algorithm is presented in Section 10.4. Section 10.5 concludes the chapter.

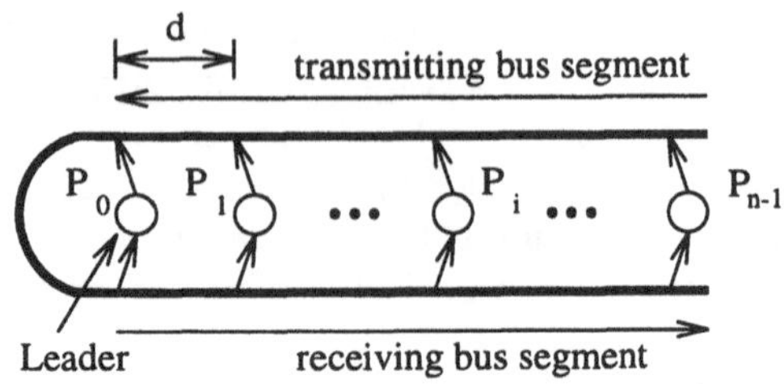

Figure 10.2 The linear array with pipelined optical buses APPB.

10.2 THE AROB MODEL

An important limitation of classical electrical buses concerns the exclusive access to the bus resources which limits throughput to a function of the end-to-end propagation time. Other fundamental constraints which bound bus interconnections in electronic systems in general are the limited bandwidth, capacitive loading, and cross-talk caused by mutual inductance. Optical communications have emerged as alternative solutions to these problems. Unlike electronic buses, on which signal propagation is bidirectional, optical channels are inherently directional and have predictable delay per unit length. As it is shown in [11, 24], this allows a pipeline of signals to be created by the synchronized directional coupling of each signal at specific locations along the channel. The possibility in optics to pipeline the transmission of signals through a channel provides an alternative to exclusive bus access. Using this kind of spatial parallelism the end-to-end propagation latency can be amortized over the number of parallel messages that are active at the same time on the bus.

10.2.1 Linear arrays with optical pipelined buses

Consider a linear array of n processors connected to an optical bus as in Fig. 10.2. Each processor is connected to the bus through two directional couplers. One coupler is used to write data on the bus and the other to read the data from the bus. Each processor sends data on the upper (*transmitting*) segment of the bus and reads messages from the lower (*receiving*) segment of the bus. The optical bus is constructed from three identical waveguides. One waveguide is used for message transmission, the *message* waveguide, and two are used for carrying address related information, the *reference* (Ref) and *select* (Sel) waveguides. During a write cycle, the data written by a processor into the bus propagate as indicated using arrows in Fig. 10.2, and may be read by any subsequent processor on the bus. Due to the directionality of the signal propagation and the predictable delay of the signal, the same bus may be used to transmit messages between other processors at the same time. Assume that each message is b bits long. Each bit is represented by an optical signal of width w seconds for a binary value of 1. The absence of such a signal represents a 0.

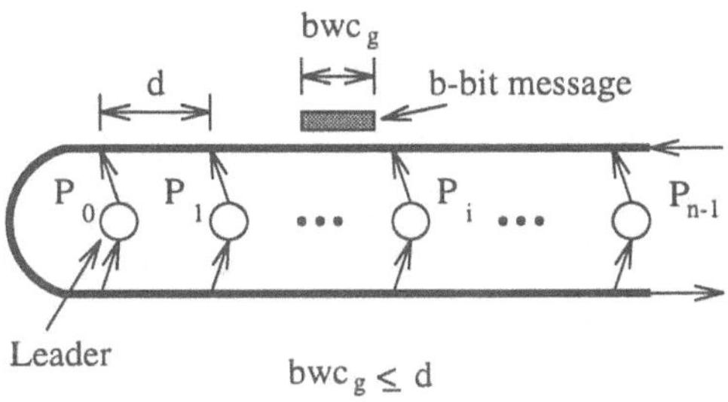

Figure 10.3 A condition to avoid message overlap.

Two conditions are essential for this kind of transmission: (1) all transmissions are synchronized and (2) the length of the optical path on the waveguide between any two adjacent processors, denoted by d in Fig. 10.3, is larger than or equal to bwc_g, where c_g is the velocity of light in the waveguide. These two conditions ensure that the signals corresponding to two different messages, which travel in a waveguide in the same direction at the same time, do not physically overlap at any point on the waveguide, that is, are space multiplexed. The *end-to-end propagation time* is $\sigma_b = 2n\tau$ seconds, where n is the number of processors in the linear array, and τ is the time taken for a message to traverse the optical distance d. This system is called the *linear array of processors with pipelined optical buses* (APPB), and its principles are introduced in [11, 24, 27].

Several approaches can be used to route messages in a linear APPB structure from one processor to another. In [11, 24] a *time waiting function* is introduced. Various message routing patterns can be realized using this function, most of which are suitable for SIMD and/or MIMD implementations. In [37] a time-division multiplexing scheme is given for communications in arrays with optical pipelined buses. This is used in [34] to implement part of the synchronous communications needed by the APPB. Two cases are considered: (1) message receiving, when the destination processor knows the index of the sender and (2) message sending, when the sender knows the destination address. It is shown that these communication schemes can be implemented in a constant number of bus cycles [34]. Arbitrary permutations can be performed using these message transmitting techniques.

The address information can also be encoded using the *coincident pulse technique* presented in [4, 21, 37, 38]. This technique can be used when the source processor knows the address (index) of the destination processor. For example, a processor can use the coincident pulse addressing mechanism in order to broadcast a value to all remaining processors [28]. Many other communication patterns are described in [4, 21, 38] using the same coincident pulse technique. The coincident pulse mechanism seems to avoid the traditional bottleneck which characterizes most of the on-line address decoding techniques [37].

A *bus cycle length* is defined as the time taken by the end-to-end propagation of messages, plus the time required to process a message at the source and

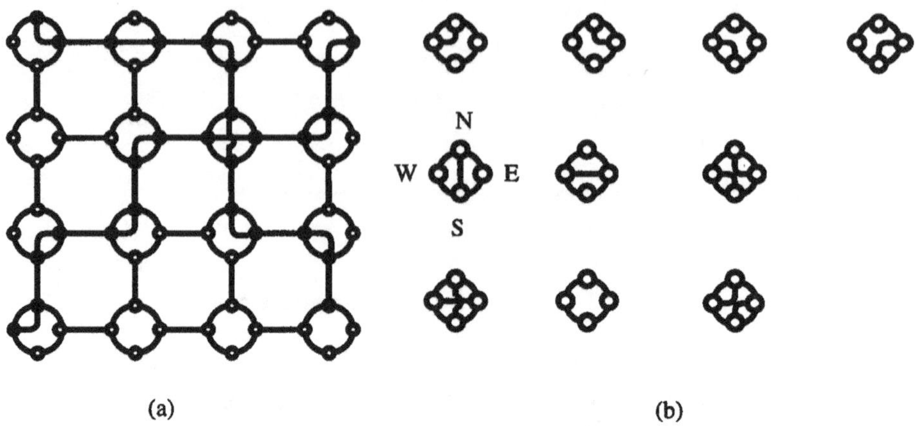

Figure 10.4 (a) Two-dimensional AROB; (b) Switch configurations.

destination processors. The latter operation includes message generation and detection, synchronization, buffering, etc. [11]. The message processing is done in parallel by all source processors, at the beginning of the bus cycle, and/or by all destination processors after the message propagation ends. The ratio, R, between the message processing time and the time for communication over distance d is estimated to be between 10 and 1000 [11]. In order for the message processing time to be of the same order of magnitude as the end-to-end propagation time σ_b, the number of processors should satisfy $n \geq R$. In this case, the bus cycle length can be considered to be $O(\sigma_b)$.

There are different factors that limit the size of an optical bus. All passive devices, as the passive couplers or optical fibers, exhibit loss. At the same time, each coupler—used by a processor to read a message from the waveguide—taps a percentage of the energy carried by the signal. One solution to these problems is to provide some means of amplitude restoration. The use of an active switch instead of a directional coupler could solve this problem [12], but it has the main disadvantage of introducing more delay. Otherwise, the number of devices connected to a bus must be restricted to an admissible value, that is, one which does not introduce errors. This second solution also reduces the value of the end-to-end propagation delay. Two-dimensional arrays with optical pipelined buses have been proposed in order to reduce the number of processors connected to a linear optical pipelined bus from $O(n)$ to $O(\sqrt{n})$ [10, 11, 36, 37].

10.2.2 The two-dimensional AROB

The two-dimensional processor array described here uses the basic architectural and functional structure of a classical reconfigurable network (RN) [3]. The communication system of this network is modified in order to allow the implementation of the pipelined optical communication mechanisms introduced in [11, 24] (and also presented in the previous subsection).

Consider n^2 processors connected in a two-dimensional reconfigurable structure as depicted in Fig. 10.4.(a). The switching system of a processor is able to connect the four ports (N, S, E, W) in one of the configurations shown in Fig. 10.4.(b). The interconnection network is assumed to use optical waveguides and optical switches for communication and reconfiguration, respectively. We call this structure the *array with reconfigurable optical buses* (AROB). The architectural and functional characteristics of the AROB are described in [1, 22, 23, 31, 32, 33, 34, 35].

It is shown in [31] that the local switch setting provides a mechanism for building arbitrary APPB-like linear structures on the AROB. A processor can be connected to up to four such buses at a time. When pipelined communications are sought, an algorithm to be implemented on the AROB is required to specify a rule in order to uniquely identify the *leader* processor of each linear array obtained by reconfiguration (similar to P_0 in Fig. 10.2). The algorithms to be simulated on the AROB are not necessarily required to specify to each processor the length of the linear buses to which it is connected or its indices relative to these buses. A method to determine the relative position of each processor in each linear bus to which it is connected, and implicitly the length of each bus, is given in [31]. Note that these values are needed in order to be able to implement the time division or coincident pulse techniques. The method of [31] is summarized in the next paragraph.

Consider the linear array depicted in Fig. 10.5.(a). For simplicity, the transmitting and receiving bus segments are not shown, and the message waveguide is entirely omitted. The leader processor is the only one in this structure which knows its index (namely, 0). Each processor, with the exception of P_0, introduces a unit delay on the receiving segment of the selection waveguide. This is followed by a bus cycle at the beginning of which P_0 sends a pair of pulses on the selection and reference waveguides. The pulse arrives at P_0 on the receiving segment at moment t_0. The difference between the arrival times of the selection and reference pulses is incremented with one time unit at each new processor. The selection pulse arrives at the jth processor with a delay of j time units relative to the reference pulse. Thus, the relative delay between the two pulses gives the index of each processor in the linear structure. This is the reversed process of the coincident pulse addressing technique. It starts with two synchronized pulses, as shown in Fig. 10.5.(b), and after a number of delays the unary-encoded information (that is, the index) is obtained at each processor, as shown in Fig. 10.5.(c). The unary-encoded information can be processed in constant time off-line at the end of the actual message transfer on the bus.

Let a *step* be either an internal operation or a communication step. It has been argued that counting the number of steps of an algorithm simulated on an array with optical pipelined buses allows for a better comparison between algorithms [9]. This method has the advantage that it abstracts from the details introduced by the technology-dependent parameters, such as σ_b, for example. Obviously, this is true only if the same assumptions regarding the bus cycle

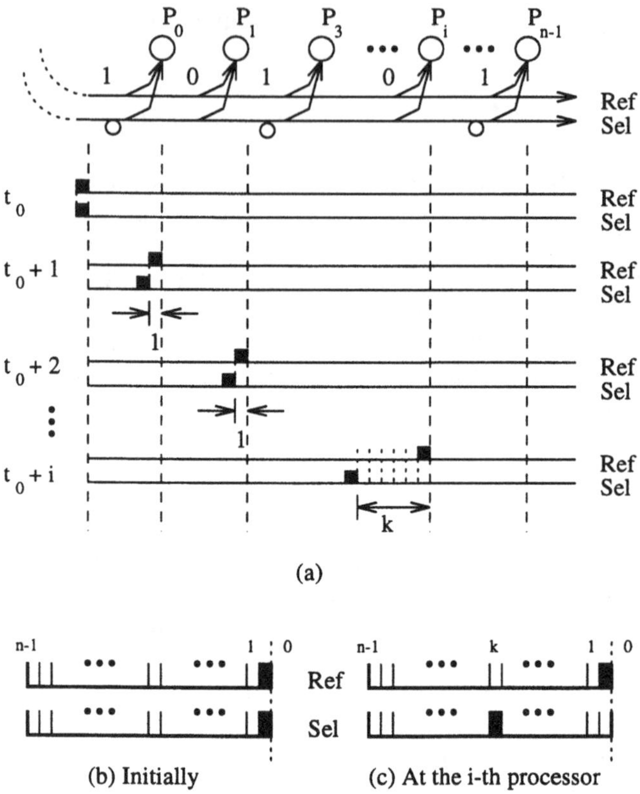

Figure 10.5 Index determination.

lengths are made. The assumption made in [24] is that the bus cycle length in a linear APPB is compatible with the computation speed of the processors. In [10, 11] and [28] the time complexity of different algorithms simulated on arrays with optical pipelined buses is also expressed in terms of the number of steps and it is shown that these results compare favorably with those obtained on different classical networks.

We adopt in this chapter the same method of specifying the time complexity of an algorithm using the number of steps it executes. However, one must be aware that a more accurate comparison of our results with those obtained on other architectures, such as the RN or the mesh with global broadcasting buses, for example, would imply the explicit use of the bus cycle length parameter, which is $O(\sigma_b)$. Obviously, the communication delays encountered in the actual implementations of the other architectures should also be accounted for.

It is shown in [34] that an AROB with n^2 processors can simulate any n^2-processor RN operation in a constant number of steps. If the propagation times on buses of the same length in the two systems are identical, then the simulation is time and cost optimal. For the RN simulation using the AROB model, the algorithms are no longer required to specify the *leader* of each broadcast-

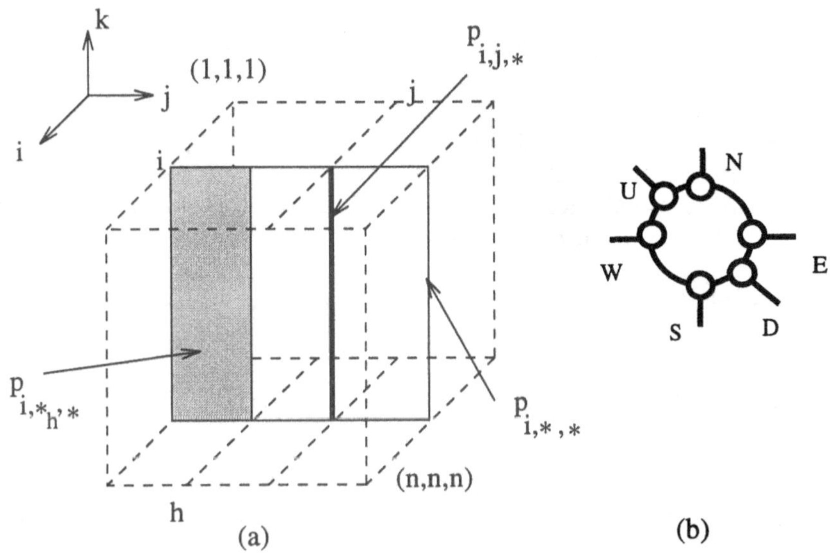

Figure 10.6 Three-dimensional AROB.

ing bus because no time division communications are needed. It is important to emphasize the differences between RN and AROB. By contrast with the classical RN, the AROB structure can use pipelining in order to allow multiple broadcasting of different messages on the same arbitrary bus, during the same bus cycle. The internal time-related operation and the coincident pulse technique, represent other important enhancements over the classical RN. The relative powers of the AROB and RN are more formally presented in [34].

10.2.3 The three-dimensional AROB

The description of two-dimensional AROBs can be extended to dimensions higher than 2. For example, a three-dimensional AROB processor has two extra ports, U and D, for communicating with neighbor processors in the third dimension. This is illustrated in Fig. 10.6. The switch configurations allow any two pairs of these six ports to be connected internally, with any port connected to only one of the other ports at the same time.

10.2.4 Binary Prefix Sums on a linear APPB

Given n one-bit values, v_i, $0 \leq i \leq n-1$, the *binary prefix sums* (BPS) problem requires the computation of $ps_i = v_0 + v_2 + \ldots + v_i$, for all $0 \leq i \leq n-1$, where "+" is the addition operation. Consider the linear APPB structure with n processors, each storing one bit of the input sequence. It is shown in [32, 34] that:

Claim 1 *The n-values BPS problem can be solved on a linear APPB with n processors in a constant number of steps.* ∎

Note that the BPS of n binary values can be computed in a constant number of steps on a $\sqrt{n} \times n$ [25], or on a $\log^2 n \times n$, RN [15]. If we assume the same propagation times for the signals in the two models, namely, RN and linear APPB, the BPS takes the same time but with considerably fewer processors on the linear APPB model. Delay lines and the coincident pulse technique are also used in [9] to compute the ***binary sum*** of n binary values on a linear array with pipelined buses enhanced with switching capabilities. In the next section we use the BPS algorithm only to obtain the binary sum of the n bits. This technique, as opposed to that in [9], has the advantage that it can determine the binary sum of the one-bit values stored by the processors connected to an arbitrary bus.

10.3 THE HT FOR LINE DETECTION ON AROB

Assume that we have an $N \times N \times p$ AROB. Assume also that we have a preprocessed (digitized) $N \times N$ image as the input data, that is, each edge (feature) pixel (i, j), for i and j in $\{0, \dots, N-1\}$, has a value of "1" and it is stored in the base array by processor $P_{i,j,0}$. All other background pixels are "0" and are also stored in the base array, one per processor.

We associate a sub-array of processors $P_{*,*,k}$ with each value of the angle parameter θ_k, $0 \le k \le p-1$. The symbol "$*$" is used in order to simplify the notation and represents all the values of the index it replaces in the notation. In the first step of the algorithm each processor $P_{i,j,0}$ in the base plane broadcasts pixel (i, j) on column $P_{i,j,*}$. Thus each sub-array $P_{*,*,k}$, $0 \le k \le p-1$, gets a copy of the input image. It is assumed also that each processor of the kth sub-array $P_{*,*,k}$ knows the value of θ_k. Usually the angles are chosen such that for all k, $\theta_k = (\pi k)/p$. Thus, each processor can compute the value of θ_k in $O(1)$ time.

The next steps of the algorithm are performed independently and in parallel by all arrays $P_{*,*,k}$, $0 \le k \le p-1$. For each pixel (i, j), $P_{i,j,k}$ computes $\rho_{i,j} = \lfloor i \cos\theta_k + j \sin\theta_k \rfloor$. All the edge pixels which have the same value $\rho_{i,j} = \rho$ describe a potential line in the image space. This line is represented by the (ρ, θ_k) pair in the parameter space. Thus, in order to compute the value of the *count*$[\rho, \theta_k]$ accumulator we need to count the number of edge pixels to which is associated the same (ρ, θ_k) pair. To simplify the presentation we restrict the angle values in the range $[0, \pi/4]$. By using symmetry properties, all the other values of θ_k, $\pi/4 < \theta_k \le \pi$, can be treated similarly.

In what follows we show how the values of *count*$[\rho, \theta_k]$ can be obtained in a constant number of steps. First we show that all the processors in a plane $P_{*,*,k}$ which have computed and stored the same values for ρ can be connected in a linear APPB-like structure. In order to do this we rely on a number of results previously obtained in [26]. The following observations are made for all k such that $0 \le \theta_k \le \pi/4$:

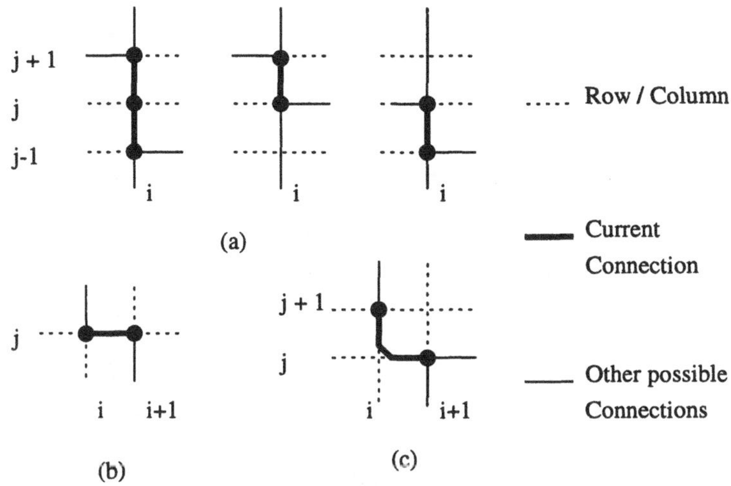

Figure 10.7 Possible connections of a processor to the bus associated to ρ, for $\theta_k \in [0, \pi/4]$.

1. For any j, with $0 \leq j \leq N-1$, the ρ-distances satisfy $\rho_{i,j} \leq \rho_{i+1,j}$ for $0 \leq i \leq N-2$. It also can be shown that no more than two consecutive values of ρ in row j can be equal.

2. The values of ρ, computed using Equation 10.1 by two consecutive processors in row j, differ by at most 1. More formally, for all i, j, $0 \leq j \leq N-1$ and $0 < i \leq N-2$, $0 \leq \rho_{i+1,j} - \rho_{i,j} \leq 1$.

3. For all values of i, j, $0 \leq i, j \leq N-2$, $\rho_{i,j} \neq \rho_{i+1,j+1}$.

4. If $\rho_{i,j} = \rho_{i,j+2}$ for $0 \leq i \leq N-1$ and $0 \leq j \leq N-3$, then $\rho_{i,j} = \rho_{i,j+1} = \rho_{i,j+2}$. If two ρ-values in a column i are equal and are placed two rows apart then the ρ-value in between should have the same value.

Similar properties can be obtained for the other values of the θ_k. An immediate consequence of these observations is that all processors which have computed the same value of the normal distance ρ, for some angle θ_k, can be connected in a linear APPB-like structure. This operation requires only local communications, each processor exchanging information with only some of its four neighbors. More specifically, the possible cases are depicted in Fig. 10.7. If $P_{i,j,k}$ computes some value ρ for the normal distance associated with pixel (i, j), and the same value is obtained for pixels $(i, j-1)$ and/or $(i, j+1)$, then processors $P_{i,j,k}$ should be connected, using the links on the ith column, to $P_{i,j-1,k}$ and/or $P_{i,j+1,k}$, as shown in Fig. 10.7.(a). When in a row j two adjacent processors have the same ρ value, the connection can be made as shown in Fig. 10.7.(b). In case processors $P_{i,j+1,k}$ and $P_{i+1,j,k}$ have to be connected, a third intermediate processor $P_{i,j,k}$ is used, as shown in Fig. 10.7.(c). If the above rules are used, a processor is connected to at most two buses at a time.

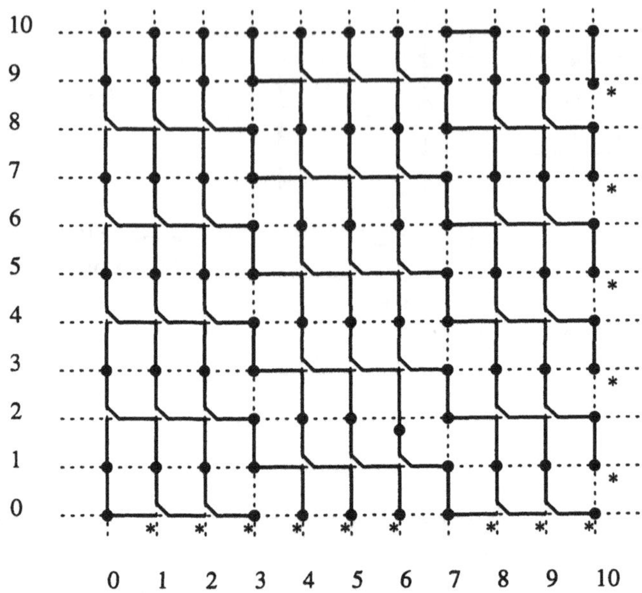

Figure 10.8 Mesh configuration for $\theta_k = \pi/6$.

No two distinct buses are connected to the same port of a processor, the actual setting of a switching system being in one of the states depicted in Fig. 10.4.(b). One end of each bus is located in the bottom row or the rightmost column of the array. A possible bus configuration is depicted in Fig. 10.8 for $\theta_k = \pi/6$. This is designated as the *leader* processor of the linear structure and is marked with a $*$ in Fig. 10.8. Leader selection requires only a constant number of computations and/or local communication.

Next we apply the BPS in each linear bus constructed in this way, and the values of $count[\rho, \theta_k]$ are obtained and stored by each leader. The last step is the thresholding operation which allows us to keep only those values of the *count* array that are larger than some given threshold value. This process is done in parallel by $P_{*,*,k}$, for all values of θ_k, $0 \leq k \leq p-1$. Thus,

Claim 2 *The Hough transform for line detection in an $N \times N$ image can be implemented in a constant number of steps on an $N \times N \times p$ AROB.* ∎

The cost of this algorithm is $O(N^2p)$. This is optimal with respect to the time complexity of the best known sequential algorithm, when the $N \times N$ image is given as input (or when $n = O(N^2)$). Since p is $O(N)$ the total number of processors is $O(N^3)$. Thus the cost of this algorithm becomes $O(N^3)$.

10.4 AN HT ALGORITHM FOR CIRCLE DETECTION ON AROB

Circle detection, like line detection, plays an important role in many computer vision applications. This section addresses the problem of circle detection using the AROB model.

The HT for circles maps each edge point in the input image to a cone in the three-dimensional Hough (parameter) space. A circle is given by the equation:

$$(x - x_0)^2 + (y - y_0)^2 = r^2, \tag{10.2}$$

where (x_0, y_0) is the center of the circle and r is its radius. In the case of circle detection one more parameter is needed, compared with line detection, and the parameter space can be described by (x_0, y_0, r). Assume that we have as input the image already preprocessed, that is, each edge pixel is marked by a "1". An edge pixel in the image can be part of many different circles in the image, and its contribution in the parameter space should affect all the associated (x_0, y_0, r) entries. All the edge pixels of the same circle in the image will of course contribute to the entry which represents the common circle. As in the case of line detection, circle detection is transformed to a more local task, namely, peak detection in the parameter space.

These operations require the quantization of the parameter space. In the case of line detection both parameters are bounded for a given $N \times N$ image. The slope θ of the perpendicular from the origin is in the interval $[0, \pi]$ and the perpendicular distance ρ from the origin lies in the interval $[0, \sqrt{2}N]$. In the case of circle detection, the parameters can be unbounded. The center of the circle can lie outside the image and at the same time only part of the circle can be visible in the image. This implies that the radius can also be much larger than $\sqrt{2}N$. For practical reasons we limit the center of the circle to lie within the image boundaries, that is, x_0 and y_0 take values in the interval $[0, N-1]$. These constraints are satisfactory for most practical applications [20]. As a consequence, any circle satisfying the above requirement has an integer radius of up to $\sqrt{2}N$. Therefore, each parameter is assumed to be quantized in p values with $p = O(N)$.

The HT algorithm for circle detection is similar to that for the line. First, the accumulator *count* associated to the parameter space is initialized to zero. Then, for each edge pixel (x, y) in the image, the algorithm increments all the parameter entries $count[x_0, y_0, r]$ for which Equation 10.2 is satisfied. In the end, the number stored by an entry $count[x_0, y_0, r]$ represents the number of pixels lying on a circle of radius r and center (x_0, y_0). The overall time complexity of this sequential algorithm is $O(N^4)$. In [20] three parallel algorithms are given for circle detection. The best of these algorithms runs in $O(N^2)$ time and uses an $N \times N$ mesh architecture. All circles of a given radius are traced in parallel, updating at the same time the parameter accumulator associated with each of these circles. The tracing technique is actually the ***midpoint circle scan-conversion algorithm***, described in [6], which is used in graphics to draw circles given in an analytic form. We present next a short description of this al-

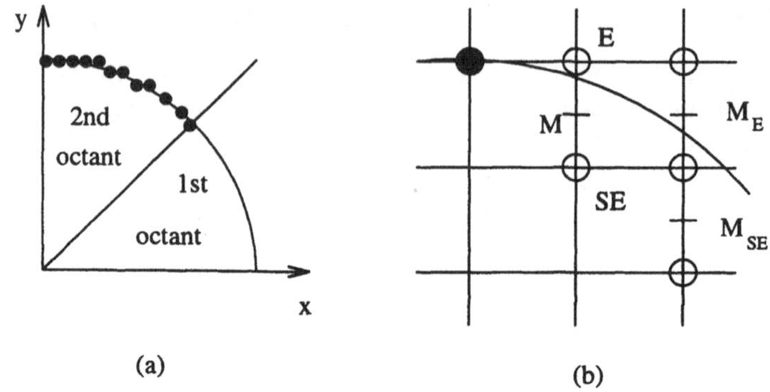

Figure 10.9 (a) The first quadrant of a circle. (b) The pixel grid.

gorithm which serves also as an intuitive introduction to our parallel algorithm for circle detection.

Consider the arc of circle depicted in Fig. 10.9.(a). We describe the technique proposed in [6, 20] for the sequential identification of the pixels lying only on the second octant of its associated circle. All other octants can be processed similarly. The idea behind this tracing technique is sketched in Fig. 10.9.(b). The shaded pixel is the initial starting point and is the topmost pixel of the circle. The pixels lying to the "east" and "south east" of the topmost pixel are marked with E and SE. Point M is the midpoint of the line segment joining E and SE. Suppose that the circle is described by Equation 10.2 and that the current pixel in the tracing sequence is the topmost one. The next pixel in a clockwise scan of the circle is either E or SE. The choice between the two can be made by determining the position of M relative to the circle. Clearly the error implied by this approximation is minimized if the next pixel is chosen to be the closest to the real circle. If M lies inside (or on) the circle then pixel E is (chosen to be) closer to the circle and if M lies outside the circle, pixel SE is closer to the real circle. The last pixel becomes the current pixel, and the above process is repeated for the entire arc in the second octant. Some modifications are necessary for the other octants [6]. This technique can be used to trace each circle in the image space. Counting the number of edge pixels encountered in this scan-conversion of a circle gives the value of an entry in the accumulator associated to the Hough space.

Assume that we have an $N \times N$ image stored on an $N \times N$ AROB, one pixel per processor. The idea behind our algorithm for circle detection is somehow similar to the one presented above and uses the same techniques employed for line detection. The first objective is to connect to the same bus all the processors which lie on the same scan-converted circle. In the next step a simple binary prefix sums algorithm can be used in order to determine the number of edge pixels associated with that circle.

As the maximum number of circle buses to which a processor can be connected at a time is two, we are no longer able to trace all circles which have the same radius in parallel as in [20]. This would require more than two buses to

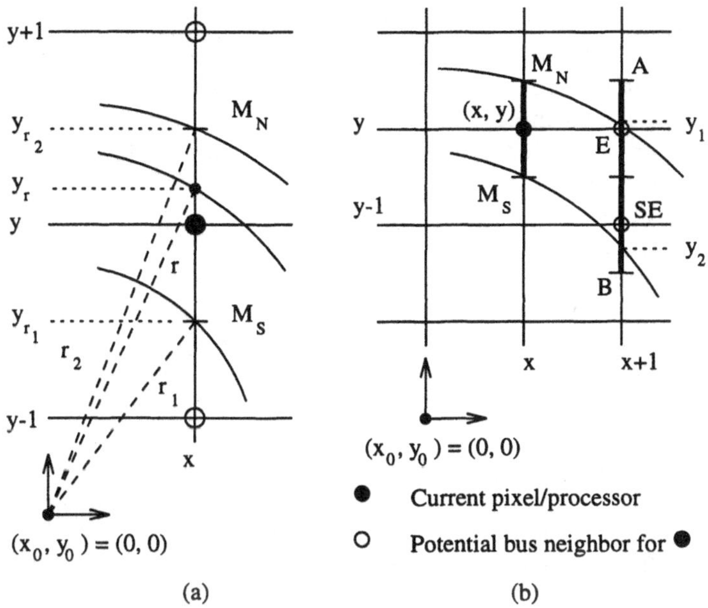

Figure 10.10 (a) Illustration for proof of Claim 3. (b) Illustration for proof of Claim 4.

pass through some processors. A better solution is to trace all the circles with the same center, say (x_0, y_0), and different radii in parallel. For simplicity and without loss of generality, the center of the coordinate system is considered to be translated to the current (x_0, y_0) center, that is, $(x_0, y_0) = (0, 0)$, a circle being described in this case by $r^2 = x^2 + y^2$. We show next that the buses associated with different circles with different radii, r, and the same center (x_0, y_0), can be constructed in parallel with no conflict.

Consider a generic pixel (x, y) in the second octant with $x \geq 1$, $y \geq 1$, and its associated processor $P_{x,y}$. A natural question is how to determine whether there is any circle with center $(x_0, y_0) = (0, 0)$ and some integer radius r, $r \geq 1$, for which (x, y) is part of that scan-converted circle.

Claim 3 *Let (x, y) be a pixel with $x \geq 1$ and $y \geq 1$.*

1. *There is a circle with center $(x_0, y_0) = (0, 0)$ and integer radius r, $r \geq 1$, such that (x, y) is a point in that circle's scan-converted representation if and only if*

$$z_1 \leq r \leq z_2, \tag{10.3}$$

where $z_1 = \lceil ((y - 0.5)^2 + x^2)^{1/2} \rceil$ and $z_2 = \lfloor ((y + 0.5)^2 + x^2)^{1/2} \rfloor$.

2. *If such an r exists then it is unique and $r = z_1 = z_2$.*

Proof. Using Fig. 10.10.(a) the proof of statement (1) is immediate. Let M_N be the point at the middle vertical distance between (x, y) and $(x, y+1)$, and M_S the point at the middle vertical distance between (x, y) and $(x, y-1)$. There is a circle with integer radius r and center $(0,0)$ such that (x, y) is a point in its scan-converted representation if and only if that circle intersects the vertical which passes through (x, y) somewhere between M_N and M_S. This condition can be written as $y_{r_1} \leq y_r < y_{r_2}$, where y_{r_1} and y_{r_2} are real values and are the radii of the two circles, centered at $(0,0)$, which pass through M_S and M_N, respectively. Since $y_{r_1} = y - 0.5$ and $y_{r_2} = y + 0.5$ one can verify that $y_{r_1} \leq y_r < y_{r_2}$ is equivalent to $((y-0.5)^2 + x^2)^{1/2} \leq r < ((y+0.5)^2 + x^2)^{1/2}$. As r must be an integer, this expression becomes $z_1 \leq r \leq z_2$.

Statement (2) can be proved by contradiction, assuming that there are two circles of integer radii r and $r+1$ such that the circles intersect the vertical at x between M_N and M_S. Let y_r and y_{r+1} be the two points of intersection in the second octant between these circles and the vertical at x. The distance between y_r and y_{r+1} is strictly greater than 1 and $y_{r_1} \leq y_r < y_{r+1} \leq y_{r_2}$. This means that the distance between y_{r_1} and y_{r_2} is strictly greater than 1, which is impossible. Thus, if such a circle exists for pixel (x, y) then it is unique. ∎

Examples : Let $x = 2$ and $y = 4$. Using Equation 10.3, $\lceil 4.03 \rceil \leq r \leq \lfloor 5.76 \rfloor$ which is equivalent to $5 \leq r \leq 5$. For $x = 9$ and $y = 10$ we obtain $14 \leq r \leq 13$ and thus there is no such r. See Fig. 10.12.

With the results in Claim 3 we have established a technique which can be used by each processor in the array to determine if its associated pixel is part of any scan-converted circle with center (x_0, y_0) and integer valued radius. Furthermore, if such a circle exists then it is unique and its radius can be determined in constant time. The entire test takes constant time.

Each processor which has determined a valid value for r must connect itself to other processors with the same r in order to establish the linear bus associated with that circle, if such processor exists. Again, we restrict our discussion to the second octant, $\pi/4 \leq \theta_k \leq \pi/2$. We describe a processor $P_{x,y}$ as *active* if it is connected to the bus associated with the corresponding circle and at the same time its pixel is part of the scan-converted circle. As we explain later, it is also possible for a processor to be connected to a bus without the corresponding pixel being part of the associated circle. This is called an *inactive* processor.

Claim 4 *Let (x, y) be a pixel which satisfies Equation 10.3, that is, there is a circle with center $(x_0, y_0) = (0,0)$ and integer radius r such that (x, y) is part of the scan-converted representation of that circle. Then, either $P_{x+1,y}$ or $P_{x+1,y-1}$ is the next active processor adjacent to $P_{x,y}$ in the clockwise scan-converted order on that circle.*

Proof. For a proof we can use Fig. 10.10.(b). The two possible extreme cases, when (x, y) is part of some scan-converted circle, are represented by the two arcs of circle which intersect the vertical passing through (x, y) in M_N and M_S.

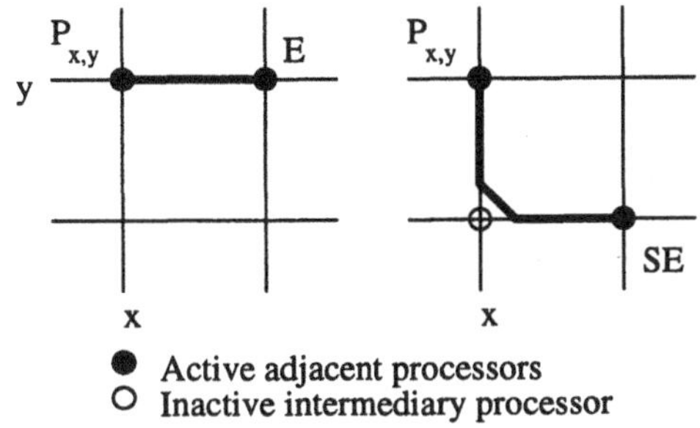

Figure 10.11 (a) The two possible connections between two neighbors on a scan-converted circle. (b) The internal switch settings used in the second octant.

One can easily verify that these circles intersect the vertical at $x+1$ somewhere between point A and point B, as illustrated in Fig. 10.10.(b). Thus, either processor E ($P_{x+1,y}$) or SE ($P_{x+1,y-1}$) is the active neighbor of $P_{x,y}$. ∎

The two possible connections are depicted in Fig. 10.11.(a). To establish the bus connection between $P_{x,y}$ and the E neighbor, the direct row link is used. To connect $P_{x,y}$ with SE, $P_{x,y-1}$ is used as an intermediate, or *inactive*, processor (by convention). Here, the processor is *inactive* as the data stored by it are irrelevant for this circle, and the processor is used only to establish a connection between $P_{x,y}$ and SE. At the same time, by reciprocity, $P_{x,y}$ must be the neighbor of either $P_{x-1,y}$ or $P_{x-1,y+1}$. The four possible internal switch connections for the second octant implied by these rules are depicted in Fig. 10.11.(b). The worst case scenario is given by the bottom-left configuration where a processor can be at the same time active and inactive but relative to two different buses. Note that no conflict arises in this case, making it possible to construct all buses in parallel. An example of parallel bus construction for the first quadrant is depicted in Fig. 10.12.(a). The different processor connections for the second octant are shown in Fig. 10.12.(b). The only exception to these rules is made for the interconnection of the four processors around P_{x_0,y_0}, that is, P_{x_0+1,y_0}, P_{x_0,y_0-1}, P_{x_0-1,y_0} and P_{x_0,y_0+1}. For this interconnection, P_{x_0-1,y_0-1}, P_{x_0-1,y_0+1}, P_{x_0+1,y_0+1} and P_{x_0+1,y_0-1} are used as intermediate, inactive processors.

A bus can correspond to either a full scan-converted circle or to an arc of a scan-converted circle. Examples of circles with center (x_0, y_0) which have at least one arc contained in the image are depicted in Fig. 10.13.(a) . The square represents the border of the $N \times N$ image. In order to apply the binary prefix sums algorithm, a *leader* processor must be specified for each constructed arc or circle bus. The leader processors are represented with shaded circles in Fig. 10.13. There are two different cases. In the first case the entire scan-converted circle with center (x_0, y_0) and radius r is contained in the image. Thus, all associated processors are connected to a linear bus and the topmost processor, that is, P_{x_0,y_0+r}, is designated as the leader of this bus. All these

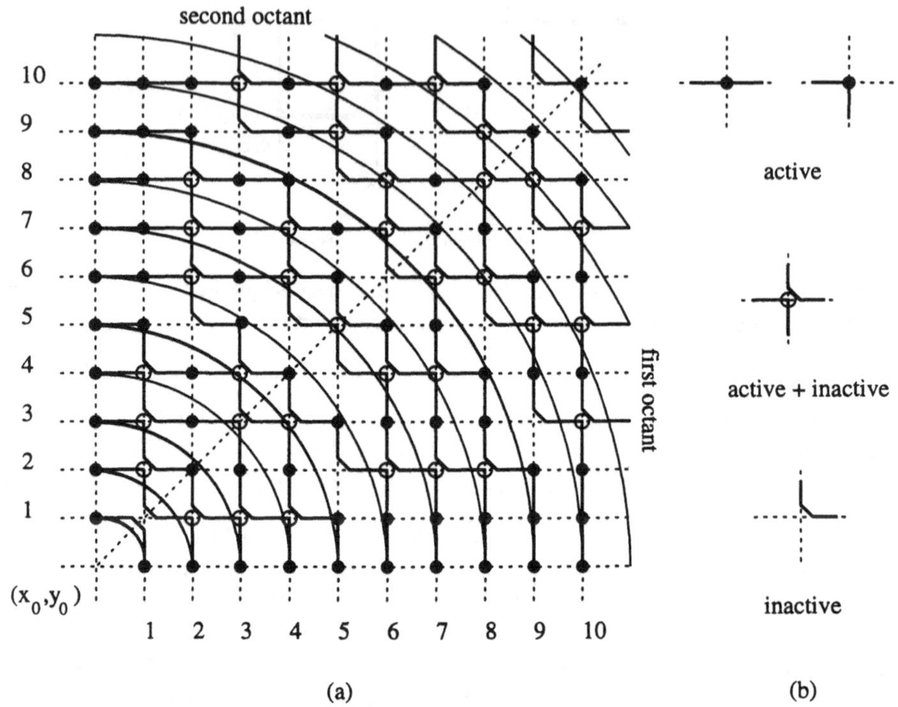

Figure 10.12 (a) The bus connections in the first quadrant. (b) Processor connections in the second octant.

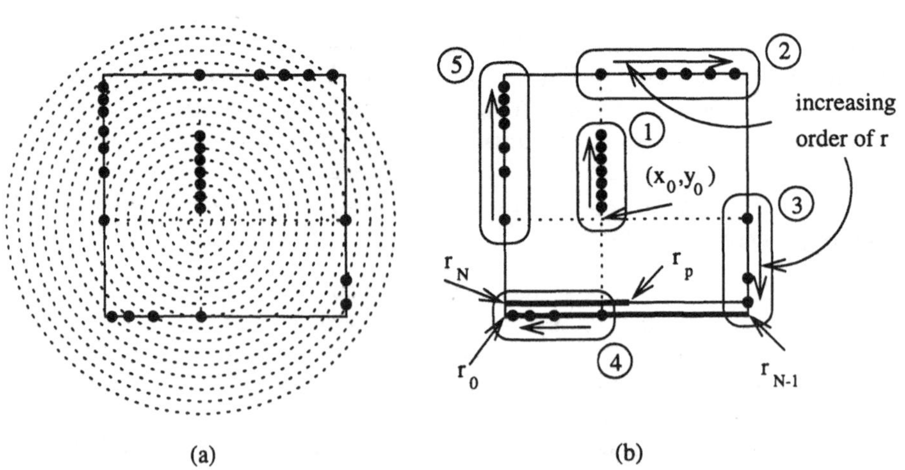

Figure 10.13 (a) Relative positions of circles within an image. (b) The leader processors and their grouping.

leaders are grouped in the rounded box marked with "1" in Fig. 10.13.(b). Note that as r has an integer value, processor P_{x_0,y_0+r} always exists when the entire circle is contained in the image. In the second case, a scan-converted circle is represented by up to four arcs, and the first processor in the clockwise order of each arc can be set as the leader. The other four rounded boxes in Fig. 10.13.(b) contain the leaders associated with these arcs. The process of leader detection uses only local computations and can be done in a constant number of steps.

The partial values of the $count[x_0, y_0, r]$ accumulators are obtained by applying the BPS algorithm in each linear bus associated with a circle or arc of circle. Each active processor participates with its pixel value in this BPS. Inactive processors use a value of 0. The leader stores the accumulator value together with the radius value r. In each of the five groups in Fig. 10.13.(b) these values are already sorted over the radii r, which are also distinct. The increasing order of r is shown in Fig. 10.13.(b) using arrows in each rounded box. The objective is to group all the accumulator values in the bottommost rows of the array in the sorted order of r. As $p \leq \sqrt{2}N$, at most two such rows are used. A possible final order of the p values of r is suggested in Fig. 10.13.(b). As no two values within each rounded box have the same destination, simple permutation routing can be used to move data to the two bottom rows. The five rounded boxes are treated sequentially because contention can occur when different partial accumulators, associated to the same entry $count[x_0, y_0, r]$, are routed to the same processor. This process is done in parallel for all values of (x_0, y_0). We therefore have:

Claim 5 *The Hough transform for circle detection in an $N \times N$ images can be implemented in a constant number of steps on an $N \times N \times p^2$ AROB.* ∎

The cost of this algorithm is $O(N^2p^2)$. This is optimal with respect to the time complexity of the best known sequential algorithm.

10.5 CONCLUSIONS

Constant time parallel HT algorithm for line detection are given in the literature for the case where the input is given as n edge pixels. The best constant time algorithm is obtained in [28] using a reconfigurable array of dimensions $n \times \log^2 n \times p \times p$. As n can be $O(N^2)$ and $p = O(N)$, the actual number of processors is $O(N^4 \log^2 N)$. The best algorithm for parallel (HT) circle detection runs in $O(N^2)$ time and uses an $N \times N$ mesh architecture [20].

In this chapter it is shown that the Hough transform can be implemented optimally in $O(1)$ steps. The costs of the two algorithms for line and circle detection are $O(N^3)$ and $O(N^4)$, respectively. These values are optimal with respect to the complexity of the best sequential algorithms and compare favorably with the best known parallel algorithms.

The number of processors can be further reduced but the price that must be paid is the increased time complexity of the algorithms. The line detection algorithm can be implemented in $O(t)$ time on an $N \times N \times \frac{p}{t}$ AROB with $t \leq p$. The circle detection algorithm takes $O(t)$ time on an $N \times N \times \frac{p^2}{t}$ AROB with $t \leq p^2$.

Acknowledgments

The authors wish to thank the Natural Sciences and Engineering Research Council of Canada for its financial support.

References

[1] S.G. Akl. *Parallel Computation: Models and Methods.* Prentice Hall, Upper Saddle River, New Jersey, 1997.

[2] S. Bhandarkar and H.R. Arabnia. The Hough transform on a reconfigurable multi-ring network. *Journal of Parallel and Distributed Computing,* 24:107–114, 1995.

[3] Y. Ben-Asher, D. Peleg, R. Ramaswami, and A. Schuster. The power of reconfiguration. *Journal of Parallel and Distributed Computing,* 13:139–153, 1991.

[4] D.M. Chiarulli, R.G. Melhem, and S.P. Levitan. Using coincident optical pulses for parallel memory addressing. *The Computer Journal,* 30:48–57, 1987.

[5] R.E. Cypher, J.L., Sanz, and L. Snyder. The Hough Transform has $O(N)$ complexity in SIMD N by N mesh array architectures. In *Proceedings IEEE Computer Society Workshop on Computer Architectures for Pattern Analysis and Machine Intelligence,* pages 115–121, 1987.

[6] J.D. Foley, A. van Dam, S.K. Feiner, and J.F. Hughes. *Computer Graphics: Principles and Practice.* Addison-Wesley, Reading, Massachusetts, 1990.

[7] A.L. Fisher and P.T. Highnam. Computing the Hough transform on a scan line array processor. *IEEE Transactions on PAMI,* 11(3):262–265, 1989.

[8] C. Guerra and S. Hambrusch. Parallel algorithms for line detection on a mesh. *Journal of Parallel and Distributed Computing,* 6:1–9, 1989.

[9] Z. Guo. Sorting on array processors with pipelined buses. In *Proceedings 1992 International Conference on Parallel Processing,* Vol. III, pages 289–292, 1992.

[10] Z. Guo. Optically Interconnected Processor Arrays with Switching Capability. *Journal of Parallel and Distributed Computing,* 23:314–329, 1994.

[11] Z. Guo, R.M. Melhem, R.W., Hall, D.M., Chiarulli, and S.P. Levitan. Pipelined communications in optically interconnected arrays. *Journal of Parallel and Distributed Computing,* 12:269–282, 1991.

[12] V.P. Heuring, H.F., Jordan, and J.P. Pratt. Bit-serial architecture for optical computing. *Applied Optics*, 31(17):3213–3224, 1992.

[13] H.A.H. Ibrahim, J.R., Kender, and D.E. Shaw. The analysis and performance of two middle-level vision tasks on a fine-grained SIMD tree machine. In *Proceedings IEEE Computer Society Conference on Computer Vision and Pattern Recognition*, pages 387-393, 1985.

[14] J. Illingworth and J. Kittler. A survey of Hough transform. *Computer Vision, Graphics, and Image Processing*, 44:87–116, 1988.

[15] J.-W Jang, H. Park, and V.K. Prasanna. A fast algorithm for computing a histogram on reconfigurable mesh. *IEEE Transactions on Pattern Analysis and Machine Intelligence*, 17(2):97–106, 1995.

[16] J.F. Jenq and S. Sahni. Reconfigurable mesh algorithms for the Hough transform. *Journal of Parallel and Distributed Computing*, 20:69–77, 1994.

[17] J. Jolion and A. Rosenfeld. An $O(\log n)$ pyramid Hough transform. *Pattern Recognition Letters*, 9:343–349, 1989.

[18] C.S. Kannan and H.Y.H. Chuang. Fast Hough transform on a mesh connected processor array. *Information Processing Letters*, 33:243–248, 1990.

[19] T.-W. Kao, S.-J., Horng, Y.-L., Wang, and K.-L., Chung. A constant time algorithm for computing Hough transform. *Pattern Recognition*, 26(2):277–286, 1993.

[20] S. Kumar, N. Ranganathan, and D. Goldof. Parallel algorithms for circle detection in images. *Pattern Recognition*, 27(8):1019–1028, 1994.

[21] S.P. Levitan, D.M. Chiarulli, and R.G. Melhem. Coincident pulse technique for multiprocessor interconnection structures. *Applied Optics*, 29(14):2024–2033, 1990.

[22] K. Li. Constant time boolean matrix multiplication on a linear array with a reconfigurable pipelined bus system. In *Proceedings 11th Annual International Symposium on High Performance Computing Systems*, pages 179–190, 1997.

[23] K. Li, Y. Pan, and S.-Q. Zheng. Simulation of parallel random access machines on a linear array with a reconfigurable pipelined bus system. In *Proceedings International Conference on Parallel and Distributed Techniques and Applications*, Vol. II, pages 590–599, 1997.

[24] R.G. Melhem, D.M. Chiarulli, and S.P. Levitan. Space multiplexing of waveguides in optically interconnected multiprocessor systems. *Computer Journal*, 32(4):362–369, 1989.

[25] S. Olariu, J.L. Schwing, and J. Zhang. Integer problems on reconfigurable meshes, with application. *Journal of Computer and Software Engineering*, 1:33–46, 1993.

[26] S. Olariu, J.L. Schwing, and J. Zhang. Computing the Hough transform on reconfigurable meshes. *Image and Vision Computing*, 11:623–628, 1993.

[27] Y. Pan. Hough transform on arrays with an optical bus. In *Proceedings 5th ISMM International Conference on Parallel and Distributed Computing and Systems*, pages 161–166, 1992.

[28] Y. Pan. A more efficient constant time algorithm for computing the Hough transform. *Parallel Processing Letters*, 4(1-2):45–52, 1994.

[29] Y. Pan and H.Y.H. Chuang. Parallel Hough transform algorithms on SIMD hypercube arrays. In *Proceedings 1990 International Conference on Parallel Processing*, pages 83–86, 1990.

[30] Y. Pan and H.Y.H. Chuang. Faster line detection algorithms on enhanced mesh connected arrays. *IEE Proceedings-E*, 2(140):95–100, 1993.

[31] S. Pavel. *Computation and Communication Aspects of Arrays with Optical Pipelined Buses.* Ph.D. thesis, Department of Computing and Information Science, Queen's University, Kingston, Ontario, 1996.

[32] S. Pavel and S.G. Akl. Matrix operations using arrays with reconfigurable optical buses. *Journal of Parallel Algorithms and Applications*, 8:223–242, 1996.

[33] S. Pavel and S.G. Akl. Area-time trade-offs in arrays with optical pipelined buses. *Applied Optics*, 35:1827–1835, 1996.

[34] S. Pavel and S.G. Akl. On the power of arrays with reconfigurable optical buses. In *Proceedings International Conference on Parallel and Distributed Processing Techniques and Applications*, pages 1443–1454, 1996.

[35] S. Pavel and S.G. Akl. Integer sorting and routing in arrays with reconfigurable optical buses. to appear in *International Journal of Foundations of Computer Science, Special Issue on Interconnection Networks*, 1998.

[36] C. Qiao. On designing communication-intensive algorithms for a spanning optical bus based array. *Parallel Processing Letters*, 5(3):499–511, 1995.

[37] C. Qiao and R.G. Melhem. Time-division communications in multiprocessor arrays. *IEEE Transactions on Computers*, 42(5):577-590, 1993.

[38] C. Qiao, R.G. Melhem, D.M. Chiarulli, and S.P. Levitan. Optical multicasting in linear arrays. *International Journal Optical Computing*, 2:31–48, 1991.

[39] A. Rosenfeld J. Ornelas, and Y. Hung. Hough transform algorithms for mesh-connected SIMD parallel processors. *Computer Vision, Graphics and Image Processing*, 41:293–305, 1988.

[40] T.M. Silberberg. The Hough transform on geometric arithmetic parallel processor. In *Proceedings IEEE Computer Society Workshop on Computer Architecture for Pattern Analysis and Image Database Management*, pages 387–393, 1985.

11 BASIC DATA MOVEMENT OPERATIONS ON THE LARPBS MODEL

Yi Pan

Department of Computer Science
University of Dayton
Dayton, Ohio 45469-2160
pan@cps.udayton.edu

Abstract: Linear array with a reconfigurable pipelined bus system (LARPBS) is a parallel computational model designed based on fiber optics. Many algorithms have been designed for the model by several research groups, and the results show that most of the algorithms can be executed efficiently on the LARPBS model. In this chapter, the main features of the LARPBS model are introduced and several basic data movement operations on the model are described. The scalability issue of the model is also addressed via demonstrating scalable algorithms on the model. Finally, its relation to other similar models is also discussed.

11.1 INTRODUCTION

Fiber optic communications offer a combination of high bandwidth, low error probability, and gigabit transmission capacity and have been used extensively in wide-area networks. Advances in optical and optoelectronic technologies indicates that they could also be used as interconnects in parallel computers. The ability to control optical delays precisely can be used in many ways to support high bandwidth multiprocessor interconnection. For example, precise delays can be used for the buffers required for temporal rearrangement of

TDM signals, collision resolution in packet switched networks, and synchronization of incoming packets on separate channels. Depending upon the material through which the signals propagate, one millimeter corresponds to three to seven picoseconds. With the precision available in mechanical layout, sub-picosecond time precision is achievable. In fact, many commercial massively parallel computers, such as the recently announced Cray T90 supercomputer system, use optical technologies in their communication subsystems. Many research papers and several journal special issues have been published on this topic [9, 17, 27, 28, 29, 42, 54]. Among them, the distributed-memory SIMD (Single Instruction Multiple Data) computer with pipelined optical buses has received much attention [10, 21, 28, 36, 37, 49, 50] due to its simplicity and low cost. In such a system, messages can be transmitted concurrently on a pipelined optical bus without partitioning the bus into several segments, while the time delay between the furthest processors is only the end-to-end propagation delay of light over a waveguided bus. This design integrates the advantages of both optical transmission and electronic computation. Several slightly different models have been proposed in the literature, including the *array processors with pipelined buses* (APPB) [11, 28], the *array processors with pipelined buses using switches* (APPBS) [12], the *array with synchronous optical switches* (ASOS) [50], and the *reconfigurable array with spanning optical buses* (RASOB) [51]. Many parallel algorithms, such as the Hough transform [36], singular value decomposition [39], order statistics [37], sorting [11, 38], matrix operations [22, 23, 24, 51], and some numerical algorithms [14], have been proposed for these models. The preliminary work indicates that arrays with pipelined buses are very efficient for parallel computation due to the high bandwidth within a pipelined bus system.

On the other hand, a lot of research has been done on reconfigurable systems. In particular, many algorithms have been designed on various arrays with reconfigurable buses [32, 33, 34, 56, 60]. In an array with reconfigurable buses, messages can be transmitted concurrently when the bus is partitioned into many segments, and the diameter problem in a point-to-point network disappears when all segments are reconfigured as a single global bus. Several journal special issues have also been dedicated to this topic [2, 26, 35, 31]. However, most of the models published there use electronic buses as their interconnections. When there is a large amount of message transfer between different sections of the network, the bus segments themselves become a potential bottleneck since they can only provide exclusive access for messages.

Based on the research in reconfigurable meshes and pipelined optical buses, arrays with reconfigurable optical buses have been proposed and studied independently by Pavel and Akl [47, 48], and Pan, Hamdi and Li [38, 41]. In these models, message can be transmitted concurrently on a bus in a pipelined fashion and the bus can be reconfigured dynamically under program control to suit communication needs. Currently, two related models have been proposed. They are: linear array with a reconfigurable pipelined bus system (LARPBS) proposed by Pan, Hamdi and Li [38, 41] and array with reconfigurable optical buses (AROB) proposed by Pavel and Akl [47]. Since then, many algorithms have been designed for these two models, including inversion number computation [18], neural network computation [8], various matrix operations [23, 24] selection [37], and sorting algorithms [38, 41] on the LARPBS model, and matrix multiplications [44], selection [52], and sorting [46, 52] on the AROB model. Although the two models are similar to some extent, a fundamental difference exists between them. In the LARPBS model, counting is not allowed during a bus cycle, while it is permitted in the AROB model. In fact, the LARPBS does not allow any processor involvement during a bus cycle, except setting switches at the beginning of a bus cycle. Unlike many theoretical models, such as PRAM (Parallel Random Access Machine), we believe that the LARPBS model is implementable and practical using current optical technologies.

In this chapter, we describe the important features and several fundamental data movement operations for the LARPBS model. In particular, efficient algorithms for computing the binary prefix sums, compression, split, maximum finding and multicasting are presented. Then, the scalability of the algorithms are discussed. Finally, the relation of this model to several recently proposed models is discussed.

11.2 THE LARPBS MODEL

A pipelined optical bus system uses optical waveguides instead of electrical buses to transfer messages among electronic processors. The advantages of using waveguides can be seen as follows: Besides the high propagation speed of light, there are two important properties of optical signal (pulse) transmission on an optical bus: unidirectional propagation and predictable propagation delay per unit length. These two properties enable synchronized concurrent access of an optical bus in a pipelined fashion [10, 28, 49, 50]. This, combined with the abilities of a bus structure to do efficient broadcasting or multicasting, makes the architecture suitable for many applications that involve intensive communication operations.

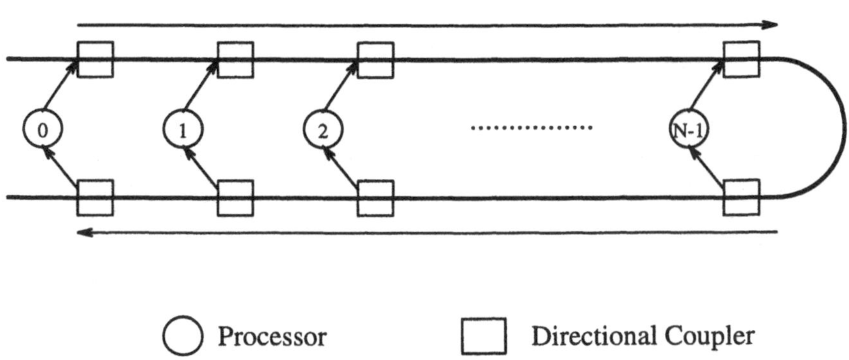

Figure 11.1 A linear optical bus system of N processors.

Figure 11.1 shows a linear array in which electronic processors are connected with an optical bus. Each processor is connected to the bus with two directional couplers, one for transmitting on the upper segment and the other for receiving from the lower segment of the bus [10, 28, 49, 50]. The optical bus contains three identical waveguides, one for carrying messages (the *message waveguides*) and the other two for carrying address information (the *reference waveguide* and the *select waveguide*), as shown in Figure 11.2. For the purpose of simplicity, the message waveguide, which resembles the reference waveguide, has been omitted from the figure. Messages are organized as fixed-length *messageframes*. Note that optical signals propagate unidirectionally from left to right on the upper segment and from right to left on the lower segment. This bus system is also referred to as the folded-bus connection in [10].

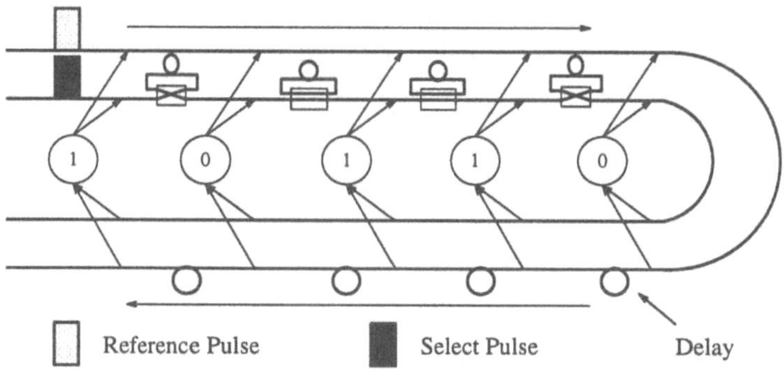

Figure 11.2 An optical bus with delays added.

Let ω be the pulse duration in seconds and c_b the velocity of light in these waveguides. Define a unit delay Δ to be the spatial length of a single optical pulse, that is $\Delta = \omega \times c_b$. Initially, processors are connected to these three waveguides such that between any two given processors, the same length of fiber is used on all three waveguides. Hence, the propagation delays between two processors are the same for all three waveguides. A bus cycle for an optical bus is defined as the end-to-end propagation delay on the bus; i.e., the time taken for an optical signal to propagate through the entire bus. If τ is the time taken for a signal to traverse the optical distance between two consecutive processors on the bus, then the length of a bus cycle for the system of Figure 11.6 is $2N\tau$. We then add one unit delay Δ (shown as a loop in Figure 11.2) between any two processors on the receiving segments of the reference waveguides and of the message waveguides. Each loop is an extra segment of a fiber and the amount of delay added can be accurately chosen based on the length of the segment. As a result, the propagation delays on the receiving segments of the select waveguide and the reference waveguides are no longer the same. Finally, we add a conditional delay Δ between any two processors i and $i+1$, where $0 \leq i \leq N-2$, on the transmitting segments of the select waveguides (Figure 11.2). The switch between processor i and $i+1$ is called $S(i+1)$ and is local to processor $i+1$. Thus, every processor has its own switch, except processor

0. The conditional delays can be implemented using 2×2 optical switches such as the Ti:LiNbO3 switches used in an optical computer [3]. Each switch can be set by the local processor to two different states: *straight* or *cross* as shown in Figures 11.3. When a switch is set to *straight*, it takes τ time for an optical signal on the transmitting segments of the select waveguides to propagate from one processor to its nearest neighbor. When a switch is set to *cross*, a delay ω is introduced and such propagation will take $\tau + \omega$ time. Clearly, the maximum delay that the switches can introduce is the duration of $N - 1$ pulses.

Figure 11.3 Conditional delays implemented using 2×2 optical switches.

Messages transmitted by different processors may overlap with each other even if they propagate unidirectionally on the bus. We call these message overlappings *transmission conflicts*. Assume each message has b binary bits, each bit represented by an optical pulse, with the existence of a pulse for 1 and the absence for 0. To ensure that there are no transmission conflicts, the following condition must be satisfied: $\tau > b\omega$, where τ is the time taken for a signal to traverse the optical distance between two consecutive processors on the bus, and ω is the pulse duration. Note that the above condition ensures that each message can fit into a pipeline cycle such that in a bus cycle, up to N messages can be transmitted by processors simultaneously without collisions on the bus. In a parallel array, messages normally have very short length; i.e., b is very small. Thus, in the following discussion, we assume that the above condition is always satisfied and that no transmission conflicts are possible as long as all processors are synchronized at the beginning of each bus cycle.

Now let us describe the LARPBS model. In the LARPBS, we insert an optical switch on each section of the transmitting bus and receiving bus. Thus, each processor has 6 more local switches, three on its three receiving segments and three on its three transmitting segments, besides its switch for conditional delay. The switches on the receiving and transmitting segments between processors i and $i+1$ are called $RSR(i)$ and $RST(i)$, respectively, and are local to processor i as shown in Figure 11.4. Here, $RSR(i), 0 \leq i < N$ are 2×1 optical switches, and $RST(i), 0 \leq i < N$ are 1×2 optical switches. In the following discussion, these switches will be called reconfigurable switches due to their function. When all switches are set to straight, the bus system operates as a regular pipelined bus system. When $RSR(i)$ and $RST(i)$ are set to cross, the whole bus system is split into two separate systems, one consisting of processors 0, 1, ..., and i and the other consisting of $i+1$, $i+2$, ..., $N-1$. The total delay for a signal passing the transmitting segment, the optical fiber between $RST(i)$ and $RSR(i)$, and the receiving segment is made to be equal to τ. Hence, the array with processors 0 to i can operate as a regular linear array with a pipelined bus system; so does the array with processors $i+1$ to $N-1$. Figure 11.4 shows the LARPBS model with 6 processors. The array is split into two subarrays, with the first one having four processors and the second one having two processors. In the figure, only one waveguide is shown. Conditional switches are omitted in the figure to avoid confusion.

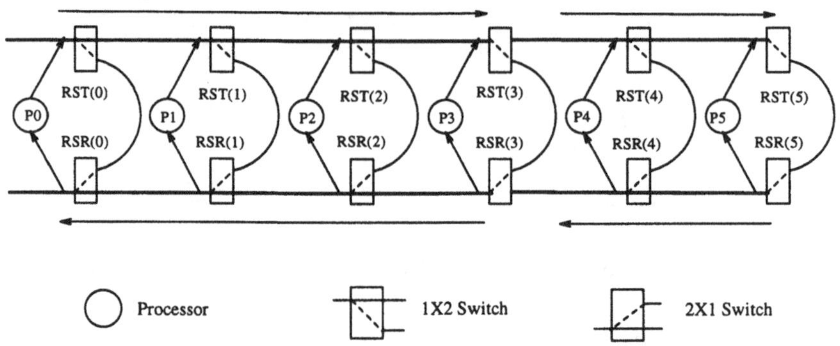

Figure 11.4 The LARPBS model of size 6 with two subarrays.

Several time-division switching methods can be applied to route messages in an optical bus system. In the first approach, each processor is assigned a fixed time slot and transmits or receives a message during that particular time slot. A sequence of time slots formed on the transmitting segment of a bus is rearranged via a time-slot interexchanger [53], and then forwarded to the receiving segment. Each time slot of the output sequence contains a message destined to the processor corresponding to that slot. In the second approach, each processor is assigned a fixed transmitting time slot. An SIMD environment is assumed in this case. Hence, each processor knows which processor is sending a message to it and knows the time slot that contains the message [10, 28]. The last approach is to use a *coincident pulse technique* [6, 11, 49]. Using this approach, the relative time delay of a select pulse and a reference pulse is determined so that they will coincide, thus producing a double-height pulse, only at receiver i. By properly adjusting the detecting threshold of the detector at processor i, this double-height pulse can be detected, thereby addressing i.

In this chapter, the coincident pulse technique is used to route messages or to broadcast messages on the bus. The switches on the sending segments are used to conditionally delay the select pulses and can be closed or opened by the local processors (see Figure 11.3). Now let us discuss in detail how to perform several basic data movement operations on the LARPBS using the coincidence pulse technique. Here we assume that the message processing time is roughly equal to the time for a bus cycle. If a basic operation takes a constant time, then a bus cycle also requires a constant time. This assumption has also been used by many other researchers [1, 11, 13, 14, 15, 18, 25, 47, 48, 50, 51, 52]. Hence, all times in the following discussion are measured in bus cycles.

Obviously, accurate delays are critical in the LARPBS model. In an practical system, there are some uncertainties in the timings. A pulse travels from one processor to the next will not take a time exactly τ. These errors may accumulate when the number of processors is large. Because the system relies on the precise timings for addressing messages, the error will eventually prohibit further scaling. This scalability problem is discussed in [7] and is called synchronization error by the authors of [7]. Some experiments have been carried out and the results indicate that large variations (on the order of one half of a pulse width) can be tolerated without significant degradation of the coincident signal [7]. For single-mode fibers with a length of a few kilometers, the synchronization error is small enough and can be tolerated. Hence, using current technology, a system using a few thousands of processors will not present any problem. For more discussion on this topic, the reader is referred to [16].

11.3 BASIC DATA MOVEMENT OPERATIONS

In this section, we present several basic data movement operations on the LARPBS model. The scalability of these operations is discussed in the next section.

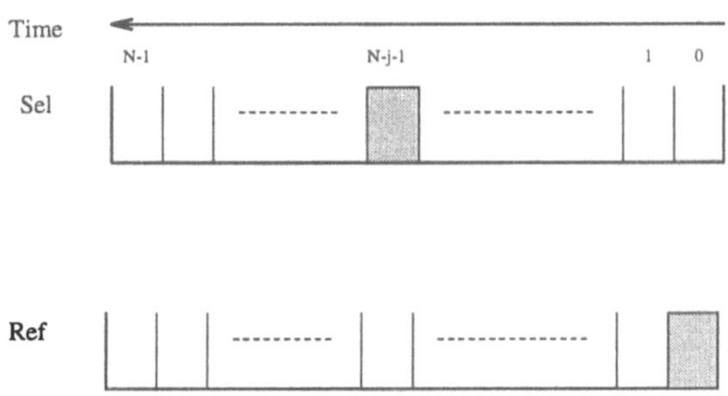

Figure 11.5 Address frames.

11.3.1 One-to-One Communication

First, set all switches on the transmitting segments to *straight* so that no delay is introduced in the switches on the transmitting segments of the select waveguides. A source processor sends a reference pulse at time t_{ref} (the beginning of a bus cycle) on the reference waveguide and a select pulse at time t_{sel} on the select waveguide. The source processor also sends a message frame, on the message waveguide, which propagates synchronously with the reference pulse. Whenever a processor detects a coincidence of a reference pulse and a select pulse, it reads the message frame. In other words, in order for processor i to send a message to processor j, we need to have the two pulses coincide at processor j. This happens if and only if $t_{sel}(j) = t_{ref} + (N - j - 1)\omega$, where $0 \leq i, j < N$. Figure 11.5 shows an address frame relative to a reference pulse for addressing processor j. Here, the select pulse is delayed for a time of $(N-j-1)\omega$ in relative to the reference pulse, and the two pulses meet at processor j after the reference pulse goes through $N-j-1$ delays.

Hence, for a given reference pulse transmitted at time t_{ref}, the presence of a select pulse at time $t_{ref}+(N-j-1)\omega$ will address processor j while the absence of a select pulse at that time will not. For example, if a processor wants to send a message to processor 0, it sends a reference pulse at time t_{ref} (the beginning of a bus cycle) and a select pulse at $t_{sel}=t_{ref}+(N-1)\omega$. Since there are $(N-1)$ unit delays in the receiving segments of the reference waveguide, these two pulses will coincide at processor 0. In order for a processor to address processor $N-1$, $t_{sel}=t_{ref}+(N-(N-1)-1)\omega=t_{ref}$; i.e., the source processor has to send a reference pulse and a select pulse at the same time. Clearly, the two pulses coincide at processor $N-1$ since there is no delay in its receiving segment of the reference waveguide.

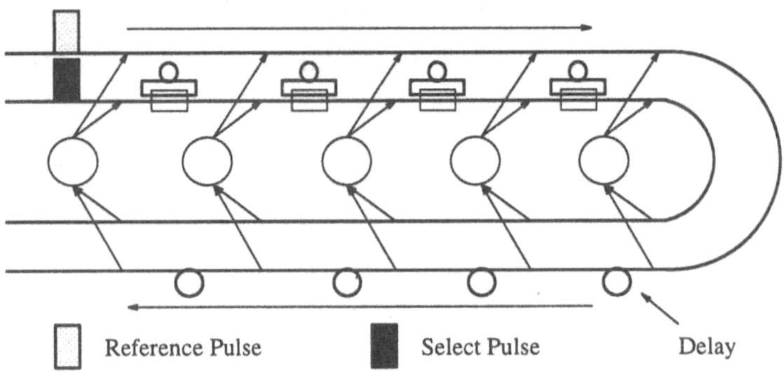

Figure 11.6 Switch settings for a broadcast operation.

11.3.2 Broadcast

All conditional switches are set to *straight*, thus introducing no delay on the transmitting segments (see Figure 11.6). The source processor sends a reference pulse at the beginning of its address frame. As described previously, the presence or absence of a select pulse determines if a processor should read the corresponding message. Thus, if the source processor sends N consecutive select pulses in its address frame on the select waveguide, as shown in Figure 11.7, every processor on the bus detects a double-height pulse and thus reads the message. This is clearly a broadcast operation. For example, when processor 0 in Figure 11.6 wants to broadcast a message, it sends a reference pulse at the beginning of its address frame on the reference bus and 5 select pulses in its address frame on the select bus. The first select pulse will meet the reference pulse at processor 4 since both pulses meet no delay on their buses. The second select pulse will meet the reference pulse at processor 3 since the select pulse meets no delay on the select bus and the reference pulse goes through one delay on the reference bus. The last select pulse will meet the reference pulse at processor 0 since the select pulse meets no delay on the select bus and the reference pulse goes through four delays on the reference bus.

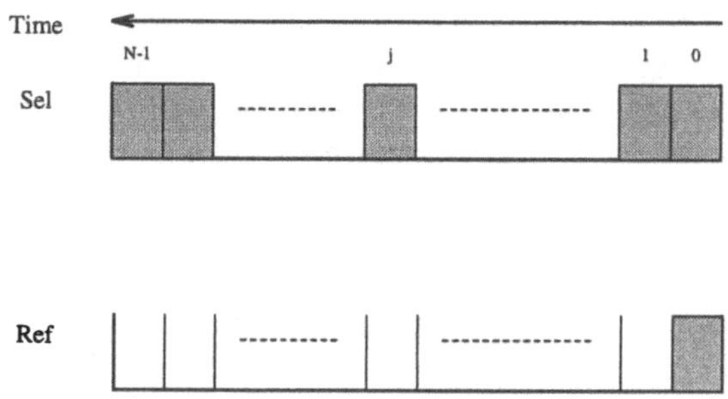

Figure 11.7 Address frames for broadcasting.

11.3.3 *Multicast*

Multicast is a one-to-many communication operation. Each processor may send a message to a group of processors in the system. Each processor receives only one message from a source processor during a bus cycle. This operation can be implemented in the following way: All conditional switches are set to straight, thus introducing no delay on the transmission segments (see Figure 11.6). Each source processor sends a reference pulse at the beginning of its address frame. If a source processor sends several select pulses in its address frame on the select waveguide as shown in Figure 11.7, the corresponding processors on the bus detect a double-height pulse and thus read the message.

11.3.4 *Compression*

Assume an array of N data elements with each processor having one data element. Also, assume that the number of active data elements in the array is s. Active elements are labeled based upon certain values of their local variables. A processor with an active element is referred to as an active processor. The compression algorithm moves these active data elements to processors $N-s-1$, $N-s$, ..., $N-1$. In other words, the compression algorithm moves all active data items to the right side of the array. In the following discussion, we assume that all active processors have their local variables $X(i)$ set to 1. In addition, we can also require the active elements remain in the same order, leading to the ordered compression problem. The ordered compression algorithm is implemented as follows: First, processor i sets its local switch $S(i)$ to cross if $X(i) = 1$, and to straight if $X(i) = 0$. Then, processor i, whose $X(i) = 1$, injects a reference pulse on the reference waveguide and a select pulse on the select waveguide at the beginning of a bus cycle. A processor also sends a message frame containing its local data in memory location D through the message waveguide during the bus cycle. Processors with $X(i) = 0$ do not put any pulse or message on the three waveguides. In other words, every active processor tries to address processor $N-1$. The select pulse sent by the processor whose index is the largest in the active set passes no delay in the transmitting segments because all the processors to its right are not in the active set and their corresponding switches are set to straight. Thus, the two pulses will meet only at processor $N-1$, and the corresponding message is picked up by processor $N-1$. Similarly, the select pulse sent by the processor whose index is the second largest in the active set passes one conditional delay in the transmitting segments because only one processor to its right is in the active set and its corresponding switch is set to cross. Since both the select and reference pulses pass one delay on the select and reference waveguides when arriving at processor $N-2$, the two pulses will meet only at processor $N-2$. Hence, processor $N-2$ receives the data item from the processor whose index is the second largest in the active set. In general, the select pulse sent by the processor whose index is the kth largest in the active set passes k conditional delays in the transmitting segments on the select waveguide because k processors to its right are in the active set and their corresponding switches are set to cross. Since both the select and reference pulses pass k delays on the select and reference waveguides when arriving at processor $N-k$, the two pulses will meet only at processor $N-k$. Clearly, this is the compression operation. Hence, the ordered compression problem can be performed on an LARPBS with N processors in one bus cycle.

A similar algorithm called *compaction* is obtained independently by Pavel and Akl [47]. However, their algorithm requires several bus cycles since it performs a binary prefix sum first followed by a permutation routing. Our algorithm requires only one bus cycle.

11.3.5 Split

In a split operation, it is desired to separate the active set from the inactive set. In other words, all data elements $D(i), 0 \leq i \leq N-1$, whose $X(i) = 1$ are moved to the upper part of the array PE($N-s$), PE($N-s+1$), ..., PE($N-1$), and all data elements $D(i), 0 \leq i \leq N-1$, whose $X(i) = 0$ are moved to the lower part of the array PE(0), PE(1), ..., PE($N-s-1$), where s is the number of elements in the active set. In other words, $D(i)$ is moved to $D(j)$ where $j = N-1-\sum_{k=i+1}^{N-1} B(k)$ if $X(i) = 1$, and $D(i)$ is moved to $D(j)$ where $j = \sum_{k=0}^{i-1} \overline{X(k)}$ if $X(i) = 0$.

The split operation is performed as follows: First, we label processor i whose $X(i) = 1$ as active; i.e., set $X(i)$ to 1 if and only if $X(i) = 1$. We call the compression algorithm to move all data elements in the active set to the upper part of the array. To avoid destroying the original data items in D, we copy them to a new memory location D_1. Second, we label all processors whose $X(i) = 0$ as active; i.e., set $X(i)$ to 1 if and only if $X(i) = 0$. Then, we call the compression algorithm to move all data elements in the set to the upper part of the array. To avoid destroying the data items in D and D_1, we copy them to a new memory location D_2. Third, move all data items in memory location

D_2 left s positions. This is a normal data transfer operation and can be carried out in one bus cycle. Finally we copy all moved data from temporary locations D_1 or D_2 to their corresponding local locations D. Hence, the split operation can be performed on an LARPBS with N processors in $O(1)$ bus cycle.

11.3.6 Binary Prefix Sums

Consider an LARPBS with N processors and N binary values v_i, $0 \leq i \leq N-1$. The binary prefix sums problem requires the computation of $psum_i = v_0 + v_1 + \cdots + v_{i-1}$, for all $0 \leq i \leq N-1$. This problem has been investigated by Pavel and Akl [47] and it is shown that $O(1)$ time is enough to solve the problem on the AROB model. The idea behind their algorithm is as follows: Processor 0 initiates a pair of pulses on the selection and reference waveguides. The time difference between the two pulses is incremented with one unit delay at each processor which has a 1 as input value. The select pulse arrives at the ith processor with k time-units delay relative to the reference pulse, where k is the number of 1-values stored by the processors with indices smaller than i, i.e., $psum_i = k$. The relative delay between two pulses is detected by each processor through a counter. The drawback with this scheme is that a high speed counter is required to count the relative delays between two pulses. In our opinion, it is impractical for an electronic processor to match the speed of an optical bus.

In this chapter, a new approach is proposed to avoid this problem. We will show that the binary prefix sums can be done in $O(1)$ bus cycle on an LARPBS without involving high speed counters. In fact, during a bus cycle, no processing is involved. Hence, as long as the processor speed matches the speed of a bus cycle, this scheme works. This also explains why a bus cycle can be considered a constant time.

First, processor i, $0 \leq i < N-1$, sets its switch $S(i)$ on the transmitting segment to *straight* if $a_i = 1$, and *cross* if $a_i = 0$. Second, processor i, $0 \leq i < N-1$, injects a reference and select pulse on the reference and select bus, respectively, at the beginning of a bus cycle. In other words, all processors try to address processor $N-1$. Thus, a train of $N-1$ address frames pass through the bus system. Now, let us consider a particular address frame sent by processor i, which will be called address frame i in the follow discussion. If processor j detects the coincidence of the reference pulse and the select pulse in address frame i, then $\sum_{s=i+1}^{N-1} x_s = N-1-j$. This can be explained as follows: Suppose that all processors to the right of processor i contains 0, all switches to the right of processor i on the select waveguide are set to straight, and thus no delay is introduced. As a result, the two pulses from processor i will coincide at processor $N-1$. Suppose that only one processor to the right of processor i contains a 1, only one switch to the right of processor i on the select waveguide is set to cross, and thus only one unit delay is introduced. As a result, the two pulses from processor i will coincide at processor $N-2$. In general, when $N-1-j$ processors to the right of processor i contain a 1, $N-1-j$ switches to the right of processor i on the select waveguide are to cross. Since there are $N-1-j$ unit delays on the select waveguide, the two pulses from processor

i will coincide at processor j. After this bus cycle, processor j simply sends a message containing its own address to processor i. Notice that it is possible that a processor may detect the coincidence of two pulses in several address frames. A naive method is to let processor j pick up all the messages and multicast a message to all those sending processors. This scheme is impractical since it requires an electronic processor to match the transmission speed of an optical bus.

To avoid this, a better method is adopted here. Assume that processors s and t contain a 1 and all processors between them contain a 0. Also assume that processor j receives a message from t. Clearly, processor j should also receive messages from processors $s+1, s+2, ..., t-1$ using the naive method. In the new scheme, we let processor j pick up only the message when detect the coincidence of two pulses the first time, and ignore the remaining coincidence of two pulses. Hence, processor j picks up a message only from t and sends a message only to t. However, since processors $s+1, s+2, ..., t-1$ should receive the same message as t does, we can simply split the bus at position t and let t broadcasts the received message, which contains an address j, to processors $s+1, s+2, ..., t-1$ in a bus cycle. In this way, all processors send (or receive) one message during a bus cycle.

Finally, when processor 0 receives a message containing an address j, it first calculates the sum of all binary numbers in the array $y = x_0 + (N-1-j)$, and broadcasts it to all processors on the bus. Processor i then gets its partial sum $x_0 + x_1 + x_2 + \cdots + x_i = y - (N-1-j)$. Thus, the binary prefix sums on the LARPBS can be computed in $O(1)$ time.

The above binary summation operation can be generalized to calculate the summation of N integers. Suppose that each integer has d bits, the sum can be calculated using d binary summations followed by a local summation in processor 0. For unbounded integer values and floating-point values, efficient algorithms also exist on the LARPBS model. The interested reader may refer to [23] for more details.

11.3.7 Selection

The selection problem is defined as finding the kth smallest in a given set of N elements. A randomized selection algorithm for the LARPBS is described in [37]. The basic idea behind the algorithm is to pick a random element as the "pivot" element and partition the input into two around the pivot; decide which partition the ith element is in; throw away the irrelevant partition and perform an appropriate selection in the relevant partition. Further, if the number of elements is much larger than the size of the LARPBS, the algorithm is still quite scalable. It is shown that the selection problem (from out of N elements) can be solved in $O(N \log N/p)$ expected number of bus cycles on an LARPBS of size p [37].

If each processor contains exactly one element, we can obtain a deterministic selection algorithm using only $O(\log N)$ bus cycles in the worst case. The algorithm is based on the one described in [25] and is an adaption of the sequential algorithm in [1]. The idea is similar to the randomized one described above. The only difference is that the pivot number is selected carefully as follows: during each iteration the N' active processors are partitioned into $N'/5$ groups with each group having exactly five active processors except for the last group, which may have fewer than five; find the median of each group, and then use the median of medians as the pivot value. In this way, we can guarantee that $O(\log N)$ bus cycles in the worst case are enough for the selection problem.

11.3.8 General Sorting

A randomized quicksort algorithm of N elements is proposed for the LARPBS model of size N in [38] and runs in $O(\log N)$ expected time. The divide-and-conquer strategy is applied to solve the sorting problem recursively. Suppose that C is the input array for a call to the sorting algorithm. In every iteration, we partition the set C into three disjoint subsets $C_1 = \{c \in C | c < v\}$, $C_2 = \{c \in C | c = v\}$ and $C_3 = \{c \in C | c > v\}$, where v is the pivot value of C selected randomly. We move all elements in C_1 to the left side of the array and move all elements in C_3 to the right side of the array. Then, we divide the array into three smaller subarrays each with a size of $|C_1|$, $|C_2|$, and $|C_3|$ respectively, and apply the same algorithm to the two subarrays containing C_1 or C_3. In this manner we can replace the given problem by two identical problems of smaller size. The main step used in each iteration is a split operation, and takes $O(1)$ bus cycles. Since the total average number of iterations of the quicksort algorithm for a set of size N is at most $O(\log N)$ [38], the total expected time of the sorting algorithm is $O(\log N)$ cycles. Further analysis shows that the quicksort algorithm is also scalable [40]. In other words, if the number of processors available is reduced to P, where $P < N$, the algorithm runs in $O((N/P)\log N)$ average time [40].

The above algorithm performs poorly in the worst case, requiring $O(N)$ bus cycles [38]. We can use a similar idea described in [25] to improve its worst case time complexity. If we use the deterministic selection algorithm described in the preceding section to find the median in the current set and use it as the pivot value, we can guarantee that the set is always divided into two subarrays of equal size. Since each iteration takes $O(\log N)$ cycles, $O(\log^2 N)$ cycles are enough in the worst case for the sorting algorithm.

11.3.9 Integer Sorting

The quicksort algorithm is suitable for general keys. If we restrict the values to integers, a radix sorting algorithm can be used. We show here that an LARPBS with N processors can sort N integers with k bits in $O(k)$ steps. The basic idea is to repeat the split algorithm k times, and each time uses the l-th bit as the mark for the active set, where $0 \leq l < k$. After k iterations, all N integers are in sorted order. Similar techniques and results are described in [25, 46, 52] for other models with optical buses.

11.3.10 Maximum Finding

Maximum finding can be carried out through the above sorting or selection algorithms. This requires either $O(\log N)$ expected time or $O(k)$ time. In this section, we show that it is possible to find the maximum of N numbers in constant time on an LARPBS of size $N^2/2$. The algorithm is not practical since it uses too many processors for such a simple problem. However, it is the basis of our improved maximum finding algorithm whose time complexity is $O(\log \log N)$ cycles using N processors.

Let us first discuss the implementation on an LARPBS with $N^2/2$ processors. Initially, processors $i, 0 \leq i < N$, hold the N numbers. We divide the array into N segments with each having $N/2$ processors. The first segment, i.e., processors $i, 0 \leq i < N/2$, multicast their data to their corresponding processors in other $N-1$ segments. The second segment, i.e., processors $i, N/2 \leq i < N$, multicast their data to their corresponding processors in other $N-1$ segments. So far, each processor receives two numbers. We reconfigure the whole array into N segments. Then processor $i/2$ in segment i, $0 \leq i < N$, broadcasts its local larger to other processors in its own segment. Each processor compares the received one with its local larger and marks its local variable $B = 0$ if the received one is larger, and $B = 1$ if smaller. A binary sum is performed over each segment. If a segment whose sum of B values is 0, the broadcast value in the segment is the largest one in the array.

Now let us turn to solving the problem of finding the maximum of N numbers on a linear array of size N. The method described below is based on the work by Kucera [19]. We partition the input into groups so that enough processors can be allocated to each group in order to find the maximum of that group in constant time by the above algorithm. As the computation progresses, the number of candidates for the maximum is reduced. This implies that the number of processors available per candidate increases, and so the group size can be increased. If the group size is s, then $s^2/2$ processors are needed to find the maximum of that group in $O(1)$ time. At the first stage, the size of each group is 2 and the maximum of each group can be found by one local comparison. At the second stage, we still have N processors, but the number of candidates has been reduced to $N/2$. We can make the size of each group 4 and have $N/8$ groups. Thus we allocate 8 processors per group which is sufficient for a group of size 4. At the third stage, the number of candidates has been reduced to $N/8$. We can make the size of each group 16 and have $N/(8 \times 16)$ groups. Thus we allocate 128 processors per group which is sufficient for a group of size 16. At the fourth stage, we have $N/(8 \times 16)$ remaining candidates. We can make the size of each group 256 and have $N/(8 \times 16 \times 256)$ groups and allocate 32768 processors per group. The above process can continue until only one candidate is left which is the maximum. Clearly, the size of the group can be squared at each stage. This implies that after $O(\log \log N)$ stages the maximum of the whole list can be found. Since each stage uses a constant number of cycles, finding the maximum of N numbers can be computed in $O(\log \log N)$ steps on an LARPBS of size N.

A randomized constant time algorithm for maximum finding is presented in [43]. Using the algorithm, the maximum of N elements can be found on an LARPBS of N processors in $O(1)$ time with high probability.

11.4 SCALABILITY ISSUES

A computer system, including hardware, system software, and application software, is called *scalable* if it can be *scaled up* to accommodate ever increasing users demand, or *scaled down* to improve cost-effectiveness. Since most algorithms on bus-based models are designed for very large arrays, most research is mainly concerned with algorithms on machines with scaled down size. If a model is scalable, then a programmer need not be so concerned with the actual size of the machine on which a program is to run. A compiler can automatically map logical processors to physical processors to run a program designed for a larger machine on a given smaller size machine. Without scalability, different problem sizes would call for different algorithms, and so different programs, to run in the best possible time on the same available machine. This is not at all a feasible approach. For a scalable model, algorithmic results have significance whether or not problem size matches machine size.

There has been some research going on to investigate the scalability issue of various models with reconfigurable buses [4, 57]. These authors investigated the ability of a model of a given size to efficiently simulate a model of the same type, but large size. In [57], two new concepts, i.e., *completely scalable* and *relatively scalable*, are proposed to address the scalability of reconfigurable meshes. Informally, a model is *completely scalable* if the *scalability factor* (overhead due to the scalability simulation) is constant and *relatively scalable* if the scalability factor is a function of the simulating (but not the simulated) model size [57]. If a model is relatively scalable, then it can run an algorithm designed for any larger size instance of the model without significant loss of efficiency on a given available machine. Limitations of existing scalability simulation of reconfigurable models encourage a focus on efficiently scaling fundamental algorithms, rather than the general model. Following this idea, Trahan *et al.* [58] studied *algorithmic scalability* of the LARPBS model. *Algorithmic scalability* refers to this ability to efficiently run an algorithm designed for a larger-sized instance of a model on a smaller-sized instance.

Trahan *et al.* [58] presented a set of algorithms that scale existing algorithms from N processors to P processors, where $P < N$. In other words, given an algorithm on an LARPBS of N processors, what is the time complexity of the algorithm on an LARPBS of P processors? To answer this question, they first need to consider the mapping of N input elements to P processors. Two obvious schemes exist. The first one, *cyclic mapping*, maps element x to processor $x \bmod P$. The second, *block mapping*, divides an array into P contiguous chunks and maps the ith chunk of $\frac{N}{P}$ elements to the ith processor. Their results [58] indicate that conversion between the block mapping and the cyclic mapping of N elements can be performed by a P-processor LARPBS in $O(\frac{N}{P})$ time. Since scaling an algorithm from N to P processors requires $\Omega(\frac{N}{P})$

time and the LARPBS can convert between mappings freely in $O(\frac{N}{P})$ time, scaling algorithms can assume either mapping.

The results described by Trahan *et al.* [58, 59] shows that all the basic algorithms described in the preceding section in this chapter are completely scalable except for the one-to-one communication algorithm which is relatively scalable on the LARPBS model. The results are very significant since scaling of algorithms (or, more generally, operations) indirectly reflects on the scaling of models. Specifically, scalability of a body of operations and a collection of communication patterns corresponds to scaling a restricted version of the model, one restricted to those operations and patterns of communication. Consequently, the restricted scalable model may still be quite general and powerful. Scalability analysis of the LARPBS in general seems to be very hard. In particular, efficiently simulating an LARPBS with a larger size by an LARPBS with a smaller size would be a very interesting research topic in the future.

11.5 CONCLUSIONS

In this chapter, we have described the important hardware features and the addressing scheme of the LARPBS model. Several basic data movement operations on the LARPBS model are also discussed. It is shown that many of these operations can be computed efficiently in the communication system (the reconfigurable pipelined optical bus) without involving any processor execution during a bus cycle. This is crucial because the processing speed of a processor is much slower than the optical bus, and it is impossible to perform even simple computations such as counting during a bus cycle. This is also the major difference between the LARPBS model described in [38, 40] and the AROB model used in [47, 52].

Since the introduction of the LARPBS model in [38], two similar models based on a linear optical bus system have been proposed. The ***linear pipelined bus*** (LPB) model proposed by Trahan *et al.* [59] is identical to the LARPBS model except it does not have the ability to reconfigure its bus; i.e., no segment switch is needed in the LPB. The segment switches allow the LARPBS to split into subsystems such that messages cannot travel from one subarray to another. Another model, called the ***pipelined optical bus*** (POB), that is similar to the LARPBS is a model proposed by Li *et al.* [25]. Both the LARPBS model and the POB model posses three waveguides and use the coincident pulse technique to receive messages. The POB is able to employ the same addressing schemes as the LARPBS. Conditional delay switches are positioned on the receiving side of the reference and message waveguides, rather than on the transmitting side of the select line. Another difference is that there are no segment switches in the model, therefore reconfiguration is not possible. Li *et al.* [25] were able to show that divide-and-conquer algorithms such as sorting and selection can still be implemented efficiently on the POB model even though it does not have the reconfiguration capability. Recently, Trahan *et al.* [59] proved that the three models are equivalent; i.e., each model can simulate any step of one of the other models in $O(1)$ bus cycles. Due to the equivalence result, we can perform an algorithm on any of the three models with only a constant slow-down factor.

Unlike many theoretical models, such as the PRAM, the LARPBS model is implementable and practical using current optical technologies. Due to its high communication bandwidth, support of versatile communication patterns, and reconfigurability, many problems can be solved more efficiently on the LARPBS model than on a PRAM model. In fact, many new results achieved so far indicate that this is true especially for many communication intensive problems [8, 18, 23, 24, 37, 38, 43]. Our model can also be extended to higher dimensional arrays. In this way, the time for a bus cycle can be reduced and complicated bus connections can be reconfigured during a computation. We believe that many graph problems and image problems can be solved more efficiently on a higher dimensional model using the results obtained for reconfigurable meshes [2, 26, 31, 35]. Further research is being carried out in these directions.

Acknowledgments

This research was supported in part by the NSF under Grants CCR-9211621 and CCR-9503882, by the U.S. Air Force under Grant F33615-C-2218, by an Ohio Board of Regents Investment Fund Competition Grant, and by an Ohio Board of Regents Research Challenge Grant.

References

[1] S. G. Akl, *Parallel Computation: Models and Methods,* Prentice-Hall, Upper Saddle River, New Jersey, 1997.

[2] S. G. Akl, J. W. Baker, and Y. Pan, eds., Special Issue on Computing on Bus-Based Architectures of *Parallel Processing Letters,* 1998.

[3] A. F. Benner, H. F. Jordan, and V. P. Heuring, "Digital optical computing with optically switched directional couplers," *Optical Engineering,* vol. 30, no. 12, pp. 1936-1941, 1991.

[4] D. Ben-Asher, A. Gordon, A. Schuster, "Efficient self-simulation algorithms for reconfigurable arrays," *Journal of Parallel and Distributed Computing,* vol. 30, no. 1, pp. 1-22, 1995.

[5] S. H. Bokhari, "Finding maximum on an array processor with a global bus," *IEEE Trans. on Computers,* vol. C-32, no. 2, pp. 133-139, 1984.

[6] D. Chiarulli, R. Melhem, and S. Levitan, "Using coincident optical pulses for parallel memory addressing," *IEEE Computer,* vol. 20, no. 12, pp. 48-58, 1987.

[7] D. M. Chiarulli, S. P. Levitan, R. G. Melhem, M. Bidnurkar, R. Ditmore, G. Gravenstreter, Z. Guo, C. Qiao, M. Sakr, and J. P. Teza, "Optoelectronic buses for high-performance computing," *Proceedings of IEEE*, vol. 82, no. 11, pp. 1701-1709, Nov. 1994.

[8] B. Cong, "Mapping of ANNs on linear array with a reconfigurable pipelined bus system," *Proc. of 1997 International Conference on Parallel and Distributed Processing Techniques and Applications,* June 30-July 3, 1997, Las Vegas, USA, pp. 522-529.

[9] P. W. Dowd, "Wavelength division multiple access channel hypercube processor interconnection," *IEEE Trans. on Computers,* vol. 41, no. 10, pp. 1223–1241, Oct. 1992.

[10] Z. Guo, R. Melhem, R. Hall, D. Chiarulli, and S. Levitan, "Pipelined communication in optically interconnected arrays," *Journal of Parallel and Distributed Computing,* vol. 12, no. 3, pp. 269-282, 1991.

[11] Z. Guo, "Sorting on array processors with pipelined buses," *1992 International Conference on Parallel Processing,* pp. 289-292, St. Charles, IL, Aug. 17-21, 1992.

[12] Z. Guo, "Optically interconnected processor arrays with switching capability," *Journal of Parallel and Distributed Computing,* vol. 23, pp. 314-329, 1994.

[13] M. Hamdi, "Communications in optically interconnected parallel computer systems," in *Interconnection Networks and Mapping and Scheduling Parallel Computations,* ed. by D.F. Hsu, A.L. Rosenberg, and D. Sotteau, pp. 181- 200, American Mathematical Society, 1995.

[14] M. Hamdi and Y. Pan, "Efficient parallel algorithms on optically interconnected arrays of processors," *IEE Proceedings - Computers and Digital Techniques,* vol. 142, no. 2, pp. 87-92, March 1995.

[15] M. Hamdi and Y. Pan, "Communication-efficient algorithms on reconfigurable array of processors with spanning optical buses," *Proc. of IEEE International Symposium on Parallel Architectures, Algorithms, and Networks,* pp. 440-446, June 12-14, 1996.

[16] H. F. Jordan, "Exploiting precise delay in optical multiprocessor interconnects," *Proc. of the 2nd International Conference on Massively Parallel Processing Using Optical Interconnections,* p. 357, October 1995.

[17] M. A. Karim and M. A. G. Abushagur, eds., Special Issue on Optical Computing of *Optics and Laser technology,* vol. 26, no. 4, 1994.

[18] H. Kimm, "Inversion number algorithm on a linear array with a reconfigurable pipelined bus system," *Proc. of 1996 International conference on Parallel and Distributed Processing Techniques and Applications,* pp. 1398-1408, Sunnyvale, CA, August 9-11, 1996.

[19] L. Kucera, "Parallel computation and conflicts in memory access," *Information Processing Letters,* vol. 14, pp. 93-96, 1982.

[20] V. Prasanna Kumar and C. S. Raghavendra, "Array processor with multiple broadcasting," *J. Parallel Distributed Computing,* vol. 4, no. 2, pp. 173-190, Apr. 1987.

[21] S. Levitan, D. Chiarulli, and R. Melhem, "Coincident pulse techniques for multiprocessor interconnection structures," *Applied Optics,* vol. 29, no. 14, pp. 2024-2039, 1990.

[22] K. Li, "Constant time boolean matrix multiplication on a linear array with a reconfigurable pipelined bus system," *Journal of Supercomputing,* vol. 11, no. 4, pp. 391-403, 1997.

[23] K. Li, Y. Pan, and S. Q. Zheng, "Fast and processor efficient parallel matrix multiplication algorithms on a linear array with a reconfigurable pipelined bus system," to appear in *IEEE Transactions on Parallel and Distributed Systems.*

[24] K. Li, Y. Pan, and S. Q. Zheng, "Fast and efficient parallel matrix computations on a linear array with a reconfigurable pipelined optical bus system," in *High Performance Computing Systems and Applications,* Jonathan Schaeffer and Ron Unrau, eds., Kluwer Academic Publishers, Boston, USA, 1998.

[25] Y. Li, Yi Pan and S.Q. Zheng, "Pipelined time-division multiplexing optical bus with conditional delays," *Optical Engineering,* vol. 36, no. 9, pp. 2417-2424, September 1997.

[26] R. Lin and S. Olariu, eds., Special Issue on High Performance Bus-Based VLSI Architectures of *VLSI Design,* 1997.

[27] A. Louri, "Three-dimensional optical architecture and data-parallel algorithms for massively parallel computing," *IEEE Micro,* vol. 11, no. 2, pp. 24-81, April 1991.

[28] R. Melhem, D. Chiarulli, and S. Levitan, "Space multiplexing of waveguides in optically interconnected multiprocessor systems," *The Computer Journal,* vol. 32, no. 4, pp. 362-369, 1989.

[29] R. Melhem and D. Chiarulli, eds., Special Issue on Optical Computing and Interconnection Systems of *Journal of Parallel and Distributed Computing,* vol. 17, no. 3, 1993.

[30] R. Miller, V. K. Prasanna-Kumar, D. Reisis, and Q.F. Stout, "Meshes with Reconfigurable Buses," *MIT Conference on Advanced Research in VLSI,* pp. 163-178, 1988.

[31] R. Miller and A. Schuster, eds., Special Issue on Dynamically Reconfigurable Architectures of *Parallel Processing Letters,* vol. 5, no. 1, pp. 1-124, 1995.

[32] K. Nakano, T. Masuzawa, and N. Tokura, "A sublogarithmic time sorting algorithm on a reconfigurable array," *IEICE Transactions,* vol. E 74, no. 11, pp. 3894-3901, November 1991.

[33] S. Olariu, J. L. Schwing, and J. Zhang, "On the power of two-dimensional processor arrays with reconfigurable bus systems," *Parallel Processing Letters,* vol. 1, no. 1, pp. 29-34, 1991.

[34] S. Olariu, J. L. Schwing, and J. Zhang, "Fast computer vision algorithms on reconfigurable meshes," *Image and Vision Computing Journal,* vol. 10, no. 9, pp. 610-616, November 1992.

[35] S. Olariu, ed., Special Issue on Algorithms for Enhanced Mesh Architectures of *Journal of Parallel Algorithms and Applications,* vol. 8, 1996.

[36] Y. Pan, "Hough transform on arrays with an optical bus," *Fifth ISMM International conference on Parallel and Distributed Computing and Systems,* pp. 161-166, Pittsburgh, PA, October 1-3, 1992.

[37] Y. Pan, "Order statistics on optically interconnected multiprocessor systems," *Proc. of the First International Workshop on Massively Parallel Processing Using Optical Interconnections,* pp. 162-169, Cancun, Mexico, April 26-27, 1994.

[38] Y. Pan and M. Hamdi, "Quicksort on a linear array with a reconfigurable pipelined bus system," *Proc. of IEEE International Symposium on Parallel Architectures, Algorithms, and Networks,* pp. 313-319, June 12-14, 1996.

[39] Y. Pan and M. Hamdi, "Singular value decomposition on processor arrays with a pipelined bus system," *Journal of Network and Computer Applications,* vol. 19, no. 3, pp. 235-248, July 1996. A preliminary version also appeared in 1993 *ACM Symposium on Applied Computing,* pp. 525-532.

[40] Y. Pan, M. Hamdi, and K. Li, "Efficient and scalable quicksort on a linear array with a reconfigurable pipelined bus system," *Future Generation Computer Systems,* vol. 13, no. 6, pp. 501-513, 1998.

[41] Y. Pan and K. Li, "Linear array with a reconfigurable pipelined bus system - concepts and applications," *Proc. of 1996 International Conference on Parallel and Distributed Processing Techniques and Applications,* August 9-11, 1996, Sunnyvale, California, USA, pp. 1431-1442.

[42] Y. Pan, K. Li, and M. Hamdi, eds., Special Issue on Parallel Computing with Optical Interconnections of *Informatica - An International Journal,* to appear in 1998.

[43] Y. Pan, K. Li, and S.-Q. Zheng, "Fast nearest neighbor algorithms on a linear array with a reconfigurable pipelined bus system," to appear in *Parallel Algorithms and Applications.*

[44] S. Pavel and S. G. Akl, "Matrix operations using arrays with reconfigurable optical buses," *Parallel Algorithms and Applications,* vol. 11, pp. 223-242, 1996.

[45] S. Pavel and S. G. Akl, "Efficient algorithms for the Hough transform on arrays with reconfigurable optical buses," *Proc. of the 10th International Parallel Processing Symposium,* pp. 697-701, April 1996.

[46] S. Pavel and S. G. Akl, "Integer sorting and routing in arrays with reconfigurable optical buses," *Proc. of 1996 International Conference on Parallel Processing,* vol. III, pp. 90-94, August 1996.

[47] S. Pavel and S. G. Akl, "On the power of arrays with optical pipelined buses," *Proc. of 1996 International Conference on Parallel and Distributed Processing Techniques and Applications,* pp. 1443-1454, Sunnyvale, California, August 9-11, 1996.

[48] S. Pavel, "Computation and communication aspects of arrays with optical pipelined buses," PhD Dissertation, Department of Computing and Information Science, Queen's University, Canada, October 1996.

[49] C. Qiao, R. Melhem, D. Chiarulli, and S. Levitan, "Optical multicasting in linear arrays," *International Journal of Optical Computing,* vol. 2, no. 1, pp. 31-48, 1991.

[50] C. Qiao and R. Melhem, "Time-division optical communications in multiprocessor arrays," *IEEE Transactions on Computers,* vol. 42, no. 5, pp. 577-590, May 1993.

[51] C. Qiao, "Efficient matrix operations in a reconfigurablc array with spanning optical buses," *Proc. 5th IEEE Symposium on the Frontiers of Massively Parallel Computations,* pp. 273-280, 1995.

[52] S. Rajasekaran and S. Sahni, "Sorting, selection and routing on the arrays with reconfigurable optical buses," *IEEE Transactions on Parallel and Distributed Systems,* vol. 8, no. 11, pp. 1123-1132, Nov. 1997.

[53] S. Ramanan and H. Jordan, "Serial array shuffle- exchange architecture for universal permutation of time slots," *SPIE Proc. of Digital Optical Computing II,* vol. 1215, pp. 330-342, Jan. 1990.

[54] M. M. Eshaghian and E. Schenfeld, ed., Special Issue on Parallel Computing with Optical Interconnects of *Journal of Parallel and Distributed Computing,* vol. 41, no. 1, 1997.

[55] Q. F. Stout, "Mesh Connected Computers with Broadcasting," *IEEE Trans. on Computers,* vol. C-32, no. 9, pp. 826-830, Sept. 1983.

[56] J. L. Trahan, R. Vaidyanathan, and C. P. Subbaraman, "Constant time graph and poset algorithms on the reconfigurable multiple bus machine," *Proc. Int. Conf. on Parallel Processing,* vol. III, pp. 214-217, 1994.

[57] J. Trahan and R. Vaidyanathan, "Relative scalability of the reconfigurable multiple bus machine," *Proc. Workshop Reconfigurable Arch. and Algs.*, 1996.

[58] J. L. Trahan, Y. Pan, R. Vaidyanathan, and A. G. Bourgeois, "Scalable basic algorithms on a linear array with a reconfigurable pipelined bus system," *Proc. of 10th International Conference on Parallel and Distributed Computing Systems,* Oct. 1-3, 1997, New Orleans, LA, USA, pp. 564-569.

[59] J. L. Trahan, Y. Pan, R. Vaidyanathan, and A. G. Bourgeois, "Scalable algorithms and simulation results for a linear array with a reconfigurable pipelined bus system," submitted for publication.

[60] B. F. Wang, G. H. Chen, and F. C. Lin, "Constant time sorting on a processor array with a reconfigurable bus system," *Information Processing Letters,* vol. 34, no. 4, pp. 187-192, April 1990.

12 FAST MATRIX MULTIPLICATION AND RELATED OPERATIONS USING RECONFIGURABLE OPTICAL BUSES

Keqin Li

Department of Mathematics and Computer Science
State University of New York
New Paltz, New York 12561-2499
li@mcs.newpaltz.edu

Abstract: We present efficient parallel matrix multiplication algorithms for linear arrays with reconfigurable pipelined bus systems (LARPBS). The main contributions are as follows. We develop five matrix multiplication algorithms with varying degrees of parallelism on the LARPBS computing model, namely, MM_1, MM_2, MM_3, and compound algorithms $\mathcal{C}_1(\epsilon)$ and $\mathcal{C}_2(\delta)$. Algorithm $\mathcal{C}_1(\epsilon)$ has adjustable time complexity in sub-linear level. Algorithm $\mathcal{C}_2(\delta)$ implies that it is feasible to achieve sub-logarithmic time using $o(N^3)$ processors for matrix multiplication on a realistic system. Algorithms MM_3, $\mathcal{C}_1(\epsilon)$, and $\mathcal{C}_2(\delta)$ all have $o(N^3)$ cost, and hence, are very processor efficient. Algorithms MM_1, MM_3, and $\mathcal{C}_1(\epsilon)$ are general-purpose matrix multiplication algorithms, where the array elements are in any ring. Algorithms MM_2 and $\mathcal{C}_2(\delta)$ are applicable to array elements that are integers of bounded magnitude, or floating-point values of bounded precision and magnitude, or boolean values. Extension of algorithms MM_2 and $\mathcal{C}_2(\delta)$ to unbounded integers and reals, as well as the application of our matrix multiplication algorithms in solving a number of important and fundamental matrix computation problems, are also discussed.

12.1 INTRODUCTION

Matrix multiplication is one of the most fundamental and important problems in science and engineering. Many other important matrix problems can be solved via matrix multiplication, e.g., finding the Nth power, the inverse, the determinant, eigenvalues, an LU-factorization, and the characteristic polynomial of a matrix, and the product of a sequence of matrices [15, 19]. Many graph theory problems are also reduced to matrix multiplication. Examples are finding the transitive closure, all-pairs shortest paths, the minimum-weight spanning tree, topological sort, and critical paths of a graph [9]. Therefore, fast and processor efficient parallel algorithms for matrix multiplication is definitely of fundamental importance.

The standard matrix multiplication algorithm takes $O(N^3)$ operations. Most existing parallel algorithms are parallelizations of the standard method. For instances, matrix multiplication can be performed in $O(N)$ time on a mesh with wraparound connections and $N \times N$ processors [5]; in $O(\log N)$ time on a three-dimensional mesh of trees with N^3 processors [15]; in $O(\log N)$ time on a hypercube or shuffle-exchange network with N^3 processors [9]. It is clear that all implementations of the standard method have *cost* , i.e., time-processor product, at least $\Omega(N^3)$. Therefore, it is interesting to develop highly parallel and processor efficient algorithms that have $o(N^3)$ cost. It turns out that such an algorithm should not be a parallelization of the standard matrix multiplication algorithm.

Extensive research has been done to develop algorithms that require $O(N^\beta)$ operations, such that $\beta < 3$. Strassen was the first one who was able to find an $O(N^{\log 7})$ time algorithm, where $\log 7 < 2.8074$, that recursively reduces the calculation of the product of two matrices of size $N \times N$ to the calculation of the products of seven submatrices of size $N/2 \times N/2$ [39]. Chandra implemented Strassen's algorithm on shared memory multiprocessors in $O(\log N)$ time using $N^{2.8074}/\log N$ processors [6]. It is well known that matrix product can be computed by a CREW PRAM in $O(\log N)$ time, using $M(N)$ processors [31], where $M(N) = O(N^\beta)$ is the number of operations done by the best sequential algorithm. Since Strassen's initial work, the value β has been reduced a number of times. The currently best matrix multiplication algorithm takes $O(N^\beta)$ operations, where $\beta < 2.375477$ [8]. However, these algorithms are highly sophisticated even for sequential execution (the constant involved in the $O(N^\beta)$ term is so large that the break-even point with the standard matrix multiplication algorithm is way above practically interesting matrix sizes); parallelization of these methods are only of theoretical interest, and far from practical (see, e.g., [30, page 624] and [15, page 312]). The reader is referred to [4] for more information on parallel matrix algorithms.

A parallel algorithm on the PRAM model only reveals the maximum parallelism of a problem. Large scale parallel computing on a shared memory system is impractical, at least using the current technology. As for distributed memory systems, to the best of the author's knowledge, even the Strassen's algorithm has not been parallelized on any popular static network with time and processor complexities $O(\log N)$ and $O(N^{2.8074})$ respectively. (Note that $\Omega(\log N)$ time is inherent in Strassen's recursive method, and is independent of any implementation on any system.) It is unlikely that the Strassen's algorithm can be parallelized on static networks with the above mentioned performance, due to its large amount of data transfer, and mismatch between its communication patterns and the structure of most of the commonly used networks.

Efficient communication is the most difficult issue in designing a high performance parallel system, which includes architecture and technology that support high communication bandwidth, and an algorithm that involves less communication. The communication capacity of static networks, such as meshes and hypercubes, is limited by their finite node degrees; while communication latency is in proportional to network diameters. Hence, increasing the size of a network may not result in further decrease in the time complexity of a parallel algorithm running on a static network. As a matter of fact, the time complexity of a nontrivial parallel computation is bounded from below by the network diameter.

One way to overcome this difficulty is to use electronic/optical buses for communication since they provide direct connection between any two processors in the system. Processor arrays with various bus systems have become increasingly interesting due to recent advances in VLSI and fiber optic technologies. Arrays of processors with global buses, multiple buses, hyperbuses, multi-dimensional bus architectures, and reconfigurable buses have been proposed for efficient parallel computations. The time taken by a datum to go from one end of an electronic bus to the other end is a function of the length L of the bus. These functions can be L^2, L, or $\log L$, depending on the physical properties of the technology used to implement the bus [2]. Similarly, the time for a signal to traverse a given section of the bus is also unpredictable. Other difficulties associated with electronic buses are low bandwidth (since they cannot support concurrent accesses), temperature sensitivity, capacitive loading, and cross talk caused by mutual inductance.

On the other hand, fiber optic communication technologies offer a combination of high bandwidth, predictable message delay, low interference and error probability, and gigabit transmission capacity, and have been used extensively in wide-area networks. Based on the characteristics of fiber optical communications, several researchers have proposed using optical interconnections to connect processors in a parallel computer system [3, 10, 12, 37, 40, 42]. In such a system, messages can be transmitted concurrently on a pipelined optical bus without partitioning the bus into several segments, while the time delay between the furthest processors is only the end-to-end propagation delay of light over a waveguided bus. This design integrates the advantages of both optical transmission and electronic computation, and leads to an entirely new realm of techniques for parallel algorithm development. Many parallel algorithms, such as the Hough transform, singular value decomposition, order statistics, sorting, and some numerical algorithms, have been proposed for systems with optical interconnections [11, 13, 24, 25, 26, 33, 34, 36, 38]. These works indicate that arrays with pipelined optical buses are very efficient for parallel computation due to the high bandwidth within a pipelined bus system.

Recently, linear arrays with reconfigurable pipelined bus systems (LARPBS) and arrays with reconfigurable optical buses (AROB) have been independently proposed by Pan, Hamdi, and Li [27, 28], and Pavel and Akl [34]. In these systems, messages can be transmitted concurrently on a bus in a pipelined fashion and the bus can be reconfigured dynamically under program control to support algorithmic requirements. Many algorithms have been designed for these two models, including selection, quicksort, matrix operations, computational geometry, and PRAM simulation on the LARPBS model [17, 18, 19, 23, 27, 29], and on the AROB model [33, 34, 35]. More recently, a linear array with a different pipelined optical bus system has been proposed in [22, 42]. This system can implement most algorithms designed for LARPBS with much less hardware.

In this chapter, we present efficient parallel matrix multiplication algorithms for linear arrays with reconfigurable pipelined bus systems. This kind of systems are able to support a large volume of parallel communication of various patterns in constant time. An LARPBS can also be reconfigured into many independent subsystems, thus, is able to support parallel implementations of divide-and-conquer computations like Strassen's algorithm. The main contributions are as follows. We develop five matrix multiplication algorithms with varying degrees of parallelism on the LARPBS computing model:

- A linear time algorithm MM_1 (applicable to an arbitrary ring or a closed semiring) using N^2 processors, and each processor needs constant amount of memory;

- A constant time algorithm MM_2 using N^3 processors, and each processor needs constant amount of memory, where we assume that array elements are integers of bounded magnitude, or floating-point values of bounded precision and magnitude, or boolean values.

- A parallelized Strassen's algorithm MM_3 (applicable to an arbitrary ring) with time complexity $O(\log N)$, using $O(N^{\log 7})$ processors, and each processor needs $O(\log N)$ amount of memory;

- A compound algorithm $\mathcal{C}_1(\epsilon)$ (applicable to an arbitrary ring) of time complexity $O(\log N + N^{1-\epsilon})$, where $0 \le \epsilon \le 1$, using $O(N^{2+0.8074\epsilon})$ processors, and each processor needs $O(\log N)$ amount of memory;

- A compound algorithm $\mathcal{C}_2(\delta)$ with time complexity $O((\log N)^{\delta})$, where $0 \le \delta \le 1$, using $O(N^3/(\frac{8}{7})^{(\log N)^{\delta}})$ processors, and each processor needs $O(\log N)$ amount of memory, and array elements are integers of bounded magnitude, or floating-point values of bounded precision and magnitude.

Algorithm $\mathcal{C}_1(\epsilon)$ has adjustable time complexity in sub-linear level. Algorithm $\mathcal{C}_2(\delta)$ implies that it is feasible to achieve sub-logarithmic time using $o(N^3)$ processors for matrix multiplication on a realistic system. Algorithms MM_3, $\mathcal{C}_1(\epsilon)$, and $\mathcal{C}_2(\delta)$ all have $o(N^3)$ cost, and hence, are very processor efficient.

Algorithms MM_1, MM_3, and $\mathcal{C}_1(\epsilon)$ are general-purpose matrix multiplication algorithms, where the array elements are in any ring. Notice that the restriction on the array elements in algorithms MM_2 and $\mathcal{C}_2(\delta)$ are actually made by virtually all sequential and parallel algorithms, since it is usually assumed that an arithmetic operation on array elements takes constant time. Otherwise, even for such a primitive operation (for integers with unbounded magnitude, or real values with unbounded precision and magnitude), one of the resources, i.e., execution time or processor hardware, should be increased in proportion to magnitude/precision. Nevertheless, extension of algorithms MM_2 and $\mathcal{C}_2(\delta)$ to unbounded integers and reals are also discussed.

In a later section of the chapter, we will briefly discuss the application of our matrix multiplication algorithms in solving a number of important and fundamental matrix computation problems on LARPBS. These problems include computing the inverse, the characteristic polynomial, the determinant, the rank, the Nth power, an LU- and a QR-factorization of a matrix, and solving linear systems of equations. Our algorithms provide a wide range of performance-cost combinations. Compared with known results, the running time of parallel solutions to all these problems can be reduced by a factor of $O(\log N)$ while maintaining costs under $o(N^4)$.

12.2 THE LARPBS COMPUTATIONAL MODEL

A linear array with a reconfigurable pipelined bus system (LARPBS) is a distributed memory system, which consists of processors linearly connected by a reconfigurable pipelined optical bus system. Each processor can perform ordinary arithmetic and logic computations, and interprocessor communication. All computations and communications are synchronized by bus cycles, which is similar to an SIMD machine. However, due to reconfigurability, an LARPBS can be divided into subsystems, and these subsystems can operate independently to solve different subproblems, which results in an MIMD machine. Input/output can be handled in a way similar to an ordinary SIMD/MIMD machine; further discussion on this issue is beyond the scope of the chapter, since we are mainly interested in algorithm development and analysis.

A pipelined optical bus system uses optical waveguides instead of electrical signals to transfer messages among (electronic) processors. In addition to the high propagation speed of light, there are two important properties of optical signal (pulse) transmission on an optical bus, namely, unidirectional propagation and predictable propagation delay. (Notice that electronic buses lack these characteristics.) These advantages of using waveguides enable synchronized concurrent accesses of an optical bus in a pipelined fashion [7, 16]. Such pipelined optical bus systems can support a massive volume of communications simultaneously, and are particularly appropriate for applications that involve intensive communication operations such as broadcasting, one-to-one communication, multicasting, global aggregation, and irregular communication patterns.

An LARPBS consists of N processors P_0, P_1, P_2, ..., P_{N-1} connected by an optical bus. In addition to the tremendous communication capabilities, an LARPBS can also be partitioned into $k \geq 2$ independent subarrays LARPBS$_1$, LARPBS$_2$, ..., LARPBS$_k$, such that LARPBS$_j$ contains processors $P_{i_{j-1}+1}$, $P_{i_{j-1}+2}$, ..., P_{i_j}, where $-1 = i_0 < i_1 < i_2 \cdots < i_k = N-1$. The subarrays can operate as regular linear arrays with pipelined optical bus systems, and all subarrays can be used independently for different computations without interference (see [21, 28] and Chapter 11 for an elaborated exposition).

As in many other synchronous parallel computing systems, an LARPBS computation is a sequence of alternate global communication and local computation steps. The time complexity of an algorithm is measured in terms of the total number of bus cycles in all the communication steps, as long as the time of the local computation steps between successive communication steps is bounded by a constant and independent of the problem size. This complexity measure implies that a *bus cycle* (i.e., the time for an optical pulse to propagate through the entire bus) takes constant time, and this assumption has been adopted widely in the literature. (Remark: To avoid controversy, let us emphasize that by "$O(f(p))$ time" we mean $O(f(p))$ bus cycles for global communication plus $O(f(p))$ number of local arithmetic/logic operations.)

For ease of algorithm development and specification, a number of basic communication, data movement, and global operations on the LARPBS model implemented using the *coincident pulse processor addressing technique* [7, 16, 37] have been developed [21, 28]. Each of these primitive operations can be performed in a constant number of bus cycles. These powerful primitive operations that support massive parallel communications, plus the reconfigurability of the LARPBS model, make the LARPBS very attractive in solving problems that are both computation and communication intensive, such as matrix manipulations. Optical buses are not only communication channels among the processors, but also active components and agents of certain computations, e.g., global data aggregations. The unique characteristics of optical buses have led to an entirely new computational model, and opened up rich avenues of algorithmic design, and novel and interesting algorithms (see Chapter 10 of [2]).

The following primitive operations on LARPBS have been used extensively in our research. Our algorithms are developed by using these operations as building blocks.

One-to-One Communication. Assume that processors P_{i_1}, P_{i_2}, ..., P_{i_m} are senders, and processors P_{j_1}, P_{j_2}, ..., P_{j_m} are receivers. In particular, processor P_{i_k} sends its value in its register $\mathtt{R}(i_k)$ to the register $\mathtt{R}(j_k)$ in P_{j_k}. The operation is represented as

```
for 1 ≤ k ≤ m do in parallel
    R(j_k) ← R(i_k)
endfor
```

(Note that we use $\mathtt{R}(i)$ to denote both the name and the content of register $\mathtt{R}(i)$.)

Broadcasting. Here, we have a source processor P_i, who sends a value in its register $\mathtt{R}(i)$ to all the N processors:

$$\mathtt{R}(0), \mathtt{R}(1), \mathtt{R}(2), ..., \mathtt{R}(N-1) \leftarrow \mathtt{R}(i)$$

Multiple Multicasting. In a multicasting operation, we have a source processor P_i, who sends a value in its register $\mathtt{R}(i)$ to a subset of the N processors $P_{j_1}, P_{j_2}, ..., P_{j_m}$:

$$\mathtt{R}(j_1), \mathtt{R}(j_2), ..., \mathtt{R}(j_m) \leftarrow \mathtt{R}(i)$$

Assume that we have g disjoint groups of destination processors, $G_k = \{P_{j_{k,1}}, P_{j_{k,2}}, P_{j_{k,3}}, ...\}$, $1 \le k \le g$, and there are g senders $P_{i_1}, P_{i_2}, ..., P_{i_g}$. Processor P_{i_k} has value $\mathtt{R}(i_k)$ to be broadcast to all the processors in G_k, where $1 \le k \le g$. Since there are g simultaneous multicasting, we have a multiple multicasting operation, which is denoted as follows:

for $1 \le k \le g$ **do in parallel**
    $\mathtt{R}(j_{k,1}), \mathtt{R}(j_{k,2}), \mathtt{R}(j_{k,3}), ... \leftarrow \mathtt{R}(i_k)$
endfor

It is important to point out that due to relative slow speed of the electronic processors, each processor can only read one message frame during one bus cycle. Thus, the g groups of destination processors must be disjoint.

Element Pair-Wise Operations. Let $P_i, P_{i+1}, ..., P_j$ be a consecutive group of processors. We use $\mathtt{R}[i..j]$ as an abbreviation of the registers $\mathtt{R}(i)$, $\mathtt{R}(i+1)$, ..., $\mathtt{R}(j)$. $\mathtt{R}[i..j]$ can be used to store a vector, or a matrix in the row-major or column-major order. Assume that A and B are two matrices of size $N \times N$. All array elements are in a domain with a binary operator $\oplus$. Elements a_{ij} and b_{ij} are stored in $\mathtt{R}[m+(i-1)N+j-1]$ and $\mathtt{R}[n+(i-1)N+j-1]$ respectively. Then $C = A \oplus B$, where $c_{ij} = a_{ij} \oplus b_{ij}$, can be done as follows, where c_{ij} is found in $\mathtt{R}[m+(i-1)N+j-1]$:

for $1 \le k \le N^2$ **do in parallel**
    $\mathtt{R}(m+k-1) \leftarrow \mathtt{R}(m+k-1) \oplus \mathtt{R}(n+k-1)$
endfor

Binary Prefix Sums. Assume that every processor P_i, where $0 \le i \le N-1$, has a register $\mathtt{R}(i)$ which holds a binary value. We need to calculate $s_j = \mathtt{R}(0)+\mathtt{R}(1)+\cdots+\mathtt{R}(j)$, and save the result in $\mathtt{R}(j)$. The operation is represented as

for $0 \le j \le N-1$ **do in parallel**
    $\mathtt{R}(j) \leftarrow \mathtt{R}(0)+\mathtt{R}(1)+\cdots+\mathtt{R}(j)$
endfor

Extraction and Compression. In this operation, every processor P_j has a value x_j, and we wish to extract those x_j's that have certain property, and compact them to the beginning of the linear array. In particular, if x_{i_1}, x_{i_2}, ..., x_{i_m} are those values that have the property, where $i_1 < i_2 < \cdots < i_m$, then we should have $\mathtt{R}(k-1) = x_{i_k}$ for all $1 \leq k \leq m$, after the extraction and compression operation.

Binary Value Aggregation. Assume that every processor P_i, where $0 \leq i \leq N-1$, has a register $\mathtt{R}(i)$ which holds a binary value. We need to calculate $\mathtt{R}(0) + \mathtt{R}(1) + \cdots + \mathtt{R}(N-1)$, and save the result in $\mathtt{R}(0)$. The operation is represented as

$$\mathtt{R}(0) \leftarrow \mathtt{R}(0) + \mathtt{R}(1) + \cdots + \mathtt{R}(N-1)$$

Boolean Value Aggregation. Assume that every processor P_i, where $0 \leq i \leq N-1$, has a register $\mathtt{R}(i)$ which holds a truth value. We need to calculate $\mathtt{R}(0)$ and $\mathtt{R}(1)$ and $\cdots$ and $\mathtt{R}(N-1)$, or, $\mathtt{R}(0)$ or $\mathtt{R}(1)$ or $\cdots$ or $\mathtt{R}(N-1)$, and save the result in $\mathtt{R}(0)$.

Integer Aggregation. The binary value aggregation operation can be generalized to calculate the summation of N integers.

Floating-Point Value Aggregation. The integer aggregation operation can be further generalized to calculate the summation of N reals.

The reader is referred to [21] for implementation details of these operations on optical buses, where the following results are established.

Theorem 0. *One-to-one communication, broadcasting, multiple multicasting, element pair-wise operation, binary prefix sums, extraction and compression, binary value aggregation, boolean value aggregation, integer (of bounded magnitude) aggregation, and real value (of bounded precision and magnitude) aggregation all take $O(1)$ time in the LARPBS computing model.* ∎

Theorem 0A. *The summation of N integers can be calculated in $O(\log\log M)$ time, using $N\phi(M)$ processors in the LARPBS computing model, where*

$$\phi(M) = \left\lceil \frac{\lceil \log M \rceil}{\lceil \log\log M \rceil} \right\rceil,$$

and M is the maximum magnitude. ∎

Theorem 0B. *The summation of N real values, where each value is represented by a p-bit mantissa and a q-bit exponent, with precision up to $2^{-(p+q-1)}$ and magnitude in the order of $M = 2^{2^q}$, can be calculated in $O(\log\log M + \log p)$ time, using $N\psi(p)$ processors, where*

$$\psi(p) = \left\lceil \frac{p}{\lceil \log p \rceil} \right\rceil,$$

in the LARPBS computing model. ∎

12.3 AN O(N) ALGORITHM USING N^2 PROCESSORS

We start with a linear time algorithm MM_1, that calculates the product of two $N \times N$ matrices by using N^2 processors in the LARPBS model. Let $A = (a_{ij})_{N\times N}$ and $B = (b_{ij})_{N\times N}$ be two matrices, where the matrix elements a_{ij} and b_{ij} are in an arbitrary ring (or, a closed semiring) with additive operator $\oplus$ and multiplicative operator $\otimes$. The product of A and B is $C = A \times B = (c_{ij})_{N\times N}$, where $c_{ij} = (a_{i1} \otimes b_{1j}) \oplus (a_{i2} \otimes b_{2j}) \oplus \cdots \oplus (a_{iN} \otimes b_{Nj})$, for all $1 \le i,j \le N$.

We find that Cannon's algorithm [5, 14] can be easily implemented on the LARPBS model. For convenience, we label the N^2 processors using pairs (i,j), $1 \le i,j \le N$, and processors are linearly ordered in terms of lexicographical order. In other words, processor $P(i,j)$ corresponds to $P_{(i-1)N+(j-1)}$ in the linear array of processors shown in Figure 1. Each processor $P(i,j)$ has three registers $\mathtt{A}(i,j)$, $\mathtt{B}(i,j)$, and $\mathtt{C}(i,j)$. Elements of arrays A and B are initially stored in A and B registers in the row-major order, i.e., a_{ij} in $\mathtt{A}(i,j)$, and b_{ij} in $\mathtt{B}(i,j)$. Finally, array C is stored in C registers in the row-major order, i.e., element c_{ij} is found in $\mathtt{C}(i,j)$, for all $1 \le i,j \le N$.

The algorithm involves an initial alignment of rows of A and columns of B, that are actually two one-to-one communications:

```
for 1 ≤ i,j ≤ N do in parallel
    A(i, cyclic-shift(j, i − 1)) ← A(i, j)
endfor
for 1 ≤ i,j ≤ N do in parallel
    B(cyclic-shift(i, j − 1), j) ← B(i, j)
endfor
```

where

$$\texttt{cyclic-shift}(i,j) = \begin{cases} i-j, & \text{if } 0 \le j \le i-1; \\ i-j+N, & \text{if } i \le j \le N-1. \end{cases}$$

After such shifts, we have the following data distribution,

$$\begin{array}{cccc} P(1,1) & P(1,2) & \cdots & P(1,N) \\ (a_{11}, b_{11}) & (a_{12}, b_{22}) & \cdots & (a_{1N}, b_{NN}) \\ \\ P(2,1) & P(2,2) & \cdots & P(2,N) \\ (a_{22}, b_{21}) & (a_{23}, b_{32}) & \cdots & (a_{21}, b_{1N}) \\ \\ \vdots & \vdots & \ddots & \vdots \\ \\ P(N,1) & P(N,2) & \cdots & P(N,N) \\ (a_{NN}, b_{N1}) & (a_{N,1}, b_{12}) & \cdots & (a_{N,N-1}, b_{N-1,N}) \end{array}$$

Then we have a sequence of N alternate local computation and row/column cyclic shifts:

```
for 1 ≤ k ≤ N do
    C(i,j) ← C(i,j) ⊕ (A(i,j) ⊗ B(i,j))
    for 1 ≤ i,j ≤ N do in parallel
        A(i, cyclic-shift(j,1)) ← A(i,j)
    endfor
    for 1 ≤ i,j ≤ N do in parallel
        B(cyclic-shift(i,1),j) ← B(i,j)
    endfor
endfor
```

(cf. Chapter 5 of [14] for a complete illustration of an implementation on a mesh with wraparound connections). All these alignments and cyclic shifts are one-to-one communications in the LARPBS model. Thus, the initial alignment and each of the subsequent cyclic shifts takes two bus cycles. During the N local computation steps, the value c_{ij} is gradually accumulated. The entire algorithm requires $O(N)$ time. Thus, we obtain the following result.

Theorem 1. *Matrix multiplication (on an arbitrary ring or a closed semiring) can be performed by using algorithm MM_1 on the LARPBS model in $O(N)$ time, with N^2 processors, and each processor needs $O(1)$ memory.* ∎

12.4 AN O(1) ALGORITHM USING N³ PROCESSORS

In this section, we present a constant time algorithm MM_2 to calculate the product of two $N \times N$ integer matrices using N^3 processors in the LARPBS model. This algorithm is applicable to integers and floating-point values of bounded precision and magnitude, and boolean values.

For convenience, we label the N^3 processors using triplets (i, j, k), where $1 \leq i, j, k \leq N$. Processors $P(i, j, k)$ are ordered in the linear array using the lexicographical order. We use $P(i, j, *)$ to denote the group of consecutive processors $P(i, j, 1)$, $P(i, j, 2)$, ..., $P(i, j, N)$. It is noticed that the original system can be reconfigured into N^2 subsystems, namely, the $P(i, j, *)$'s. Each processor $P(i, j, k)$ has three registers $\mathtt{A}(i, j, k)$, $\mathtt{B}(i, j, k)$, and $\mathtt{C}(i, j, k)$.

The input and output of algorithm MM_2 are specified as follows. Initially, elements a_{ij} and b_{ji} are stored in registers $\mathtt{A}(1, i, j)$ and $\mathtt{B}(1, i, j)$ respectively, for all $1 \leq i, j \leq N$. If we partition A into N row vectors A_1, A_2, ..., A_N, and B into N column vectors B_1, B_2, ..., B_N, then the initial data distribution is as follows:

$$\begin{array}{cccc} P(1,1,*) & P(1,2,*) & \cdots & P(1,N,*) \\ (A_1, B_1) & (A_2, B_2) & \cdots & (A_N, B_N) \end{array}$$

(Other processors are not shown here for simplicity, since they do not carry any data initially.) In other words, the two input matrices A and B are put in the beginning of the linear array in the row-major and the column-major orders respectively. This is a quite simple and easy to manage way of preparing the input data. When the computation is done, c_{ij} is found in $\mathtt{C}(1, i, j)$. If we divide C into N row vectors C_1, C_2, ..., C_N, then the output layout is

$$\begin{array}{cccc} P(1,1,*) & P(1,2,*) & \cdots & P(1,N,*) \\ (C_1) & (C_2) & \cdots & (C_N) \end{array}$$

Again, such an arrangement makes it easy for C to be used as an input to another computation.

The algorithm proceeds as follows. First, we change the placement of matrix A in such a way that element a_{ij} is stored in $\mathtt{A}(i,1,j)$, for all $1 \le i,j \le N$. This can be accomplished by a one-to-one communication:

```
for 1 ≤ i,j ≤ N do in parallel
    A(i,1,j) ← A(1,i,j)
endfor
```

After such replacement, we have the following data distribution,

$$\begin{array}{cccc} P(1,1,*) & P(1,2,*) & \cdots & P(1,N,*) \\ (A_1,B_1) & (A_2,B_2) & \cdots & (A_N,B_N) \\ \\ P(2,1,*) & P(2,2,*) & \cdots & P(2,N,*) \\ (A_2,-) & (-,-) & \cdots & (-,-) \\ \\ \vdots & \vdots & \ddots & \vdots \\ \\ P(N,1,*) & P(N,2,*) & \cdots & P(N,N,*) \\ (A_N,-) & (-,-) & \cdots & (-,-) \end{array}$$

where the linear array of N^3 processors are logically arranged as a two dimensional $N \times N$ array, and each element in the array stands for a group of N processors $P(i,j,*)$. The symbol "$-$" means that the A and B registers are still undefined.

Next, we distribute the rows of A and columns of B to the right processors, such that processors $P(i,j,*)$ hold A_i and B_j, for all $1 \le i,j \le N$. This can be performed using multiple multicasting operations:

```
for 1 ≤ i,k ≤ N do in parallel
    A(i,2,k), A(i,3,k), ..., A(i,N,k) ← A(i,1,k)
endfor
for 1 ≤ j,k ≤ N do in parallel
    B(2,j,k), B(3,j,k), ..., B(N,j,k) ← B(1,j,k)
endfor
```

After multicasting, the data distribution is as follows:

$$\begin{array}{cccc} P(1,1,*) & P(1,2,*) & \cdots & P(1,N,*) \\ (A_1,B_1) & (A_1,B_2) & \cdots & (A_1,B_N) \\ \\ P(2,1,*) & P(2,2,*) & \cdots & P(2,N,*) \\ (A_2,B_1) & (A_2,B_2) & \cdots & (A_2,B_N) \\ \\ \vdots & \vdots & \ddots & \vdots \\ \\ P(N,1,*) & P(N,2,*) & \cdots & P(N,N,*) \\ (A_N,B_1) & (A_N,B_2) & \cdots & (A_N,B_N) \end{array}$$

At this point, processors are ready to compute. In particular, $P(i,j,k)$ calculates $\mathtt{A}(i,j,k) \times \mathtt{B}(i,j,k)$:

```
for 1 ≤ i,j,k ≤ N do in parallel
    C(i,j,k) ← A(i,j,k) × B(i,j,k)
endfor
```

Then, the values $\mathtt{C}(i,j,k)$, $1 \le k \le N$, are assembled, using integer aggregation operations, and the result c_{ij} is in $\mathtt{C}(i,j,1)$, as shown below,

```
for 1 ≤ i,j ≤ N do in parallel
    C(i,j,1) ← C(i,j,1) + C(i,j,2) + ··· + C(i,j,N)
endfor
```

Here, for the purpose of multiple aggregations, the original system is reconfigured into N^2 subsystems $P(i,j,*)$'s, for $1 \le i,j \le N$. After calculation, the data distribution is as follows:

$$\begin{array}{cccc} P(1,1,1) & P(1,2,1) & \cdots & P(1,N,1) \\ (c_{11}) & (c_{12}) & \cdots & (c_{1N}) \\ \\ P(2,1,1) & P(2,2,1) & \cdots & P(2,N,1) \\ (c_{21}) & (c_{22}) & \cdots & (c_{2N}) \\ \\ \vdots & \vdots & \ddots & \vdots \\ \\ P(N,1,1) & P(N,2,1) & \cdots & P(N,N,1) \\ (c_{N1}) & (c_{N2}) & \cdots & (c_{NN}) \end{array}$$

Finally, one more data movement via one-to-one communication brings the c_{ij}'s to the right processors:

```
for 1 ≤ i ≤ N, 2 ≤ j ≤ N, do in parallel
    C(1,i,j) ← C(i,j,1)
endfor
```

It is clear that each step in the algorithm involves a simple local computation, or a primitive communication operation, and hence, takes $O(1)$ time. Thus, the entire MM_2 algorithm can be executed in $O(1)$ time. This gives rises to the following claims.

Theorem 2. *For integers of bounded magnitude, or floating-point values of bounded precision and magnitude, or boolean values, matrix multiplication can be performed by using algorithm MM_2 on the LARPBS model in $O(1)$ time, with N^3 processors, and each processor needs $O(1)$ memory. (For boolean values, $+$ is replaced by* `or`*, and $\times$ by* `and`*.)* ∎

It is observed that in virtually all sequential and parallel algorithm design and analysis, it is assumed that an arithmetic operation can be performed by a processor in $O(1)$ time, regardless of the magnitude/precision of the operands. In other words, the maximum magnitude M of integers, and p and q for floating-point number representations are all fixed. Therefore, Theorem 2 is already enough for most applications, and is theoretically very reasonable. However, algorithm MM_2 can also be extended to handle matrices containing integers/reals with unbounded magnitude and precision. We present the following results and leave out the details.

Theorem 2A. *For integers with unbounded magnitude, matrix multiplication can be performed in $O(\log\log M)$ time on the LARPBS model, with $N^3\phi(M)$ processors, and each processor needs $O(\log\log M)$ memory, where M is the largest magnitude.* ∎

Theorem 2B. *For reals with precision up to $2^{-(p+q-1)}$ and magnitude in the order of $M = 2^{2^q}$, matrix multiplication can be performed in $O(\log\log M + \log p)$ time on the LARPBS model, with $N^3\psi(p) \approx N^3 p/\log p$ processors, and each processor needs $O(\log p)$ memory.* ∎

A constant time matrix multiplication algorithm has been obtained on a reconfigurable meshes with $O(N^4)$ processors [32], and on the AROB model using $N^3 \log N$ processors [33], assuming that all integers have $O(\log N)$ bits. It is clear that algorithm MM_2 uses fewer processors, and Theorems 2A and 2B are applicable to more general cases.

12.5 AN O(log N) ALGORITHM USING O($N^{2.8074}$) PROCESSORS

The matrix multiplication algorithm MM_3 described in the present section is a parallelization of Strassen's recursive method, which is applicable to an arbitrary ring. First of all, let us describe Strassen's strategy. Without loss of generality, we assume that N is a power of two, i.e., $N = 2^n$, for some $n > 1$. Given two matrices,

$$A = \begin{bmatrix} A_{11} & A_{12} \\ A_{21} & A_{22} \end{bmatrix}, \qquad \text{and} \qquad B = \begin{bmatrix} B_{11} & B_{12} \\ B_{21} & B_{22} \end{bmatrix},$$

where A_{ij} and B_{ij} are matrices of size $N/2$, the Strassen's algorithm to calculate

$$C = A \times B = \begin{bmatrix} C_{11} & C_{12} \\ C_{21} & C_{22} \end{bmatrix}$$

is shown below:

$$\begin{array}{ll}
X_1 = A_{11} + A_{22}, & Y_1 = B_{11} + B_{22}, \\
X_2 = A_{21} + A_{22}, & Y_2 = B_{11}, \\
X_3 = A_{11}, & Y_3 = B_{12} - B_{22}, \\
X_4 = A_{22}, & Y_4 = B_{21} - B_{11}, \\
X_5 = A_{11} + A_{12}, & Y_5 = B_{22}, \\
X_6 = A_{21} - A_{11}, & Y_6 = B_{11} + B_{22}, \\
X_7 = A_{12} - A_{22}, & Y_7 = B_{21} + B_{22};
\end{array}$$

$$\begin{array}{l}
M_1 = X_1 \times Y_1, \\
M_2 = X_2 \times Y_2, \\
M_3 = X_3 \times Y_3, \\
M_4 = X_4 \times Y_4, \\
M_5 = X_5 \times Y_5, \\
M_6 = X_6 \times Y_6, \\
M_7 = X_7 \times Y_7;
\end{array}$$

$$\begin{array}{l}
C_{11} = M_1 + M_4 - M_5 + M_7, \\
C_{12} = M_3 + M_5, \\
C_{21} = M_2 + M_4, \\
C_{22} = M_1 - M_2 + M_3 + M_6.
\end{array}$$

There are three major steps in Strassen's algorithm. First, matrices X_g and Y_g, $1 \leq g \leq 7$, are calculated using matrix addition and subtraction. Second, matrices M_g, $1 \leq g \leq 7$, are calculated using matrix multiplication of the X_g's and Y_g's. Third, matrices C_{11}, C_{12}, C_{21}, C_{22}, are calculated using matrix addition and subtraction of the M_g's. The original motivation of the algorithm is to reduce the number of element multiplications from N^3 to $N^{2.8074}$, at the cost of more element additions and subtractions. However, the total number of element level operations is only $O(N^{2.8074})$.

The key point of Strassen's algorithm is that the product C of two matrices A and B of size N can be calculated based on seven products M_1, M_2, ..., M_7 of matrices of size $N/2$. Such a reduction of matrix size can be continued until the size becomes $N = 1$, i.e., $n = 0$. Due to this recursive nature, Strassen's algorithm takes $\Omega(\log N)$ time. We are going to show that each level of the recursion in Strassen's algorithm can be implemented in $O(1)$ time, and the entire algorithm takes $O(\log N)$ time. Since each level of the recursion increases the number of processors by a factor 7, the total number of processors in this parallel implementation is $7^n = 7^{\log N} = N^{\log 7} = N^{2.8074}$. The reconfigurability of the LARPBS model is crucial in such an implementation. Also, a pipelined optical bus architecture can efficiently support all necessary data transfer in the parallelized Strassen's algorithm.

Let the $p = N^{\log 7}$ processors be labeled as P_0, P_1, ..., P_{p-1}. Each processor P_i has six registers A(i), B(i), C(i), X(i), Y(i), and M(i), for each level of the recursion. Hence, $O(\log N)$ amount of memory are required in each processor. Let $\Pi(i, j)$ denote a partition of an LARPBS system consisting of consecutive processors P_i, P_{i+1}, ..., P_j. The original system is $\Pi(0, p-1)$. When we mention a partition $\Pi(i, j)$, it is assumed that the original system has been reconfigured such that $\Pi(i, j)$ is a complete and independent LARPBS system. We use R[$i..j$] as an abbreviation of the registers R(i), R($i+1$), ..., R(j).

Algorithm MM$_3$ is recursively defined. Our first call to the algorithm is

$$\text{MM}_3(A, B, C, N, \Pi(0, N^{\log 7} - 1)),$$

which means calculating the product C of matrices A and B of size N using all the $N^{\log 7}$ processors. Later on, algorithm MM$_3$ will be invoked to calculate sub-matrices of C of smaller sizes using partitions of the system. Thus, our description of MM$_3(A, B, C, N, \Pi(i, j))$ is general, where $j - i + 1 = N^{\log 7}$.

The input arrays A and B are stored in the A and B registers of the first N^2 processors of the partition, both in the row-major order. The output array C is also stored in the C registers of the first N^2 processors of the partition, in the row-major order.

The processors P_i, P_{i+1}, ..., P_j are divided into seven groups of equal size, or, the partition $\Pi(i, j)$ is reconfigured into seven independent sub-partitions during recursion: $\Pi_g = \Pi(i + (g-1)\pi, i + g\pi - 1)$, where $1 \le g \le 7$, and $\pi = (j - i + 1)/7$. We wish the input arrays being stored within the range of Π_1. This is possible only when $N > 8$.

Thus, if $N \le 8$, we reach the base case of the recursion, and calculate C directly using any method, e.g., the sequential algorithm. This takes constant time.

In general, when $N > 8$, we calculate X_g, Y_g, and M_g using the partition Π_g, and all the seven computations are done in parallel. To this end, we first make A and B available to all the seven partitions via multiple multicasting, though a partition may only need parts of A and B:

```
for 0 ≤ k ≤ N^2 − 1 do in parallel
    A[i + π + k], A[i + 2π + k], ..., A[i + 6π + k] ← A[i + k]
endfor
for 0 ≤ k ≤ N^2 − 1 do in parallel
    B[i + π + k], B[i + 2π + k], ..., B[i + 6π + k] ← B[i + k]
endfor
```

Next, X_g and Y_g are calculated on $\Pi(i + (g-1)\pi, i + g\pi - 1)$, and saved in X[$i + (g-1)\pi..i + (g-1)\pi + N^2 - 1$] and Y[$i + (g-1)\pi..i + (g-1)\pi + N^2 - 1$] respectively, for all $1 \le g \le 7$, in parallel:

```
for 1 ≤ g ≤ 7 do in parallel
    calculate X_g on Π(i + (g − 1)π, i + gπ − 1),
    and save the results in X[i + (g − 1)π..i + (g − 1)π + N^2 − 1]
endfor
```

```
for 1 ≤ g ≤ 7 do in parallel
    calculate Y_g on Π(i + (g − 1)π, i + gπ − 1),
    and save the results in Y[i + (g − 1)π..i + (g − 1)π + N^2 − 1]
endfor
```

Different operations are performed for different Π_g's. However, only one element pair-wise matrix addition and subtraction is involved in each partition Π_g. Then, the M_g's are obtained recursively on Π_g, for all $1 \leq g \leq 7$, in parallel:

```
for 1 ≤ g ≤ 7 do in parallel
    recursively call MM3(X, Y, M, N/2, Π(i + (g − 1)π, i + gπ − 1))
endfor
```

Finally, the C_{ij}'s are calculated in constant time, using element pair-wise matrix additions and subtractions, and the results are moved to the right processors.

It is clear that except recursion, all steps can be conducted in constant time. Hence, the entire MM_3 algorithm takes $O(\log N)$ time. In general, when N is not a power of two, we can round N up to $2^{\lceil \log N \rceil}$, and the total number of processors is $7^{\lceil \log N \rceil} < 7^{\log N + 1} = 7N^{2.8074} = O(N^{2.8074})$.

We would like to point out that since no global aggregation (which is useful only for certain types of data) is conducted, our implementation of Strassen's algorithm is applicable to all rings.

Theorem 3. *Matrix multiplication (on an arbitrary ring) can be performed by using algorithm MM_3 on the LARPBS model in $O(\log N)$ time, with $O(N^{2.8074})$ processors, and each processor needs $O(\log N)$ memory.* ∎

The time complexity $O(\log N)$ is due to the nature of Strassen's algorithm, not the physical limitation of LARPBS systems. It is unlikely that the Strassen's algorithm can be easily parallelized on static networks such as meshes and hypercubes with the same performance as that of MM_3, nor on ordinary bus systems with limited bandwidth.

12.6 COMPOUND ALGORITHMS

A *compound algorithm* is a combination of multiple existing algorithms solving the same problem. Two compound algorithms for matrix multiplication on LARPBS, i.e., $C_1(\epsilon)$ and $C_2(\delta)$, are given in this section. As a matter of fact, $C_1(\epsilon)$ and $C_2(\delta)$ are algorithm schemes with a parameter, such that they have adjustable time complexity in different levels. Without loss of generality, we assume that N is a power of two in the following discussion.

We first consider algorithm $C_1(\epsilon)$, the combination of MM_3 and MM_1, with a parameter ϵ. Algorithm $C_1(\epsilon)$ proceeds in the same way as MM_3 for the first m levels of the recursion. Then the submatrix sizes are reduced to $N/2^m$. At this point, $C_1(\epsilon)$ employs algorithm MM_1 to calculate the remaining products of submatrices of size $N/2^m$, using $(N/2^m)^2$ processors for each product. Thus, the time complexity is $O(m + N/2^m)$, and the total number of processors is $O(7^m(N/2^m)^2)$. If we let $m = \epsilon \log N$, for some constant ϵ, where $0 \leq \epsilon \leq 1$, then the time complexity becomes $O(\log N + N^{1-\epsilon})$, and the total number of processors is $O(N^{2+\epsilon\log(7/4)})$. The above discussion can be summarized as

Theorem 4. *Matrix multiplication (on an arbitrary ring) can be performed by using algorithm $C_1(\epsilon)$ on the LARPBS model in $O(\log N + N^{1-\epsilon})$ time, with $O(N^{2+0.8074\epsilon})$ processors, where $0 \le \epsilon \le 1$, and each processor needs $O(\epsilon \log N)$ memory.* ∎

The time and processor complexities of algorithm $C_1(\epsilon)$ are illustrated in Table 12.1. Algorithm $C_1(\epsilon)$ has adjustable time complexity in sub-linear level. For example, algorithm $C_1(0.5)$ has time complexity $O(\sqrt{N})$ and uses $O(N^{2.4037})$ processors. It is noticed that $\text{MM}_1 = C_1(0)$, and $\text{MM}_3 = C_1(1)$. The total cost of algorithm $C_1(\epsilon)$ i.e., the time-processor product, is $O(N^{3-\epsilon\log(8/7)}) = O(N^{3-0.1926\epsilon})$, for $0 \le \epsilon < 1$, and the cost is $o(N^3)$ for $0 < \epsilon \le 1$.

In a similar way, we can obtain the second compound algorithm $C_2(\delta)$, where $0 \le \delta \le 1$, which is a combination of algorithms MM_3 with MM_2. When the submatrix sizes are reduced to $N/2^m$, $C_2(\delta)$ invokes algorithm MM_2 to calculate the remaining products of submatrices of size $N/2^m$, with $(N/2^m)^3$ processors for each product. Thus, the time complexity is $O(m)$, and the total number of processors is $O(7^m(N/2^m)^3)$. If $m = (\log N)^\delta$, for some constant δ, where $0 \le \delta \le 1$, then the time complexity is $O((\log N)^\delta)$, and the total number of processors is $O(N^3(7/8)^{(\log N)^\delta})$. This gives rises to the following claim.

Theorem 5. *For integers of bounded magnitude, or floating-point values of bounded precision and magnitude, matrix multiplication can be performed using algorithm $C_2(\delta)$ in $O((\log N)^\delta)$ time on the LARPBS model, with*

$$O\left(\frac{N^3}{1.1428^{(\log N)^\delta}}\right)$$

Table 12.1 Time and Processor Complexities of $C_1(\epsilon)$.

ϵ	time	processor	cost
0.0	$O(N)$	$O(N^2)$	$O(N^3)$
0.1	$O(N^{0.9})$	$O(N^{2.0808})$	$O(N^{2.9808})$
0.2	$O(N^{0.8})$	$O(N^{2.1615})$	$O(N^{2.9815})$
0.3	$O(N^{0.7})$	$O(N^{2.2423})$	$O(N^{2.9423})$
0.4	$O(N^{0.6})$	$O(N^{2.3230})$	$O(N^{2.9230})$
0.5	$O(N^{0.5})$	$O(N^{2.4037})$	$O(N^{2.9037})$
0.6	$O(N^{0.4})$	$O(N^{2.4845})$	$O(N^{2.8845})$
0.7	$O(N^{0.3})$	$O(N^{2.5652})$	$O(N^{2.8652})$
0.8	$O(N^{0.2})$	$O(N^{2.6459})$	$O(N^{2.8459})$
0.9	$O(N^{0.1})$	$O(N^{2.7267})$	$O(N^{2.8267})$
1.0	$O(\log N)$	$O(N^{2.8074})$	$O(N^{2.8074}\log N)$

processors, where $0 \le \delta \le 1$, *and each processor needs* $O((\log N)^\delta)$ *memory.* ∎

It can be verified that for all $x > 0$, and $\delta \ge 0$,

$$x^{(\log N)^\delta} = N^{\log x/(\log N)^{1-\delta}}.$$

Hence, we have

$$\frac{N^3}{1.1428^{(\log N)^\delta}} = N^{3-\log(8/7)/(\log N)^{1-\delta}} = o(N^3), \quad 0 < \delta \le 1.$$

Theorem 5 implies that it is feasible to achieve sub-logarithmic time using $o(N^3)$ processors for matrix multiplication on a realistic system.

We notice that algorithm $C_2(\delta)$ can also be extended for arbitrary integers and reals.

Theorem 5A. *For integers with unbounded magnitude, matrix multiplication can be performed in* $O((\log N)^\delta + \log\log M)$ *time on the LARPBS model, with*

$$O\left(\frac{N^3\phi(M)}{1.1428^{(\log N)^\delta}}\right) = O\left(\frac{N^3}{1.1428^{(\log N)^\delta}} \cdot \frac{\log M}{\log\log M}\right)$$

processors, where $0 \le \delta \le 1$, M *is the largest magnitude, and each processor needs* $O((\log N)^\delta + \log\log M)$ *memory.* ∎

Theorem 5B. *For reals with precision up to* $2^{-(p+q-1)}$ *and magnitude in the order of* $M = 2^{2^q}$, *matrix multiplication can be performed in* $O((\log N)^\delta + \log\log M + \log p)$ *time on the LARPBS model, with*

$$O\left(\frac{N^3\psi(p)}{1.1428^{(\log N)^\delta}}\right) = O\left(\frac{N^3}{1.1428^{(\log N)^\delta}} \cdot \frac{p}{\log p}\right)$$

processors, where $0 \le \delta \le 1$, M *is the largest magnitude, and each processor needs* $O((\log N)^\delta + \log p)$ *memory.* ∎

Since the set of boolean values with operations **and/or** is a closed semiring (which is not a ring), Strassen's algorithm is not directly applicable. However, boolean matrix multiplication can be treated as integer matrix multiplication over the ring $\mathbf{Z}_{N+1}$ (cf. Section 6.6 in [1]), we have the following result based on Theorem 5A.

Theorem 5C. *Boolean matrix multiplication can be performed in* $O((\log N)^\delta + \log\log N)$ *time on the LARPBS model, with*

$$O\left(\frac{N^3\phi(N)}{1.1428^{(\log N)^\delta}}\right) = O\left(\frac{N^3}{1.1428^{(\log N)^\delta}} \cdot \frac{\log N}{\log\log N}\right)$$

processors, where $0 \le \delta \le 1$, *and each processor needs* $O((\log N)^\delta + \log\log N)$ *memory.* ∎

Recently, the well known Four Russians' algorithm has been parallelized on the LARPBS model, which has constant execution time, and uses $O(N^3/\log N)$ processors [17].

12.7 APPLICATIONS TO OTHER MATRIX OPERATIONS

It has long been recognized that a number of important matrix problems are quite closely related, in the sense that up to polynomial changes in the size of the matrices, and constant factor changes in the running time of the algorithms, the amount of parallel time needed to solve these problems are the same. In particular, it is well known that finding the inverse, the characteristic polynomial, the determinant, the rank, the Nth power, and an LU-factorization of a matrix, and the product of a matrix chain, etc., can be reduced to each other, such that all of them have the same parallel computing complexity (cf. Section 2.4.5 of [15] for proofs of the above claim). In addition to the parallel time complexity, the cost of an algorithm, which is the product of the execution time and the number of processors employed by the algorithm, is widely used to measure the efficiency of the algorithm. Currently the fastest parallel algorithms for these problems (except LU-factorization) run in $O((\log N)^2)$ time, by using $O(N^4)$ processors connected by some commonly used static networks such as meshes of trees and hypercubes [15]. Since the reduction among these problems may significantly increase the problem size, and hence, the number of processors, the costs are not preserved under these reductions. To keep the processor complexity at $O(N^4)$, the best algorithm for LU-factorization has time complexity $O((\log N)^3)$ on static networks.

While the problem of finding tight time and cost bounds of each of these problems remains open, answering the following two questions appears to be challenging:

Q1. Can the parallel time complexities of these problems (except LU-factorization) be reduced to $o((\log N)^2)$, and to $o((\log N)^3)$ for LU-factorization?

Q2. Can the costs of parallel algorithms for these problems be reduced to $o(N^4)$?

In answering question Q1, we notice that the major hurdle is that all algorithms for these problems ultimately use a matrix multiplication algorithm as a subroutine, and the best matrix multiplication algorithm has time complexity $O(\log N)$, even on theoretical models like parallel random access machines. The parallelism in matrix multiplication can be explored to such extend that the N^2 vector products can be calculated simultaneously. The $O(\log N)$ bottleneck is due to the summation of the N values in an inner vector product, which cannot be done any faster on existing computing models. Theoretically, the answer to question Q2 is affirmative. It is well known that there exist matrix multiplication algorithms on PRAMs that run in $O(\log N)$ time by using $O(N^\beta)$ processors, where $O(N^\beta)$, $2 < \beta < 3$, is the time complexity of the best sequential algorithm for matrix multiplication [31]. The fact that the known smallest value of β is less than 2.3755 [8] implies that the costs of parallel PRAM algorithms for many matrix problems are less than $O(N^4)$. However, such a sequential algorithm is quite sophisticated, and its implementation is not considered practical even on sequential machines. Consequently, its parallelization on realistic parallel systems is far from feasible.

With results like Theorems 2 and 5, we are able to give affirmative answers to questions Q1 and Q2 on realistic parallel computing systems. In [19], we solved seventeen fundamental matrix problems on LARPBS. These problems and the time and processor complexities of our solutions are listed in Table 12.2. Problems (1)-(7) are basic matrix operations, whose algorithms are repeatedly used in other algorithms. Problems (8)-(17) include a number of fundamental matrix operations. Our results provide the following answers to questions Q1 and Q2.

A1. Problems (6)-(15) can all be solved in $O(\log N)$ time, and problems (16)-(17) in $O((\log N)^2)$ time, using no more than $O(N^4)$ processors. Thus, compared with known results, the running time of parallel solutions to all these problems mentioned in Table 12.2 can be reduced by a factor of $O(\log N)$.

A2. Problems (6)-(15) can be solved in $o((\log N)^2)$ time, and problems (16)-(17) in $o((\log N)^3)$ time, with cost $o(N^4)$ on LARPBS.

Furthermore, as shown in Table 12.2, our results provide a wide range of cost/per- formance combinations of our algorithms, where $0 \leq \delta \leq 1$.

Table 12.2 The Problems and Time/Processor Complexities.

No.	Problem	Time Comp.	Processor Comp.
(1)	matrix transposition	$O(1)$	$O(N^2)$
(2)	vector chain addition	$O(1)$	$O(N^2)$
(3)	matrix chain addition	$O(1)$	$O(N^3)$
(4)	matrix-vector multiplication	$O(1)$	$O(N^2)$
(5)	matrix multiplication	$O((\log N)^\delta)$	$O(N^3/(\frac{8}{7})^{(\log N)^\delta})$
(6)	matrix powers	$O((\log N)^{1+\delta})$	$O(N^4/(\frac{8}{7})^{(\log N)^\delta})$
(7)	matrix chain product	$O((\log N)^{1+\delta})$	$O(N^4/(\frac{8}{7})^{(\log N)^\delta})$
(8)	lower triangular matrix inversion	$O((\log N)^{1+\delta})$	$O(N^3/(\frac{8}{7})^{(\log N)^\delta})$
(9)	upper triangular matrix inversion	$O((\log N)^{1+\delta})$	$O(N^3/(\frac{8}{7})^{(\log N)^\delta})$
(10)	the characteristic poly. of a matrix	$O((\log N)^{1+\delta})$	$O(N^4/(\frac{8}{7})^{(\log N)^\delta})$
(11)	the determinant of a matrix	$O((\log N)^{1+\delta})$	$O(N^4/(\frac{8}{7})^{(\log N)^\delta})$
(12)	the rank of a matrix	$O((\log N)^{1+\delta})$	$O(N^4/(\frac{8}{7})^{(\log N)^\delta})$
(13)	arbitrary matrix inversion	$O((\log N)^{1+\delta})$	$O(N^4/(\frac{8}{7})^{(\log N)^\delta})$
(14)	solving linear systems of equations	$O((\log N)^{1+\delta})$	$O(N^4/(\frac{8}{7})^{(\log N)^\delta})$
(15)	Krylov matrix	$O((\log N)^{1+\delta})$	$O(N^4/(\frac{8}{7})^{(\log N)^\delta})$
(16)	LU-factorization of a matrix	$O((\log N)^{2+\delta})$	$O(N^4/(\frac{8}{7})^{(\log N)^\delta})$
(17)	QR-factorization of a matrix	$O((\log N)^{2+\delta})$	$O(N^4/(\frac{8}{7})^{(\log N)^\delta})$

12.8 CONCLUDING REMARKS

We developed a number of matrix multiplication algorithms on linear arrays with reconfigurable pipelined bus systems. Algorithm $\mathcal{C}_1(\epsilon)$ has adjustable time complexity in sub-linear level. Algorithm $\mathcal{C}_2(\delta)$ implies that it is feasible to achieve sub-logarithmic time using $o(N^3)$ processors for matrix multiplication on a realistic system. Algorithms MM_3, $\mathcal{C}_1(\epsilon)$, and $\mathcal{C}_2(\delta)$ all have $o(N^3)$ cost, and hence, are very processor efficient. Algorithms MM_1, MM_3, and $\mathcal{C}_1(\epsilon)$ are general-purpose matrix multiplication algorithms, where the array elements are in any ring. Extension of algorithms MM_2 and $\mathcal{C}_2(\delta)$ to unbounded integers and real values were also discussed. Many other important matrix problems and graph theory problems can be solved using a matrix multiplication algorithm as a subroutine [19]. Hence, the algorithms developed in this chapter can be used as subroutines for solving a large class of problems.

We have not seen parallel computing systems with conventional interconnection networks that solve the matrix multiplication problem in constant running time, or support efficient implementation of Strassen's algorithm, and/or provide such a wide range of cost and performance combinations. Our results have the following important implication, namely, due to its high communication bandwidth, the versatile communication patterns it supports, and its ability of utilizing communication reconfigurability as an integral part of a parallel computation, the LARPBS is a powerful architecture for exploiting large degree of parallelism in a computational problem that most other machine models cannot achieve.

To conclude the chapter, we would like to briefly discuss the scalability of our algorithms. While the algorithms presented in this chapter are fast and processor efficient in the theoretical sense, it is less practical to build an LARPBS system with $O(N^c)$ processors, where $c > 1$, for large scale matrix computations. Typically, the number of processors available is far less than what is required. Therefore, it is necessary to design algorithms that can be executed on an LARPBS with p processors, where p is arbitrarily chosen in the range $[1..N^3]$ for the standard method, and in the range $[1..N^{2.8074}]$ for Strassen's method. For a system with fixed size, a large matrix should be partitioned into submatrices, such that processors can perform computations and communications at the submatrix level. In other words, our primitive operations should be extended to handle blocks of data, not just a single data item. Some initial work in this direction has been reported in [41].

An important issue in parallel computing is the scalability of a parallel algorithm on certain parallel system. Though there is no unified definition, scalability essentially measures the ability to maintain speedup that is linearly proportional to the number of processors. To be more specific, when the number of processors is less than the maximum required, the parallel time complexity can generally be represented as $O(T(N)/p + T_{\mathrm{comm}}(N,p))$, where N is the problem size, p is the number of processors available, $T(N)$ is the time complexity of the sequential algorithm being parallelized, and $T_{\mathrm{comm}}(N,p))$ is the overall communication overhead of a parallel implementation. We say that a parallel implementation is *scalable* in the range $[1..p^*]$ if linear speedup can be achieved for all $1 \leq p \leq p^*$. The implementation is *highly scalable* if p^* is as large as

$$\Theta\left(\frac{T(N)}{T^*(N)(\log N)^k}\right)$$

for some constant $k \geq 0$, where $T^*(N)$ is the best possible parallel time. The implementation is *fully scalable* if $k = 0$. A fully scalable parallel implementation means that the sequential algorithm can be fully parallelized, and communication overhead in parallelization is negligible.

We have obtained the following results, namely, there is a fully scalable implementation of the standard matrix multiplication algorithm on LARPBS, and there is a highly scalable parallelization of Strassen's algorithm on LARPBS with $k = 2.4771\ldots$.

Theorem 6. *There is a fully scalable parallelization of the standard method (where we assume that array elements are integers of bounded magnitude, or floating-point values of bounded precision and magnitude, or boolean values), i.e., for all $1 \leq p \leq N^3$, multiplying two $N \times N$ matrices can be performed on a p-processor LARPBS in $O(N^3/p)$ time.* ∎

Theorem 7. *There is a highly scalable parallelization of Strassen's method (applicable to an arbitrary ring) with $k = 2.4771$. In particular, for all $1 \leq p \leq N^{2.8074}$, multiplying two $N \times N$ matrices can be performed on a p-processor LARPBS in*

$$O\left(\frac{N^{2.8074}}{p} + \left(\frac{N^2}{p^{0.7124}}\right)\log p\right)$$

time, which implies that when

$$p = O\left(\frac{N^{2.8074}}{(\log N)^{3.4771}}\right),$$

linear speedup and cost-optimality can be achieved. Note that $T^*(N) = O(\log N)$ is the best possible parallel time for Strassen's algorithm. ∎

Due to space limitation, we will not present the details of this part of the research, and refer the reader to [20].

Acknowledgments

The author's research was supported by National Aeronautics and Space Administration (NASA) and the Research Foundation of State University of New York through the NASA/University Joint Venture in Space Science Program (1996-1999) under Grant NAG8-1313, the 1996 NASA/ASEE Summer Faculty Fellowship Program granted by NASA and American Society for Engineering Education, and 1998 New York State/United University Professions Professional Development and Quality of Life Committee's Individual Development Award.

References

[1] A.V. Aho, J.E. Hopcroft, and J.D. Ullman, *The Design and Analysis of Computer Algorithms*, Addison-Wesley, Reading, Massachusetts, 1974.

[2] S.G. Akl, *Parallel Computation: Models and Methods*, Prentice-Hall, Upper Saddle River, New Jersey, 1997.

[3] A.F. Benner, H.F. Jordan, and V.P. Heuring, "Digital optical computing with optically switched directional couplers," *Optical Engineering*, vol.30, pp.1936-1941, 1991.

[4] D. Bini and V.Y. Pan, *Polynomial and Matrix Computations, Vol.1, Fundamental Algorithms*, Birkhäuser, Boston, 1994.

[5] L.E. Cannon, *A Cellular Computer to Implement the Kalman Filter Algorithm*, Ph.D. Thesis, Montana State University, Bozman, Montana, 1969.

[6] A.K. Chandra, "Maximal parallelism in matrix multiplication," Report RC-6193, IBM T.J. Watson Research Center, Yorktown Heights, New York, October 1979.

[7] D. Chiarulli, R. Melhem, and S. Levitan, "Using coincident optical pulses for parallel memory addressing," *IEEE Computer*, vol.30, pp.48-57, 1987.

[8] D. Coppersmith and S. Winograd, "Matrix multiplication via arithmetic progressions," *Journal of Symbolic Computation*, vol.9, pp.251-280, 1990.

[9] E. Dekel, D. Nassimi, and S. Sahni, "Parallel matrix and graph algorithms," *SIAM Journal on Computing*, vol.10, pp.657-673, 1981.

[10] P.W. Dowd, "Wavelength division multiple access channel hypercube processor interconnection," *IEEE Transactions on Computers*, vol.41, pp.1223-1241, 1992.

[11] Z. Guo, "Sorting on array processors with pipelined buses," *Proceedings of International Conference on Parallel Processing*, pp.289-292, August 1992.

[12] Z. Guo, R. Melhem, R. Hall, D. Chiarulli, and S. Levitan, "Pipelined communications in optically interconnected arrays," *Journal of Parallel and Distributed Computing*, vol.12, pp.269-282, 1991.

[13] M. Hamdi and Y. Pan, "Efficient parallel algorithms on optically interconnected arrays of processors," *IEE Proceedings - Computers and Digital Techniques*, vol.142, pp.87-92, March 1995.

[14] V. Kumar, A. Grama, A. Gupta, and G. Karypis, *Introduction to Parallel Computing: Design and Analysis of Algorithms*, Benjamin/Cummings, Redwood City, California, 1994.

[15] T. Leighton, *Introduction to Parallel Algorithms and Architectures: Arrays · Trees · Hypercubes*, Morgan Kaufmann, San Mateo, California, 1992.

[16] S. Levitan, D. Chiarulli, and R. Melhem, "Coincident pulse techniques for multiprocessor interconnection structures," *Applied Optics*, vol.29, pp.2024-2039, 1990.

[17] K. Li, "Constant time boolean matrix multiplication on a linear array with a reconfigurable pipelined bus system," *Journal of Supercomputing*, vol.11, no.4, pp.391-403, 1997.

[18] K. Li, Y. Pan, and S.-Q. Zheng, "Simulation of parallel random access machines on a linear array with a reconfigurable pipelined bus system," *Proceedings of International Conference on Parallel and Distributed Processing Techniques and Applications*, vol. II, pp.590-599, July 1997.

[19] K. Li, Y. Pan, and S.-Q. Zheng, "Fast and efficient parallel matrix computations on a linear array with a reconfigurable pipelined optical bus system," in *High Performance Computing Systems and Applications*, J. Schaeffer and R. Unrau, eds., Kluwer Academic Press, 1998.

[20] K. Li, Y. Pan, and S.-Q. Zheng, "Scalable parallel matrix multiplication using reconfigurable pipelined optical bus systems," *Proceedings of 10th International Conference on Parallel and Distributed Computing and Systems*, October 1998.

[21] K. Li, Y. Pan, and S.-Q. Zheng, "Fast and processor efficient parallel matrix multiplication algorithms on a linear array with a reconfigurable pipelined bus system," to appear in *IEEE Transactions on Parallel and Distributed Systems.*

[22] Y. Li, Y. Pan, and S.-Q. Zheng, "Pipelined TDM optical bus with conditional delays," *Optical Engineering*, vol.36, no.9, pp.2417-2424, 1997.

[23] Y. Li and S.-Q. Zheng, "Parallel selection on a pipelined TDM optical buses," *Proceedings of International Conference on Parallel and Distributed Computing Systems*, pp.69-73, Dijon, France, September 1996.

[24] Y. Pan, "Hough transform on arrays with an optical bus," *Proceedings of 5th International conference on Parallel and Distributed Computing and Systems*, pp.161-166, October 1992.

[25] Y. Pan, "Order statistics on optically interconnected multiprocessor systems," *Proceedings of the 1st International Workshop on Massively Parallel Processing Using Optical Interconnections*, pp.162-169, April 1994.

[26] Y. Pan and M. Hamdi, "Efficient computation of singular value decomposition on arrays with pipelined optical buses," *Journal of Network and Computer Applications*, vol.19, pp.235-248, July 1996.

[27] Y. Pan, M. Hamdi, and K. Li, "Efficient and scalable quicksort on a linear array with a reconfigurable pipelined bus system," *Future Generation Computer Systems*, vol.13, no.6, pp.501-513, 1998.

[28] Y. Pan and K. Li, "Linear array with a reconfigurable pipelined bus system – concepts and applications," *Information Sciences – An International Journal*, vol.106, no.3-4, pp.237-258, 1998.

[29] Y. Pan, K. Li, and S.-Q. Zheng, "Fast nearest neighbor algorithms on a linear array with a reconfigurable pipelined bus system," to appear in *Parallel Algorithms and Applications.*

[30] V. Pan, "Parallel solution of sparse linear and path systems," in *Synthesis of Parallel Algorithms*, J.H. Reif ed., pp.621-678, Morgan Kaufmann, San Mateo, California, 1993.

[31] V. Pan and J. Reif, "Efficient parallel solution of linear systems," *Proceedings of 7th ACM Symposium on Theory of Computing*, pp.143-152, May 1985.

[32] H. Park, H.J. Kim, and V.K. Prasanna, "An $O(1)$ time optimal algorithm for multiplying matrices on reconfigurable mesh," *Information Processing Letters*, vol.47, pp.109-113, 1993.

[33] S. Pavel, *Computation and Communication Aspects of Arrays with Optical Pipelined Buses*, Ph.D. Thesis, Dept. of Computing and Information Science, Queen's University, Ontario, Canada, 1996.

[34] S. Pavel and S.G. Akl, "Matrix operations using arrays with reconfigurable optical buses," *Journal of Parallel Algorithms and Applications*, vol.8, pp.223-242, 1996.

[35] S. Pavel and S.G. Akl, "On the power of arrays with reconfigurable optical buses," *Proceedings of International Conference on Parallel and Distributed Processing Techniques and Applications*, vol.III, pp.1443-1454, August 1996.

[36] C. Qiao, "Efficient matrix operations in a reconfigurable array with spanning optical buses," *Proceedings of the 5th Symposium on the Frontiers of Parallel Computation*, pp. 273-280, 1995.

[37] C. Qiao and R. Melhem, "Time-division optical communications in multiprocessor arrays," *IEEE Transactions on Computers*, vol.42, pp.577-590, 1993.

[38] S. Rajasekaran and S. Sahni, "Sorting, selection and routing on the array with reconfigurable optical buses," *IEEE Transactions on Parallel and Distributed Systems*, vol.8, no.11, pp.1123-1131, 1997.

[39] V. Strassen, "Gaussian elimination is not optimal," *Numerische Mathematik*, vol.13, pp. 354-356, 1969.

[40] C. Tocci and H.J. Caulfield, eds., *Optical Interconnection – Foundations and Applications*, Artech Houce, Inc., 1994.

[41] J.L. Trahan, Y. Pan, R. Vaidyanathan, and A.G. Bourgeois, "Scalable basic algorithms on a linear array with a reconfigurable pipelined bus system," *Proceedings of 10th International Conference on Parallel and Distributed Computing Systems*, pp. 564-569, October 1997.

[42] S.-Q. Zheng and Y. Li, "Pipelined asynchronous time-division multiplexing optical bus," *Optical Engineering*, vol.36, no.12, pp.3392-3400, 1997.

Index

SPRINGER NATURE

GPSR Compliance

The European Union's (EU) General Product Safety Regulation (GPSR) is a set of rules that requires consumer products to be safe and our obligations to ensure this.

If you have any concerns about our products, you can contact us on ProductSafety@springernature.com

In case Publisher is established outside the EU, the EU authorized representative is:

Springer Nature Customer Service Center GmbH
Europaplatz 3
69115 Heidelberg, Germany

FSC
www.fsc.org
MIX
Papier | Fördert
gute Waldnutzung
FSC® C083411

Kolibri 360 GmbH
Ferdinand-Jühlke-Straße 7
99095 Erfurt, Deutschland
produktsicherheit@kolibri360.de